C0-ACM-921

Systems engineering series

A. Terry Bahill, Editor-in-Chief

The Road Map to Repeatable Success: Using QFD to Implement Change
Barbara A. Bicknell, Bicknell Consulting, Inc.
Kris D. Bicknell, Bicknell Consulting, Inc.

Linear Systems Theory
Ferenc Szidarovszky, University of Arizona
A. Terry Bahill, University of Arizona

Engineering Modeling and Design
William L. Chapman, Hughes Aircraft Company
A. Terry Bahill, University of Arizona
A. Wayne Wymore, Systems Analysis and Design Systems

Model-Based Systems Engineering
A. Wayne Wymore, Systems Analysis and Design Systems

The Theory and Applications of Iteration Methods
Ioannis K. Argyros, Cameron University
Ferenc Szidarovszky, University of Arizona

System Integration
Jeffrey O. Grady, JOG System Engineering

System Engineering Planning and Enterprise Identity
Jeffrey O. Grady, JOG System Engineering

System Engineering Planning and Enterprise Identity

Jeffrey O. Grady

CRC Press
Boca Raton Ann Arbor London Tokyo

LIMITED WARRANTY

CRC Press warrants the physical diskette(s) enclosed herein to be free of defects in materials and workmanship for a period of thirty days from the date of purchase. If within the warranty period CRC Press receives written notification of defects in materials or workmanship, and such notification is determined by CRC Press to be correct, CRC Press will replace the defective diskette(s).

The entire and exclusive liability and remedy for breach of this Limited Warranty shall be limited to replacement of defective diskette(s) and shall not include or extend to any claim for or right to cover any other damages, including but not limited to, loss of profit, data, or use of the software, or special, incidental, or consequential damages or other similar claims, even if CRC Press has been specifically advised of the possibility of such damages. In no event will the liability of CRC Press for any damages to you or any other person ever exceed the lower suggested list price or actual price paid for the software, regardless of any form of the claim.

CRC Press SPECIFICALLY DISCLAIMS ALL OTHER WARRANTIES, EXPRESS OR IMPLIED, INCLUDING BUT NOT LIMITED TO, ANY IMPLIED WARRANTY OF MERCHANTABILITY OR FITNESS FOR A PARTICULAR PURPOSE. Specifically, CRC Press makes no representation or warranty that the software is fit for any particular purpose and any implied warranty of merchantability is limited to the thirty-day duration of the Limited Warranty covering the physical diskette(s) only (and not the software) and is otherwise expressly and specifically disclaimed.

Since some states do not allow the exclusion of incidental or consequential damages, or the limitation on how long an implied warranty lasts, some of the above may not apply to you.

Library of Congress Cataloging-in-Publication Data

Grady, Jeffrey O.
System engineering and enterprise identity / Jeffrey O. Grady.
p. cm.— (Systems engineering series)
Includes bibliographical references and index.
ISBN O-8493-7832-X
1. Industrial engineering. 2. System engineering. 3. Engineering economy. I. Title. II. Series.
T56.24.G73 1994
620′001′1—dc20 94-31150
CIP

Direct all inquiries to CRC Press, Inc., 2000 Corporate Blvd. N.W., Boca Raton, Florida 33431.

International Standard Book Number 0-8493-7832-X
Library of Congress Card Number 94-31150
Printed in the United States of America 1 2 3 4 5 6 7 8 9 0
Printed on acid-free paper

Contents: Part I

Following is a condensed Contents listing for Part II of this book, the SEM/SEMP section. A complete Contents listing immediately precedes the SEM/SEMP section.

Contents: Part II — SEM/SEMP

Foreword

This book materialized as a result of war stories from peers in industry who expressed concern for the ability of their companies to compete for Department of Defense (DoD) contracts under the terms defined in MIL-STD-499B. How could it possibly be true that seasoned system engineers and managers might feel this way? These same people had worked on programs for a decade in response to MIL-STD-499A. Why should a revision to this document cause such a pessimistic response?

While employed in industry, I had been involved from time to time in writing system engineering management plans (SEMP) compliant with MIL-STD-499A and found it to be a difficult and frustrating job. My theory was that we should write the plan for us to follow when we won the contract. A SEMP written this way would, however, almost never pass review. During proposal efforts it is very difficult to suppress the urge to use sales language and all too often when we did write a SEMP as part of a proposal effort it was written as part of the proposal. And like the proposal, it was seldom opened again after a win. I confess that I don't recall ever having had to work strictly in accordance with a SEMP on a contract in industry. This is what MIL-STD-499B attempted to change in its short and unfulfilled life and thus carried with it a lot of concern on the part of those who do not now have an effective system engineering process.

I wondered just how difficult it could be to write a company system engineering management plan compliant with MIL-STD-499B. Government insistence that the SEMP be a contractual document under the new standard appeared to me to encourage the writing of a plan that would actually be followed after a contract win and that had a certain appeal. At the same time, I was mindful that the defense sector is contracting relative to the commercial sector so an effort should be made to broaden the application of the product to encourage wider readership than might be the case for a narrow DoD market appeal. As a result, the planned book would have to be responsive to not only MIL-STD-499B but also to other authoritative system engineering process requirement sources to which companies might have to respond. Would it be possible to write a generic plan that many different companies would find useful as a basis for their own plan that could be used across their customer base that might include commercial interests?

Having recently accepted early retirement from a large aerospace company and started my own system engineering consulting business, I had the time to devote to this activity. It also appeared that the results might yield a service and product of use to my new company. Following a year of work and shortly after sending the manuscript to the publisher, it became evident that DoD was not going to approve MIL-STD-499B because it was running at cross purposes to another trend within DoD to appeal to commercial standards. As a result, I switched the basic structure from 499B to 499A recognizing full well that 499A was never more than a U.S. Air Force standard. And, while few had ever actually adhered to a SEMP prepared in accordance with this standard and it fails to accommodate the latest systems engineering techniques such as concurrent development, it is still not a bad way to organize the information that must be contained in a SEMP.

You, the reader, will find the body of this book, Part I, to be fairly short for it is only a way to convey to you the attached System Engineering Manual (SEM)/SEMP, Part II. The book has two planned applications. First, it may be used by a person in industry as a source of a model for a SEMP. You will find it relatively easy to edit this document for your own purposes compared to building one from scratch. Some system engineering managers had told me that they had a lot of good system engineers to do specific tasks but that most of their people would be intimidated by the task of writing from scratch a SEMP. A computer disk is sold with the book to provide the attached document in a form that you can begin to use immediately as a basis for your own plan. Simply put your company name on the cover or edit in some detail as you see fit.

Second, the book is intended to be used in system engineering certificate programs in a system engineering management course as a supplement to a more general book like *Systems Engineering Management* by Ben Blanchard. In such a class, this book could be used as a basis for a class team project to write a SEM or SEMP for a real or imaginary company or program. My consulting company offers a week long course based on these two books in which an on-site class prepares two or more SEM/SEMP using the enclosed version as a basis and recognizing a brief list of requirements provided by a red team selected from company management. Upon completion of the class, the red team reviews the manuals and selects one, or some combination of parts, as the basis for further work to create a company SEM/SEMP.

The central theme throughout this book is that an enterprise should have a well-defined identity in terms of its written internal procedures. It should have only one standard set of procedures (one identity) and these procedures should be repeated from program to program taking advantage of repetition as a means of improving personnel performance. The enterprise should actually use these procedures in performing its work and work to incrementally improve them over time. The attached SEM/SEMP attempts to blend the best of MIL-STD-499A, MIL-STD-499B, the U.S. Air Force integrated management system concept, the draft IEEE Standard for Systems Engineering

P1220, Electronic Industries Association Systems Engineering Standard SYSB-1, and ISO 9001 coupled with an internal generic planning and transform engine into a description of a powerful, integrated system development environment. My experiences in industry as a system engineer and engineering manager in systems engineering acted as the selection and integration engine for the final result. I also gained benefit from conversations over the years with several members of the National Council on Systems Engineering and other practicing system engineers.

I find it surprising that many people in commercial practice have such hostile feelings towards MIL-STD-499 and anything derived from a process looked upon with favor by DoD. There is apparently a feeling that such processes are fundamentally at cross purposes to rapid time to market, innovative and creative thinking, and good business sense. When pressed for their views of an effective systems approach, however, you find that few of these people have a real, comprehensive alternative and when you look at attempts to date to create commercial systems engineering standards they contain essentially the same process embodied in MIL-STD-499A. There is a good reason for this that is exposed in Chapter 1 of this book. The challenge is to apply this organized process to best effect in your particular circumstances. To do so you need to understand the whole and how to compromise on parts of the process so as to balance your process between risks: the risk, for example, of being late to market vs. the risk of missing something of importance on the first development pass.

These decisions are focused on the need for and cost of perfection. If you are going to Mars, you will seek perfection. If you are marketing a new bicycle you may be more interested in market forces and design adequacy. Regardless, the same fundamental process is very hard to beat in encouraging sound solutions to complex problems. It is problem complexity that drives the need for a systems approach. If you are in a business that solves simple problems, you may not need the systems approach at all.

DoD happens to have had more experience than any organization with the development of systems to solve very complex problems within an environment promising very dire consequences for mistakes. We may debate how well DoD spent our money or how efficiently programs were managed, but all who value their freedom in late 1994 owe a debt of gratitude to the availability of an effective systems development approach for DoD weapons systems through the period between the end of WW II and the fall of communism. Rather than dismissing the possibility of any utility in this process for commercial purposes, even a cursory review with an open mind might find value in these methods before one's competition does.

I hope you are able to use this prescription in your enterprise to great benefit. I confess that it may not be easy to understand the broad and deep sweep of the intended process. I have tried to cover a lot of territory within the confines of these covers. Even at that, there are many specialized areas in the SEM/SEMP that are only briefly touched upon. In these areas you may wish to expand to suit your company's needs. There is a great deal of interest in a simple version of the system engineering process that unfortunately will be very elusive if those searching for it also target a comprehensive definition. If you seek both broad and deep knowledge simultaneously, it may be difficult to master immediately. That is the bad news.

The good news is that, while this book offers a comprehensive window into the planning and execution process for a system engineering work effort, not everyone working on a program should be exposed to the full sweep of this work. Those responsible for system engineering in companies and on programs must partition this knowledge into parts that are needed by specific persons and teams and not try to burden everyone with the complete picture. If you attempt to educate everyone in matters they need not understand it will likely result in discouragement and a conclusion that the process is too complex for normal human beings. The offered planning process coordinates the planning work of many specialists each of whom must understand their specialty but need not master the complete framework within which the program functions. Someone must understand it, however, to make the whole process sing.

Jeffrey O. Grady
San Diego, CA

Author

Jeffrey O. Grady is the Principal of JOG System Engineering, a consulting firm focusing on helping firms achieve excellence in the systematic development of complex products through methodological improvements and education. He previously worked in industry for over 25 years in the capacities of engineering manager, system engineer, project engineer, field engineer, and customer training instructor. He has worked in several aerospace companies on a wide range of systems including space transport, cruise missiles, unmanned aircraft, underwater fire control, and superconducting magnets with a customer base that included the DoD (USAF and USN), NASA, DoE, and private corporations. Mr. Grady also served in the U.S. Marines in the communications field. He received a M.S. degree in systems management from the University of Southern California and a B.A. in mathematics from San Diego State University. In addition to this book, Mr. Grady has authored *System Requirements Analysis,* McGraw-Hill, 1993, and *System Integration,* CRC Press, 1994. He was the first elected Secretary of the National Council on Systems Engineering (NCOSE) and the founding Editor-in-Chief of *Systems Engineering, The Journal of the NCOSE.* Mr. Grady teaches system engineering courses at the University of California, San Diego, and the University of California, Irvine, in the system engineering certificate programs he helped to found.

Acknowledgments

CRC Press has allowed me to use excerpts from the book *System Integration* (SI), which I had earlier written, in preparing this book and that is appreciated.

I confess that my initial reaction to MIL-STD-499B, around which much of this book evolved, was negative and it was only through the repeated friendly but firm encouragement of Dr. Jerry Lake, the first President of the National Council on Systems Engineering, at the time a professor at the Defense System Management College, and now a consultant, that I actually read the document on the way to becoming informed of its content. I very much appreciate his encouragement for otherwise I may have fallen into the trap of ignorant hostility. I regret that a lot of work that Jerry and many other fine system engineers put into this standard was not immediately accepted by DoD.

There are many military and industry persons unknown to the author who contributed to what the U.S. Air Force has called the integrated management system which was applied to the F-22 Program by Air Force Systems Command and several contractors working on that program. The person who awoke my interest in this work was U.S. Air Force Col. Tom Bucher (now retired) and I appreciate that new knowledge. I regret that many people see this approach as very complex, as will some readers of this book. It may be very complex at the global level but it is not necessary that everyone master the method in its entirety. It actually provides a simple framework within which a large mass of information from a large number of specialists can be coordinated. The important point to understand about this method is that none of the specialists need understand all of the details of this process, only their part of the job. Hopefully, I have not damaged the concept in applying it in this book.

I would also like to thank Michael Dick, Director of Techno-Tiger, for awakening my interest in the standards prepared by the International Standards Organization, ISO 9001 in particular. I used ISO 9001 as a quality cross check of the attached SEM/SEMP after it was created responsive to the MIL-STD-499B model and found several things not well addressed. You will find the traceabilty matrix from this cross check study as Table A-3 in the attached SEM/SEMP. The resultant changes made this book a better product.

Thanks again to my wife Jane for enduring yet another book campaign and for her sound business sense throughout them all.

System Engineering Planning and Enterprise Identity

Part I

chapter one

Introduction

1.1 *What is the purpose?*

The fundamental premise of this book is that an enterprise that develops and manufactures complex products should be guided by a standard, documented process for performing work that, in turn, defines a unique enterprise identity. We should all know by now that a successful enterprise must also be in a constant state of change toward a goal of excellence in a changing world. These two goals may appear to many readers to be mutually exclusive because in the past it has, in many firms, required months to gain approval of process changes. So, how can an enterprise possibly document its process while constantly changing it to keep up with or stay ahead of the competition? One alternative is to simply give up on process definition documentation and allow it to be captured only in the corporate memory of the current employee base. Some companies with this attitude that have also endured severe downsizing may find that important elements of their process definition have been dismissed with the staff.

We will find that just as a successful firm must speed up its product development process to remain competitive, it must also speed up its identity definition and maintenance process. The same object that is driving the rate of change in industry, the computer, can also be used to accommodate this identity definition acceleration. The same machinery used to allow us to capture our identity can also become an integral part of the enterprise's ability to speed up its program definition process whether it be for the purpose of serving commercial customers or for preparation of a proposal in response to a request for proposal from a government agency.

A key element in our enterprise identity is our process of integrating the many different specialized engineers together into an effective force to develop and produce products involving complex relationships. This key element is called system engineering or concurrent engineering as you prefer. The author believes that concurrent engineering or development is only a variation on the original purpose of the system engineering process but one that has re-vitalized interest in the process for many. This book focuses on developing and maintaining an effective system engineering process through an optimum mixture of generic and program-specific planning and the documentation of the system engineering portion of your identity. This portion can serve as a model for your completion of the remaining necessary planning data for your whole enterprise or simply act as a basis for your development and maintenance of an enterprise system engineering process.

This book conveys to you an attached, fully prepared document (Part II) that may be used for several purposes. The attached document is termed an integrated System Engineering Manual (SEM) and generic System Engineering Management Plan (SEMP) or SEM/SEMP. A system engineering manual is thought of as a functional department plan created by functional management in a company managed using the matrix approach. This kind of document provides guidance for the implementation of the system engineering process on a program. There should be only one of these documents in a business unit because a standard process has certain advantages that will be covered in some detail in this book. A SEMP is thought of as a program system engineering plan created in response to a customer standard such as the U.S. Air Force MIL-STD-499. In the past it has been common for a company to write a unique SEMP for each unique program.

This book may be used to satisfy two needs. First, the attached SEM/SEMP may be used as the basis for a company functional SEM, program SEMP, or integrated SEM/SEMP as intended. Secondly, it may be used as a text book in a system engineering management course in combination with a more general systems engineering management book.

Let us first discuss the independent SEM application. A functional system engineering manual should define a repeatable process and provide information about how to perform each task comprising the process. This need could be satisfied in at least two different ways:

a. The SEM provides all system engineering process requirements as well as detailed instructions on performing each task.
b. The SEM provides all system engineering process requirements only supplemented by separate, more detailed functional procedures describing in detail how to perform each task required by the SEM.

The author chose to follow the first alternative and to use the MIL-STD-499 model as the principal basis for the attached SEM/SEMP. The A revision of this standard had been in use on U.S. Air Force programs for several years when an attempt was made to update it for important improvements and in the process gain Department of Defense (DoD) approval of the standard. Unfortunately, after several years of work, the B revision appeared ready for signature within DoD at a time when DoD was working to move toward use of commercial standards. The work on 499B will very likely be taken over by the National Council on Systems Engineering or some other society; so, the content of this standard may yet see the light of day, hopefully merged with some good ideas missed on the first pass. The author has tried to respect the basic structure of MIL-STD-499A while integrating into that structure the latest ideas expressed in the B revision, the U.S. Air Force integrated management system, ISO 9001, and other standards.

An attempt was made in the appended SEM/SEMP to augment the raw process requirements with some descriptive material such that a company could get by without writing additional procedures if they chose not to do so. Alternatively, the SEM/SEMP could be augmented by a series of detailed procedures describing in some detail how to perform some or all of the tasks defined.

The attached document could also be used as a basis for preparing an independent program-peculiar SEMP in a company that does or does not have an existing functional SEM. In the past, companies have commonly thought it necessary to prepare a program-unique SEMP focused on the individual customer's needs. Since the attached document is based on the MIL-STD-499A model, it should be adequate for this purpose on a U.S. Air Force program as a minimum, perhaps with some tailoring of the standard and/or the accompanying data item description (DID). Since there is no other systems engineering standard in DoD, if another service requires a systems approach, the one addressed here will likely be adequate.

A third alternative is offered in this book. If there is value in a company only having one system engineering process, and there is, then why should we re-design ourselves for each contract and prepare a program-unique SEMP? In the past, companies in the weapons business have been encouraged in this direction by a short-sighted DoD.

In recent years there have been signs that DoD has seen the error in this approach and is now leaning toward favoring companies with a sound process that is well documented and traceable to DoD requirements. If this be the case, then why can we not merge the functional SEM with a generic SEMP and apply that standard process on every program? In the military standard MIL-STD-499A and more recent Government documents, the Government intends to require that the SEMP be written by the contractor and placed under contract, suggesting that the contractor would have to actually follow the content during the program. The one impediment to successful system engineering planning in this concept under the MIL-STD-499B format was that the SEMP would have included program-peculiar material in it. This would have made it impossible for a company to develop a generic SEM/SEMP. MIL-STD-499A had no such content provision; so, a generic approach is in keeping with the standard.

In parallel with the development of ill-fated MIL-STD-499B, another U.S. Air Force initiative was developed at Wright Patterson Air Force Base called the integrated management system. Unfortunately these two teams could not find a way to work together and neither approach includes all of the best elements of the other. That is a pity because they can be blended together to produce a much better planning product than possible without one or the other. Perhaps those who pick up the 499B work will see the good sense in this position and move to merge the best of both.

In this book, and the attached integrated SEM/SEMP, we will follow this third course. We seek to create a single document that will provide a functional system engineering manual for our enterprise that will also satisfy any customer's requirements for a system engineering management plan appropriate for their contract. We approach this goal by applying the following principles:

a. There shall be a single SEM/SEMP for our enterprise and it shall be applied to all contracts enabling task repetition and the resultant improving personnel task performance.
b. Where the enterprise must respond to a customer requiring application of the U.S. Air Force integrated management system, the SEM/SEMP will provide the narrative system engineering process material called for in the Integrated Master Plan (IMP) and be referenced therein as the source of the system engineering process description narrative.
c. The proposed SEM/SEMP should not be a contractual document as encouraged in MIL-STD-499B because the program-peculiar requirements for each contract would defeat our need for a common process definition. The contractor should be required to map its systems engineering process description to a standard placed on contract, such as MIL-STD-499A or a systems engineering standard developed by the National Council on Systems Engineering, IEEE, or other society.
d. The program-peculiar content that MIL-STD-499B called for (such as a trade study list or specific risks anticipated) will not be included in the SEM/SEMP. Rather, it will be included in a program peculiar plan such as the IMP noted above or a Program Plan. As a result, the SEM/SEMP should not have to be changed for a given program because it has no program-peculiar content.
e. On a DoD program, a program-peculiar plan (such as an IMP) will require compliance with the internal SEM/SEMP either directly or through a customer standard tailored for identity to the internal SEM/SEMP. Customer concerns for inability to control the content of the internal SEM/SEMP will be satisfied by mapping the content of the SEM/SEMP to the tailored customer standard and a contractual obligation to satisfy the requirements of the tailored customer standard. Thus, any change to the internal SEM/SEMP subsequent to accepting a contract obligation must respect any existing contractual requirements or run the risk of being non-compliant with customer requirements. Some changes to the internal SEM/SEMP may require a no-cost contract change proposal for one or more existing contracts at the time they are implemented. The contractor needs to make these points clearly in any proposal to a customer that requires a particular standard or requires that the contractor write a program-peculiar system engineering document.

f. Program peculiar plans will answer five of the traditional reporter's questions: who, what, where, when, and why. The SEM/SEMP will answer the question, "How shall system engineering tasks be done?" We may choose to augment the SEM/SEMP with more detailed process description documents. One could, for example, write a detailed process description for technical performance measurement or risk management referenced in the SEM/SEMP. This same model can be applied to all functional areas of an enterprise.
g. Program planning work should have two components: (1) that planning work which we can do in a generic way, and (2) that which must be done for a specific program. The more work we can push into the former category the better because we only have to do that planning work once and then apply it on each program. By minimizing the program-peculiar planning we will also be able to respond more quickly to customer needs.

The central theme throughout the book and attached SEM/SEMP is that your enterprise should have a clear identity defined in its generic planning and practices. You should apply this standard process to the benefit of all of your customers, quickly and incrementally improving it through lessons learned on programs.

1.2 *Where are we going?*

For reasons made clear in this chapter, we must plan the complex work that will require the application of system engineering techniques. Most of us would accept that planning is an inseparable part of good management. Another way to say this is that management is the effective control of the faithful execution of good planning. In this chapter we seek to specifically answer the question, "Why must we plan?" There are, as you may already know, some very fundamental forces at work within us and in our environment that drive the need to plan our work.

This book grew out of the author's experiences and research into the application of system engineering in industry. The content focuses on planning work associated with the application of system engineering techniques on programs but, since system engineering planning and work cuts across everything we do on programs, a great deal of the work that must be done on a program properly falls under the influence of the system engineering planning effort. Therefore, the content of this book may be useful to persons other than system engineers and those who would wish to understand the systems approach to problem solving. We need not think of system engineering as an organization; you may not have such an organization. System engineering is a process for solving large and complex problems efficiently. A fundamental part of that process is to clearly understand the goal we seek, plan the needed work very carefully to reach that goal, and to execute that plan, faithfully tracking the current situation against the plan.

In Chapter 2 we find out that it is okay for us, as an enterprise, to have a defining identity and that we can use that identity as a means to become better and better at what we do on each program through small incremental changes in a consciously applied continuous process improvement program. This identity will also provide our employees with a solid foundation upon which to grow in their chosen careers.

In Chapter 3 we explore some sources of inspiration for our identity and fix on MIL-STD-499A and a U.S. Air Force initiative called the integrated management system, all intertwined with the integrated product development notion. Readers who are interested in commercial enterprises should not immediately close the covers of this book because our planning machinery draws upon DoD experience. The facts are that DoD has experienced the development of the greatest number of the most complex systems under the pressure of the most urgent timing, with the most seriously adverse consequences of failure than any other customer. DoD procurement offices have shown a lot of interest in the systems approach for reasons rooted in bad experiences in this process and there have been many of those.

We must plan for a specific organization; so, Chapter 4 defines an organizational structure for use in this book and provides rationale for this selection. The author encourages a matrix form for enterprises involved with multiple programs or customers. Nearly everyone who has worked within a matrix management environment has at least one horror story supporting the notion that it is a bad structure. The author believes that most of these problems stem from poor management at the top that permits the strong personalities in an organization to shape the way the matrix is applied rather than these persons being forced to function within the bounds of a predefined relationship. The functional axis of the proposed matrix is very lean but necessary to preserve and build the enterprise's long term capabilities in specialized areas.

In Chapter 5 you will find the core of the book in the form of a planning process that you can apply in your own enterprise that builds on a clear definition of your company's identity by mapping that identity to your customer's needs. Appended, you will find an example, or model, of the results of having accomplished the work explained in Chapter 5 in the form of a company SEM and generic SEMP. The content of this document is formally traced to MIL-STD-499A and applicable to a wide range of situations. A computer disk is included with the book containing a copy of this document in Microsoft Word for Macintosh on a disk formatted on an IBM PC. It may be opened on an IBM PC and converted in the process to a Word for Windows or DOS document or the disk placed in a Macintosh super drive and opened in Word For Macintosh. The graphics were created in Claris MacDraw II and saved as PICT files on the included disk. The disk copy does not precisely match the pagination of the paper copy included in the book because the user will generally prefer an $8^1/_2'' \times 11''$ document. The content, however, is the same. Additional computer products are available from the author's consulting firm to capture the definition of your identity and to map that identity to your customer's needs.

The process defined here is compliant with the ISO 9000 series, as would be almost any process description. The difficult part of being ISO 9000 compliant is in having the good sense to actually follow the procedures that you have. There are several options here:

a. Bad procedures that are not followed.
b. Bad procedures faithfully implemented.
c. Good procedures that are not followed.
d. Good procedures that are faithfully implemented.

Clearly, only option "d" is worthwhile as a goal. Many firms will find themselves in condition "a" at the present where their procedures prepared some time ago have not been maintained. They may have been effective procedures at some time past but conditions have changed and they have not. There was a lot of ownership volatility in industry through the 1980s and one division System Engineering Director with whom the author talked had procedures that still bore the identification of the company that had owned the division two owners previous. These may still be very good procedures, of course. But, there have been a lot of good ideas come to the surface in this field in the late 80s and early 90s that should have found their way into a company's methods.

There will be readers who believe that the goal of good procedures, even if momentarily attained, consumes resources better spent in improving an enterprise's capital base or investor profit. They will contend that it is more important that the team work together in a creative way in response to ad hoc management direction motivated by a sense of the current situation and needs in accordance with a simple policy statement. The author wishes that a great story from management literature would apply to every situation. In that story, the CEO asks the staff to bring their department manuals to the next meeting. At that meeting the CEO throws them all in a 55-gallon drum and torches them. He then issues a two-page memo that explains the company mission and related policy. There are businesses that can function this way very well. But, those who would develop complex systems will require a more organized environment within which to conduct their customer's business.

If you do not have such an environment now and cannot succeed using a two-page memo, you may find the content of this book and the attached SEM/SEMP useful.

We seek a common or generic process for our enterprise that can be transformed into a range of possible program processes acceptable by the customers of those programs. There will be a great attraction in those programs to diverge from the common approach; so, we must have the ability to audit programs against our preferred processes description. Customers are more concerned that an enterprise have a repeatable process that they actually follow than a process that perfectly matches their prescription. Some of these customers may even insist that you be able to demonstrate that you are following your own procedures. Chapter 6 offers an audit program with two axes to allow us to compare a program's process to our generic enterprise model and to compare our enterprise model to some reasonably universally accepted model. Ideally, we should find on all programs that our process conforms to our generic process and that our generic process compares favorably with some world-class standard.

Chapter 7 provides a guide for moving from a current condition of an ad hoc development company dependent on serendipity between several functional departments and rework during first article fabrication and test into a powerhouse systems organization. Your organization may have already begun this trip and might even have taken some correct turns in the roadway, so much the better. If so, you will know that it is a very difficult road. Hopefully, the content of Chapter 7 will encourage a more rapid approach to your goals.

1.3 Why plan?

Planning is something we believe everyone else should do well before starting a complicated task. We don't always feel compelled to do so for our own affairs, however. Why is this the case? Well, planning is certainly hard work and it does not appear, at the time we have to do it, to be leading to a constructive result on the path we know we must travel. We engineers are action and solution oriented people. We want results now. We especially shun the planning work others would do for us. What can they know about our needs anyway? These may be acceptable attitudes where the goal is to buy a new car for personal use, build a room addition on our home, or go on a camping or fishing trip alone. Customers of complex products insist on and deserve better.

The fundamental motivation for planning is that time is important. It is a unidirectional phenomenon which we cannot get back if we use it badly. Also, the painful rules of economics apply to much of what we do. We will ever have to deal with fewer resources (including money) than we would like to have available to us. If time and money mean nothing to us, then planning is not a real necessity. Few readers of this book will ever find themselves in this situation.

Even if we were to accomplish a needed task all by ourselves, it would be helpful in terms of our time and resources to list a series of steps we will have to accomplish, and in what order, as a means of assuring that we thoroughly understand the work necessary to reach our goal and that we have accounted for all of the foreseeable impediments to success. Where two or more people must work together to achieve a common goal, it becomes more important to plan the work required by each to reach the common goal. The larger the number of people involved, the more resources that have to be committed, the more demanding or complex the need, the more adverse the consequences of failure to finish at a particular time, the greater the schedule difficulty, and the more physical separation between the members of our total team, the more necessary good planning becomes.

1.4 *When plan?*

Unfortunately, we must plan work prior to accomplishing the work in order for it to have any beneficial effect. It is true that at the time we must plan work, we seldom have command of all of the knowledge needed to plan perfectly. We must plan future work under conditions of imperfect knowledge of the future and, therefore, there is some risk of failure that must be mitigated, also through good planning. This means that our plans will almost never match perfectly the conditions that unfold during the time we implement the plan so we must recognize that we will have to change our plan to some degree to reflect unforeseen events as they become apparent to us.

Because we have little hope of planning perfectly, does that mean that planning is a waste of time? No! Good planning will result in a closer match with future realities and smaller risk exposure. Good planning teamed with good implementation management flexibility in making changes to resolve problems offers the best chance of satisfying program goals of cost and schedule. We evade the message encouraging good planning at our peril. Our competitors may discover the secret to success through system engineering planning while we are still in denial.

1.5 *One man's axioms of system engineering*

Our efforts to plan future work must recognize some fundamentals of human nature and the nature of complexity. First, we are gifted with a marvelous thinking instrument in our brain. Sad to say that brain is limited in its capability to make efficient use of information. We are knowledge limited. At the same time, because of that marvelous brain, men and women over the centuries have amassed a tremendous amount of useful information and found ways of storing and communicating that information in books, audio and video instruments, computers, mail, telephones, FAX, and so on.

Many centuries ago man arrived at the point where one person could not know everything known by man. Our civilization has long since passed beyond that marker. Today the pace of increased knowledge is tremendous, driven by economic advantage to those that have it, an unquenchable appetite to know that which we do not, and the tremendous power of computers to help create, store, and make that information available.

Man has worked out a method for taking advantage of the full storehouse of knowledge called specialization. Our enterprises are often functionally organized this way into engineering, finance, manufacturing, procurement, and other departments focused on particular slices of knowledge. Our college degree programs are based on a particular breakdown of knowledge into science, business, engineering, and so forth. Within engineering we have mechanical, electrical, chemical, and other disciplines. It is possible to specialize even more finely than these narrow disciplines. The result is that when confronted by a complex task that must involve a wide range of knowledge and technology, we are forced to team together to get that task accomplished. No one person can master all of the knowledge needed to solve a complex problem well in a reasonable period of time.

Even if any one of us could master all of the knowledge needed for a particular project, that person would not do as well in developing a solution as a team of specialists well led. The reason for this is that we all have different experiences whatever our specialized knowledge base. Two or more people will conceive a wider range of solution options than any one person because of these experience differences. And, no matter how marvelous our procedures for implementing a systems approach, it all should be focused on providing a safe space within which to conceive good ideas that satisfy our customer's needs. The more possibilities considered thoughtfully, the better the chance that the best solution is among them.

These ideas have been blended into a set of unifying axioms for system engineering that the author has found useful as a foundation for the content of this and his other books on system engineering, teaching work, and consulting work. They can be summarized succinctly as follows:

1. Man is limited in his capacity to master knowledge.
2. Knowledge expands monotonically.
3. In the modern era, available knowledge exceeds the amount of knowledge any one human can master.
4. Competition rewards those who can command the greatest access and most efficient and effective use to knowledge pertaining to particular problems.
5. Specialization solves the problem of coverage of the knowledge space applicable to a given problem.
6. Decomposition of a large problem into a series of related smaller problems produces problems with a scope upon which relatively small teams of specialists can effectively work together in a cooperative way. These smaller problems and their assignment to teams should be characterized by minimized cross-organizational interface, minimizing the necessity for the teams to cooperate. The opportunities for intra- and inter-team cooperation and communication must be maximized.
7. Traceable requirements identification assures that all specialists are working to solve the same large problem.
8. Small problem teams will not always solve their problems optimally at the large problem level requiring integration and optimization of the work of the small problem teams at the system level.
9. Pure analytical thought is insufficient to assure that our work (composed of the fruits of our thoughts) has satisfied product requirements leading to a need for convincing proof in the form of demonstration, inspection, and testing. That is, we must verify that we have solved the original problem with convincing evidence to that effect.

chapter two

The importance of enterprise identity

2.1 It works for people

We would all agree that it is okay and even important that a human being have a unique identity permitting others to identify them as a particular, unique individual. We would certainly find life less enjoyable in a rigorously planned society like that some countries tried to make work in the 20th century with standard people providing society with standard work units. No matter how efficient we might become by rigorous adherence to some standard personality, few of us would voluntarily submit to this prescription. One wonders whether the drive for competitiveness felt throughout the world in 1994 carries with it dangers of unreasonable emphasis on one's work and forced worker conformity. How do you increase production and reduce costs when you have improved everything else conceivable in your plant, procedures, and work force? While one's work is an important element of one's life, few of us would accept that it is our complete life.

Everyone we know is a unique human being, adding to everyone's enjoyment of life most often. We are defined by our physical being, our position, our family and friendships, the material things we own, our personality, and our experience and knowledge. Philosophers like Locke and Hume had difficulty defining identity but believed that one's identity had two components: body and mind. We begin life physically unique and most of our experiences increase that uniqueness by influencing the content of our mind. Our resultant identity differentiates us from all other people.

Even if it were possible to improve the process of developing systems by reducing the diversity among us, it would be no service to mankind. There are those who believe that system engineering can only thrive in an environment of powerfully controlled order. But, there are limits within which we must remain in designing our system engineering process. An eleventh commandment is proposed to protect us from overzealous planners as follows: "Thou shall not interfere or tamper with the unique identity of your employees." These differences are a fundamental part of being alive and should not be tampered with no matter if a benefit were derived by the enterprise in the process. These differences identify us as individual human beings.

We take advantage of this uniqueness of the people in our work force in the integrated product development teams encouraged in this book. The author once thought that the only advantage in teaming to solve problems was that you were able to cover a larger knowledge scope to counter the effects of specialization. The reality is that there is another effect every bit as important and that is that we also benefit from the differences in team member experiences that influence them to think differently than their teammates. As a result, a team will conceive a richer mix of alternative solutions to a problem than a single human being and in that mix will be a best solution that might be missed by a single person with limited experiences.

The author would draw an analogy between an individual and a single business unit. We as individuals would have difficulty functioning internally if forced to reflect two or more external identities. We would be internally inefficient. The same is true of a single business unit that permits its programs to reflect a different character or identity to different customers or the same customer on different programs. Personnel assigned to these programs from a common pool would have to master multiple methods for the same tasks rather than mastering a single universal process.

2.2 What constitutes an identity for an enterprise?

Philosophers have difficulty agreeing upon what defines or constitutes an individual's identity. They would likely also disagree on how to define an enterprise's identity. The author, being no philosopher, will avoid an appeal to a philosopher's logic to prove a particular postulation in this matter and attempt to use a simple persuasive argument. An enterprise depends on a positive perception of its uniqueness in the minds of its customers and its good works for those customers that encourages that positive perception. Customers do discriminate in favor of those enterprises they think have the better product or will provide the best service for them. Customers are also interested in firms that focus on solving the problems they believe they face. Customers relate to the identity of a firm defined by its physical plant; location; product and/or services quality, price, performance, and availability; the characteristics of their human interface with that enterprise; and, of course, by the advertising offered by that enterprise.

Behind this outer layer exposed to the public, how do we internally define our identity as an enterprise? How do we ensure that our enterprise is presenting an enduring, consistent image to our customers? In a small enterprise the identity is very likely embodied, for better or worse, in the personality of the owner. In a large enterprise the identity is properly embodied in its mission statement, policy, and procedures. If an enterprise does not have these in written form, such that they can be communicated to employees, or it fails to follow the standards it has, then its employees may have difficulty understanding the internal expression of the enterprise identity.

The result can be internal friction that detracts from a singleness of purpose oriented toward satisfying the enterprise mission. This difficulty will be coupled to the external identity expression in a form that leads to customer concerns.

Our enterprise needs a common process, defined in our mission statement, policy, and practices, that can be applied efficiently to its customer's needs. This common process forms an internal identity responsible for how our customer's view us through the products we develop and produce or services we provide.

Now a corporation may find it advantageous to permit its business units (divisions or whatever) to develop a unique identity based on their customer base and product line. The corporation can use these differences to evaluate the advantages and disadvantages of particular processes but should be aware that because a particular process may work well in one business unit does not mean it will work well under all circumstances. We do not seek to limit a differentiated enterprise's options. The identity argument offered only applies to a single business unit and its common personnel pool. At higher levels of organization, the same arguments can be turned around in support of separate identities (possibly equivalent but not necessarily so) for each business unit in a large enterprise.

We can avoid some of these problems by insisting that the work force for any new product will be partitioned into a separate business unit such that every element of our large enterprise is a separate business unit with a dedicated work force. In this case every business unit will have a single process description corresponding to its single program process. There are other problems in this configuration dealing with instability of the work force that the matrix organizational structure solves that encourage the author to support a matrix structure. But this is one alternative approach with some merit.

If a business unit, currently organized as a matrix, perceives strains in trying to force all programs into a common process mold, it should evaluate whether it should partition its business into two or more business units. There are certainly limits beyond which a common process identity is hard to maintain. Very different customer requirements, product types, and production methods encourage separation into multiple units.

Given that you now use a matrix or wish to do so, then you have to deal with the problem of a common process for the common labor pool.

2.3 *Customer standards, tailoring, and enterprise identity*

Given that we have a vision of an ideal enterprise process definition, how might we be successful in gaining acceptance of that process by our customers? If we are producing products for the commercial market place, we need only concern ourselves with the customer's view of our external identity. If, on the other hand, we are interested in serving a large customer of large and complex systems, like NASA, DoD, FAA, and DoE, we will be forced to comply with standards insisted upon by those customers. There are some 50,000 standards that might be invoked by these customers and we can be sure that they do not all correspond perfectly to our preferred process description. Perhaps, our interest in a generic process is dead on arrival.

It has not always been so but today most large customers, like all of those mentioned above, recognize that their desired product or service will cost more if they force you to comply precisely with their standards than it will for you to comply with your own standards. The fact that your internal standards are different will give rise to concern on the part of your customer. Is there a way to bridge this gap? Yes, and that way is through tailoring your customer's standards for identity with your own and showing the traceability between these standards set.

Tailoring is an editing process where you change the content of a document. These changes are commonly listed as a series of changes made to the document and these statements may include deletions, changes, or even additions to the document content. The combination of the original document adjusted as defined in the tailoring produces a modified document. In the attached SEM/SEMP you will find an example of these statements against MIL-STD-499A to cause it to agree perfectly with the attached SEM/SEMP.

As part of the definition of your identity, your enterprise should understand the relationship between its process and the process described in the standards commonly called by your customer base. For example, let us say that the U.S. Air Force (USAF) is a member of your customer base and that USAF commonly applies MIL-STD-499A, Systems Engineering, as a requirement for its development programs. You should understand the requirements of that standard and its relationship to your internal process. If there are differences, you should have identified the tailoring that will have to be applied to MIL-STD-499A to allow you to apply your internal process to a DoD program with no changes. When you bid a USAF program calling MIL-STD-499A, you simply include that tailoring in the statement of work that you include with your proposal.

The fact that you can tell your customer how you will perform their work in the context of their standards is very important. It will give your customer confidence in your process. The understanding is, of course, that you have a documented process and that you are continuously working to improve that process by reflecting on your program experiences. Your internal procedures should at any time reflect your conclusions about the very best way your current enterprise can function to produce the highest quality product for the lowest possible cost.

Refer to Chapter 5 of the author's book *System Requirements Analysis* (McGraw-Hill, 1993) for a detailed discussion of applicable documents analysis and tailoring toward the goals outlined above.

chapter three

Sources of inspiration

3.1 Help!

We need a good model of a system engineering process upon which to base our planning efforts. Where do we turn? There are many sources of input on what constitutes a good system engineering process. We can choose from several Department of Defense (DoD) models, any number of specialized computer software development models, or some text and reference books that have been published. We need not select a single existing model. We can pick and choose components from several existing models or invent an entirely new model. Let's first look at some that exist or have existed.

3.2 AFSCM 375 series

The U.S. Air Force Systems Command published a series of manuals (AFSCM) numbered 375 in the early 1960s. These manuals provided a detailed description of what Systems Command expected of contractors in planning and implementing a system engineering process. The rigorously defined process involved completing a lot of specially designed forms with information derived from the application of the process.

Some people believe that this prescription resulted in a backlash hostile to the systems approach that survives today in design group attitudes. Some companies, fearing that the constraining effects of this organized process would damage the creativity of their design engineers, invented system engineering organizations to shield their "real engineers" from contamination. System engineers thus became typecast as documentation specialists rather than system architects. Despite these faults (as perceived by the author), this process was involved in the development of some very successful systems.

The author has benefited a lot from studying the 375 series but has concluded that, while this series identified much of the information that must be captured and mastered in the development of a complex system, this completeness carries with it a message of complexity that is not appreciated. Systems engineering practices must be built in layers of complexity, just like product systems, that respect the attention or knowledge span of people. If you try to educate everyone in all of the details, you will fail. You need the simple overview that everyone involved in the process must understand supported by details that those who need them can selectively master.

At the time the 375 series was first introduced, computer applications were not well developed to capture and relate the tremendous amount of information called for in the many forms required; so, the process entailed a lot of bookkeeping work to maintain the current baseline definition. More recently, computer tool companies have developed many excellent tools useful in parts of the process defined in the 375 series, particularly the systems requirements analysis area.

3.3 SAMSO and BMO standards

Each ballistic missile program that the Ballistic Missile Office (BMO) or Space and Missile System Office (SAMSO) of the U.S. Air Force Systems Command developed had a system requirements analysis standard that had clear ancestry from the 375 series. These documents commonly were more focused on the computer system that would serve the process than descriptive of the process itself and included the following:

a. Exhibit 68-62, System Requirements Analysis Program For the Minuteman ICBM, SAMSO, 1968
b. SAMSO STD-77-6, System Requirements Analysis Program for the MX Weapons System, SAMSO, 1977
c. BMO STD-77-6A, System Requirements Analysis Program For SICBM, BMO, 1986

3.4 U.S. Army field manual

We should not be surprised that the U.S. Army published a field manual on System Engineering. That document, FM 770-78, was released in April 1979 and revised in 1986. It included an effective system engineering process description with less reliance on forms and rigor than the 375 series. It sketched out an overall process and provided considerable detail about the process.

3.5 MIL-STD-499

Dissatisfaction with the 375 series led to efforts to create a standard that would be more tolerant of tailoring to the specific development situation and of different but effective company processes. MIL-STD-XXX was one attempt to create such a document but it was essentially a condensed version of the 375 series focused on the computer systems that would serve the process rather than the process itself.

The first version of the surviving military standard, MIL-STD-499, was released in July 1969. It specifically stated that it did "...not prescribe or imply a specific contractor system engineering process or management methodology, organizational structure or form of contractor internal documentation." The Government's initial intent was, however, that firms that would do business with DoD would have their system engineering and system management processes validated by government inspectors as, at the time, they would already have to have their cost and schedule control systems validated in accordance with a criteria.

Mr. Elmer Peterson, now President of a system engineering consulting firm, Intec Associates Inc., was involved in the early development and use of this standard. He told the author that of three trial program applications (B-1, an airborne weather reconnaissance system, and the F-15) only the F-15 program retained it throughout the development program. The principal criticism was that while the intent was to validate the contractor's system engineering practices, they inevitably were actually validating the capability of specific people doing the work. When the standard was updated to the A revision, the validation intent was dropped, the process description generalized, and the intent to certify dropped.

MIL-STD-499A, released in May 1984, required that a contractor prepare a Systems Engineering Management Plan (SEMP) in accordance with a data item description. Specific system engineering activity statements were included in an appendix that could be selectively tailored and included in a program statement of work. The intent was that the SEMP would be approved by the customer and followed by the contractor during the program implementation. The SEMP was generally not made a contractual document and often ignored. Often the plan was prepared as a proposal volume with promises not always kept during execution.

MIL-STD-499B resulted from an intensive and extensive review of DoD needs and ways to define an effective systems approach. Like its predecessors, it was principally a U.S. Air Force initiative, but real efforts were made to coordinate with the other services and industry. It was scheduled for release in early 1994, but in May of 1994 it was rejected because it ran counter to the plan for DoD to appeal to commercial standards.

Unfortunately the 499B development team did not work in a coordinated way with the Air Force people developing, at much the same time, the integrated management system approach for the F-22 Program. The combination of these two initiatives offers many advantages of which this book tries to take advantage. This book attempts to merge these ideas into one process description.

3.6 NASA system engineering manual

NASA released a systems engineering manual from the JPL organization in 1992. It represented a great advance in NASA acceptance of the systems approach. NASA and its contractors have produced some very spectacular results but not always as a result of a clearly defined and respected systems approach.

3.7 Commercial and societal standards

The Electronic Industries Association (EIA) offers a system engineering standard numbered SYSB-1. It does not provide any how-to information but does cover the fundamental notions upon which a successful systems approach must be based.

International Electrical and Electronic Engineering (IEEE), at the time this book was written, was circulating standard P1220 on system engineering in preparation for release. Some of the people involved in MIL-STD-499B development also worked on this standard with the intent to satisfy the needs of the commercial market place.

The National Council on Systems Engineering (NCOSE) was a relatively new organization at the time this book was written, only having started in 1989. But, NCOSE will likely evolve as the accepted source of systems engineering process information. The membership, approaching 2000 in 1993, includes respected practitioners from many companies, campuses, and Government agencies who collectively certainly have the knowledge to provide a useful standard. At the time this book was written, the author and some other NCOSE members were discussing among themselves and with publishers the possibilities of publishing a system engineering manual under the auspices of NCOSE. Upon the death of MIL-STD-499B, the NCOSE Board undertook efforts to acquire the electronic media to continue work on that standard under the NCOSE banner.

3.8 International Organization For Standardization

The International Organization For Standardization, based in Switzerland, has developed a series of standards to which many commercial firms feel obligated to respond. While the series does not prescribe a specific system engineering process, the ISO 9000 series on quality does encourage that an enterprise have a written process description and that an enterprise follow their own procedures. ISO 9001 outlines the requirements for an effective product development and manufacturing capability. It addresses some very fundamental notions, but so many firms would find it difficult to prove their compliance with these simple requirements that few of us would argue with their inclusion in the standard.

3.9 DSMC Systems Engineering Management Guide

Several versions of the Defense System Management College (DSMC) Systems Engineering Management Guide have been published for use in a DSMC education program for future DoD program managers. The author has a preference for the original issue that grew out of a Lockheed Missiles and Space Company Space Systems Division systems engineering manual and a San Jose State University system engineering program under the leadership of Mr. Bernard Morais, now the President of Synergistic

Applications in Sunnyvale, California. The guide focuses on how to perform a particular necessary set of system engineering tasks and as such is not a bad model for a generic SEM. Much of the author's knowledge of the system engineering process was initially derived or improved from the content of this document.

3.10 Books in print and otherwise

Many good books have been published covering both the grand design of a systems approach and the detailed how-to methods appropriate to that process. The author has found not one of them that adequately covers the complete story in a way that we would all accept as the most desirable generic process. There are useful fragments in all of those listed below and some common themes accepted in all or most.

Blanchard, B.S., *System Engineering Management*, New York, John Wiley, 1991.
Blanchard, B.S. and Fabrycky, W.J., *Systems Engineering and Analysis*, 2nd Edition, Englewood Cliffs, Prentice Hall, 1990.
Chase, W.P., *Management of Systems Engineering*, New York, John Wiley, 1974.
Chestnut, H., *Systems Engineering Methods*, New York, John Wiley, 1967.
Grady, J.O., *System Integration*, Boca Raton, FL, CRC Press, 1994.
Grady, J.O., *System Requirements Analysis*, New York, McGraw-Hill, 1993.
Lacy, J.A., *Systems Engineering Management*, New York, McGraw-Hill, 1992.
Quade, E.S. and Boucher, W.I., *Systems Analysis and Policy Planning, Applications in Defense*, New York, Elsevier.
Rechtin, E., *Systems Architecting*, Englewood Cliffs, NJ, Prentice Hall, 1991.

3.11 Selection

All of these sources, and others, were studied as a prerequisite to selection of a generic process definition for use in this book and as a basis for continuing work. The author chose to follow the model encouraged by MIL-STD-499A blended with 499B content and the integrated management system process modified to allow the enterprise to evolve over time an excellent system engineering process through common process repetition and continuous process improvement based on incremental improvements stimulated through program experience with that common process. This approach encourages the enterprise to develop a system engineering identity respected by their customers. The author would encourage enterprises to extend this model to the other parts of their company beyond the system engineering function.

The attached SEM/SEMP includes appended to it a generic process diagram that the author created over a period of several years while an employee of General Dynamics Space Systems Division. This process description has been through many changes since the author departed GD to adapt to new knowledge and system engineering technology initiatives. This process diagram is believed to be consistent with any of the system engineering descriptions described in the documents discussed above.

chapter four

Organizational structures

4.1 Updating matrix management

In order to discuss system engineering planning, we must make some assumptions about the organization through which we will deal. After all, the object of our planning work will have to be executed by the organization. There are three principal organizational structures common in industry: (1) projectized, (2) functional, and (3) matrix. In the projectized organization, the personnel are assigned to each program which has hiring and termination decision-making authority. In the functional organization, all personnel are assigned to departments that specialize in particular kinds of work such as: engineering, production, quality assurance, etc. The matrix organization approach imperfectly attempts to marry the positive aspects of each of these approaches and avoid the negative aspects.

The author supports the matrix management arrangement for large companies with multiple programs. The matrix is characterized by: (1) functional departments led by a supervisory hierarchy, possibly including Chiefs, Managers, Directors, and Vice Presidents, which provide qualified personnel, tools, and standard procedures, to (2) project organizations responsible for blending these resources into a set of effective product-oriented cross-functional teams called integrated Product Development Teams (PDT) and managing these teams to achieve program success measured in customer terms.

The functional departments provide a pool of qualified specialists trained to apply department-approved standard best practices and tools in the development and production of products appropriate to the company's product line and customer base. The functional departments are charged with the responsibility to continuously improve the company's capability through small improvements in training, tools, and procedures based on lessons learned from program experiences and continuing study of available technology, tools, methods, company needs, and the capabilities of competitors. Company programs are internal customers of the functional departments. Note that this division of labor should result in a very lean functional management structure that is in tune with the times.

Program PDTs are organized about the product architecture reflected in the product architecture diagram overlaid by the program Work Breakdown Structure (WBS). The teams are selected and formed by program management. They are led by persons selected by program management or by the teams from the personnel assigned with the approval of program management. Once identified, the team leader is responsible for: (1) molding assigned personnel into an effective team, (2) concurrently developing the assigned product requirements, followed by (3) concurrently developing a responsive product design, test, manufacturing (tooling, material, facilitization, and production), operations and logistics, and quality concepts.

PDT personnel must first focus on team building matters in order to form an effective cross-functional group. When the PDT approach is first begun, there is a tendency to place an exaggerated emphasis on this step. You have to get through this phase as fast as possible to get your focus back on the work that these teams must do.

Next, the team must focus on product and process requirements analysis. The product of this work should be documented and approved by program management. Ideally, the teams would apply a structured, top-down requirements analysis process such as that described in the author's book *System Requirements Analysis*. The team must, concurrently with requirements work, develop alternative concepts responsive to the requirements as a way of validating and demonstrating understanding of the maturing requirements. Where there is no clear single solution, the team should trade off the relative merits of alternative concepts. The team must be very careful not to influence the requirements identification too strongly by concept work (that is, it must avoid leaping to familiar past point designs) but should take advantage of valid concept-driven requirements identified. This requires a delicate balance that should be assertively monitored by a systems PDT, called the Program Integration Team (PIT), and by program management at internal reviews. The requirements must also drive manufacturing, quality, logistics, material, test, and operational design work as well as product design.

Only after approval of the requirements and complying preferred concepts at an internal design review, and at appropriate customer and/or Independent Verification & Validation (IV&V) reviews, if required, teams should be authorized to proceed with concurrent product detailed design work and test, manufacturing, tooling, material, operations, logistics, and quality process design work. The team must implement all of the assigned tasks within budget and schedule constraints producing documentation that clearly defines what must be produced and how it shall be produced, tested, verified, and used to comply with the pre-defined requirements.

Functional management staffs programs with qualified personnel appropriate to the tasks identified on the project. Personnel must be assigned to programs with a reasonable degree of longevity because PDTs require personnel assignment stability to be effective. The reason for this is that the principal work of teams is communications, and communications networks are fractured by personnel turnovers and these fractures require time to heal.

Day-to-day leadership of personnel assigned to programs should be through the PDTs and their leaders rather than the functional departments. Personnel assigned to one team throughout a complete quarterly personnel evaluation cycle should receive an anonymous evaluation from their fellow team members coordinated by the team leader. Personnel assigned to more than one team in one evaluation period can receive evaluations from all teams within which they served. Each functional Chief should review all of the program team quarterly evaluations for persons from his/her department and integrate this data into department ranking and rating lists used as a basis for all administrative actions (training needs, compensation adjustments, promotion/status quo/setback decisions, and program assignment considerations).

Functional department supervisory personnel, and senior working personnel under their guidance, should monitor the performance of the personnel assigned to programs from their department and provide coaching and on-the-job training where warranted. Functional management also should provide programs with a source of project red team personnel used by the Program Manager to review the quality of program performance in the development of key products. Functional department Chiefs can then use this experience as a source of feedback on improvement needs appropriate to current standard procedures, tools, and training programs.

Functional Chiefs, Managers, and Directors should be encouraged to follow the performance of personnel from their department on programs, but should be forbidden to provide program work direction for those personnel. That is, functional management may provide help and advice for their program personnel in how to do their program tasks but should not direct them in terms of what tasks to do nor when to do them. Only the program-oriented PDT leaders and lower tier supervision, if any, should be allowed to direct the work of team members.

Functional management personnel, in this environment, are rewarded based on the aggregate performance of their personnel on programs (in terms of CDRL submittals, major review results, budget and schedule performance, and noteworthy personal efforts recognized by project management) and the condition and status of their department metrics, which may include depth, breadth, and quality of standard procedure coverage for the department charter, toolbox excellence, and personnel training program effectiveness.

Program budgets are assigned to PDTs by Program Finance. The PDT leaders are not only responsible for the technical product development tasks defined in our program planning documentation, but the team budget and schedule as well. PDT leaders report to the Program Manager. Teams are generally, but not necessarily always, led by engineering personnel during the early project phases (when the principal problem is product concept development and design) and later by production personnel (when the principal problem is factory oriented). Some companies have solved this leadership problem with product-oriented engineering-led development teams through first article inspection and factory- or facility-oriented production teams thereafter.

4.2 A model program organization structure

In our model of the perfect world, each project includes two or more PDTs and one PIT. Some people prefer the team name system engineering and integration (SEI) team in place of PIT and these readers can insert that term or any other in place of PIT. The author avoided the term SEI because it has a special application in large contracts where the customer contracts with a firm to perform system engineering work for them.

The PIT should be led by someone qualified to serve as the deputy Program Manager drawn from available senior personnel. During the early phases, this person should be someone with extensive engineering experience. In later program phases, this position should be reassigned to someone with extensive production experience. The senior engineering person assigned to the PIT in some companies would be called the Program Chief Engineer and that is the case in this book. This person is responsible for monitoring all engineering work on the program and coordinating changes in engineering personnel team assignments through the PIT and PDT Leaders. The Chief Engineer is the principal product technical decision-maker for a program. Other major functional departments will have persons assigned to the PIT with similar responsibilities. The Chief Engineer is the Engineering leader of the PIT and could be the PIT Leader as well in early program phases.

The PIT is responsible for technical direction and integration of the work products of the PDTs toward development of a complete product. This includes: (1) performance of initial system analysis and development of program level documentation (such as the system specification and other high level specifications, program plans, system architecture and interface block diagrams and dictionaries, and system level analyses); (2) mapping of the evolving system architecture to PDTs and formation and staffing of the teams, (3) review and approval of PDT requirements documentation; (4) granting authorization for PDTs to begin design work based on an approved set of requirements, concepts, schedules, and budget; (5) monitoring the development of interfaces between team items; and (6) development of interfaces between the complete product and external elements (system environment and associate contractor items).

The PIT, like the PDTs, reports to the Program Manager as illustrated in Figure 4-1. A project Business Team includes all of the project level administration functions such as scheduling, finance, configuration management, data management, personnel, program procedures, project level meeting management and facilitation, action item management, program reference document libraries, information systems services, and the program calendar of events.

PDTs organized at the system level may have to further decompose items for which they are responsible in order to reduce the problem space to workable proportions. In this event, the system level PDTs (those illustrated on the first tier of Figure 4-1 which report to the Program Manager) will create their own sub-teams, like PDT 2 in Figure 4-1, and acquire any additional

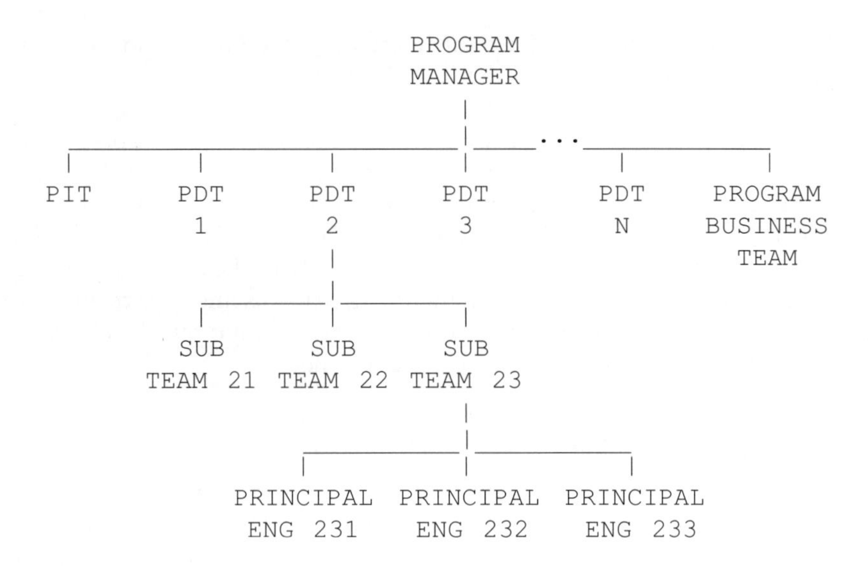

Figure 4-1 Program team structure.

personnel needed to accomplish sub-team tasks. Sub-teams may be fully staffed for independent work or rely heavily on the parent level team for specialists. Sub-team leaders are referred to either as PDT Leaders or Item Principal Engineers as a function of the magnitude of the task.

At the lower system levels, such as an on-board computer or valve assembly, the person assigned development responsibility is referred to as a Principal Engineer (as under sub-team 23 in Figure 4-1). The Principal Engineers draw upon the specialty engineers assigned to the parent PDT (or sub-team) for related specialty work. PDTs are responsible for integration of the work of all sub-teams created under their authority. The PIT has this same responsibility over all PDTs at the system level.

During the early development phases up through Functional Configuration Audit (FCA), the Chief Engineer should normally lead the PIT and be the principal product system level technical decision-maker subject to management by exception by the Program Manager. The PIT is the internal reviewing and approving authority for the system PDTs. PIT must review each system team's requirements and concept sets before the team is allowed to proceed to concurrent design.

Each PDT Leader is the principal technical decision-maker for the PDT and should review and approve subordinate team requirements, concepts, and designs. Each PDT is responsible for development of all of its internal interfaces and jointly responsible, with the team responsible for the opposing interface terminal, for all of their external interfaces. This responsibility extends from PIT at the system level down through PDTs, sub-teams, and principal engineers as a function of the scope of their responsibility.

Phantom PDTs can be formed for items to be developed by associate contractors and work accomplished with this team through what is commonly called an Interface Control Working Group (ICWG). The senior member of this team will be selected for their interface development and inter-personal skills. On small programs the ICWG may be a subordinate task of the PIT.

Teams or sub-teams may also be formed around major supplier items such as a rocket engine, guidance set, or other high-dollar or schedule-critical item. Leaders of these major supplier items may report to the Program Manager like other system level teams or be organized as sub-teams to a propulsion or guidance system level team. The Team Leader of a supplier team might be selected from procurement. These teams might be treated like principal engineer items as a function of the dollar amount, development complexity, schedule criticality, or program policy.

4.3 Physical collocation options

The preceding encourages a matrix structure with a very thin functional organization and a strong program organization through which the enterprise gets its contract work accomplished using the process, personnel, and tools provided by the functional organization. The understanding is that the personnel will be physically collocated in product teams in program work spaces. Many companies that have tried to implement this structure from a starting point as a matrix organization with a strong functional management structure have found it very difficult because it de-emphasizes the functional organization where the power resided at the time the change was attempted.

4.3.1 The virtual organization

William H. Davidow and Michael S. Malone in their book *The Virtual Corporation* published in paperback by Harper Business in 1992 sketched an interesting alternative to physical collocation of the people that may find adherents in the management of strong functional matrix organizations. A virtual team could be formed of members not necessarily physically collocated. The members would be connected via networked multi-media computers capable of simultaneous real time video, audio, and computer data. With this equipment, a team could virtually function as if it were physically collocated.

The author's first thought after reading Davidow and Malone was that people could cooperate as a team from their homes and between different business units using the virtual concept. But a company could conceivably function under the PDT concept while remaining collocated by function and gain some benefit from economies of scale in personnel assignment across several programs. This is an example of how you can become out of step with the advancing waves of technology and knowledge. Let us say that you conclude that you must de-emphasize your functional structure and team on programs. You succeed at this after a year of in-fighting with your functional managers. Then you find that you have the computer resources to enable your workers to cooperate via virtual teams from their original functional organizations.

Please read on before you conclude that this is the best alternative in organizing your personnel. This argument is only made to emphasize the point that you should be very careful, without being paralyzed, before committing to the latest management technology. The waves of change are striking the beach at a much faster pace than in times past.

4.3.2 *Virtual functional organization*

We might argue that if we can form virtual teams for programs, we could form virtual functional organizations as well using the same resources leaving the personnel physically collocated by program. This is true. The thin functional management structure could periodically gather with its personnel collocated by program on a network version of a functional department meeting. In these brief meetings, the program people could discuss their problems, new approaches to their common process, and ways to use their tools. These meetings would be an important part of the continuous process improvement plan. So, we have our choice between the virtual product team (people physically collocated by function working on product teams) or virtual functional departments (people physically collocated by program team cooperating about the functional axis via networked computers).

4.3.3 *Flexible optimum collocation policy*

Between these two choices, it would seem sensible to place the emphasis on virtual functional organizations and real product teams. The relationships are more complex between the product team personnel than they are between the members of the same functional department. Given that face-to-face communications is more effective than even the most effective virtual capability, we should choose to align the physical collocation with the most complex relationship. So the author's recommendation is to physically collocate where ever possible by product team on programs.

There will be particular specialties in every company and on every program that cannot be supported by a single team or even a single program. For example, it may be difficult to justify a full time reliability engineer, mass properties engineer, or nuclear survivability and vulnerability engineer on every team on a program. In these cases, the isolated specialists can effectively contribute to many teams via the virtual team concept and the company gain some cost benefit from the resultant economy of scale.

4.4 *Resistance to PDT*

There are many reasons why the organizational structure described here is very difficult to put in place in a company currently organized to perform autonomously within their functional organizations. These reasons include: (1) it disturbs existing organizational power relationships, (2) a conclusion that it is in conflict with the Department of Defense (DoD) cost/schedule control system (C/SCS) criteria defined in DoD Instruction 5000.2, Part 11, Section B, Attachment 1 and the work breakdown structure concepts defined in MIL-STD-881B, (3) PDT creates staffing problems for the functional organization, and (4) PDT interferes with an efficient personnel evaluation process. Let us look at each of these reasons.

4.4.1 Human resistance

Many of those currently in functional management roles will perceive this prescription for distributed power as a personal, professional, and career threat. If they remain in functional management, they will see their relative power in the organization decline as it must to implement concurrent development. This will commonly result in resistance from those now in functional management roles. This resistance is inevitable. There is no best way to make the transition other than to announce it as far in advance as possible and openly discuss opportunities for those currently in power under the new system. It is, of course, entirely possible that at least some of those in functional management should not be allowed to survive the change.

The long range view of this shift, which many will not be open to see, is that a shift to the kind of organizational structure described above will attract a different collection of people to functional management positions. This position should focus on providing programs with qualified people trained in the standard techniques defined for a function and skilled with the standard tools supportive of those methods. Personnel selection and training, continuous process improvement, and tool building are the principal roles of this new kind of functional manager.

Many of those now in functional management positions will not wish to continue working in the new environment. Some will prefer to migrate to program centers of power (program office, PDT leadership, or program functional leadership roles). This is not a disaster, though some may perceive it to be; it is simply re-balancing of the power centers. Those for whom power is a motivator will migrate to it wherever it resides.

Those who have worked in a matrix management structure for many years have observed many shifts in the balance of power between the functional organization and program organizations. These are not uncommonly driven by the personalities of those in power rather than by some sound rationale for improvement in the company's capability. The change we are discussing here requires an enlightened management whose first priority is to satisfy the company's customers and needs for company profitability. They must have a willingness to make changes toward these ends despite a possible temporary set back in attainment of personal goals. This is a lot to ask of perfectly normal human beings populating management positions and it will not often be satisfied in the real world. On the other hand, the change may be a necessary prerequisite to the company's survival in a very competitive world and this can be a very good motivator.

Obviously, movement into PDT requires a high level of support and leadership in an organization. Anything short of total commitment at the top will prolong the transition or destroy best efforts from any courageous and selfless people at the lower level. If your employees have endured a few aborted crusades in the past, it will require an even stronger attachment to your goal and some early success to build the perception of the inevitability of that goal.

4.4.2 C/SCS criteria conflict

4.4.2.1 What is the criteria?

The Department of Defense has prepared a criteria, called the cost/schedule control system (C/SCS) criteria, for accounting and reporting systems that contractors must support in order to be considered for award of a government contract. The C/SCS criteria listed in DoD Instruction 5000.2, Part 11, Section B, Attachment 1 (formally in DoD 7000.2) includes five criteria: (1) organization, (2) planning and budgeting, (3) accounting, (4) analysis, and (5) revisions and access to data.

Some companies following the matrix management model have concluded that there is a fundamental conflict between the C/SCS criteria and integrated product development. The conclusion is that they must relate all program cost in a matrix between their functional organization and the contract work breakdown structure (WBS), which is a hierarchical organization of all products and services covered by the contract. Given that they wish to manage programs using integrated product development teams focused on WBS items, they cannot give the leaders of these teams control of the budget allocated to those items because they feel the criteria requires them to allocate it to cost account managers oriented toward their functional organization. Without budget control, the PDT structure is doomed to failure.

If there is a conflict between these criteria and PDT, it resides in criteria 1 or 3 as a function of how we interpret the phrases "functional organizational elements", "organizational structure", and "organizational elements". None of these terms are specifically defined in DoD I 5000.2, Part 11, Section B, Attachment 1. Definitions for the terms "performing organization" and "responsible organization" do not refer to either the functional or program structures of a matrix organization only to a "...defined unit within the contractor's organization structure...." DoD does not require a contractor to organize under any particular model (matrix, projectized, or functionally). So, the criteria must be implementable by companies following any of the three kinds of organizational structures.

The DoD I 5000.2 criteria elements that relate to direct costs and organizational structures are listed below. Note that the numbering and lettering of these steps is different between DoD I 5000.2 and 7000.2 but the content is the same except that the contract WBS acronym used in 7000.2 is spelled out in the newer version.

"1. *Organization*
   b. Identify the internal organizational elements and the major subcontractors responsible for accomplishing the authorized work.

c. Provide for the integration of the contractor's planning, scheduling, budgeting, work authorization and cost accumulation systems with each other, the contract work breakdown structure, and the organizational structure.

e. Provide for the integration of the contract work breakdown structure with the contractor's functional organization structure in a manner that permits cost and schedule performance measurement for contract work breakdown structure and organizational elements.

3. *Accounting*

   c. Summarize direct costs from the cost accounts into the contractor's functional organizational elements without allocation of a single cost account to two or more organizational elements."

4.4.2.2 *Alternative approaches*

The premise is that there is a conflict between this criteria and the use of PDT, or what the DoD finance community calls a work team concept. Given that this is a true premise, there are several alternative arrangements we should consider in reaching a condition of compatibility between this criteria and PDT.

4.4.2.2.1 Status quo. Traditionally in a matrix organization, we allow the cost account manager (CAM), assigned from a functional perspective, to manage a functional department's budget and we hold him/her accountable for variances to a Program Manager responsible for one or more product-oriented budget segments, a role referred to in some companies (and in this book) as a WBS Manager. Total program cost is decomposed into intersections between functional departments and the WBS. Program Office WBS managers manage the WBS columns of the matrix and CAMs manage the organizational rows of the matrix. The PDT structure seldom overlays this matrix in total alignment with either matrix axis. PDT leaders are cut out of the budget management process and cannot possibly manage their team without budget control.

4.4.2.2.2 Change the criteria. The DoD finance community and many other customer organizations are wedded to the criteria till the end and maybe properly so. It was born out of contractor abuse and it will require decades of highly ethical behavior on the part of contractors to erase the stain of past excesses. Let us assume that we cannot change the criteria in any substantial way nor can we change it in any way in the near term.

4.4.2.2.3 Functional organization suppression. The criteria does not attempt to define what a functional organization is so we could suppress our functional organization in a matrix organization and establish a matrix between the WBS and PDTs on programs. Probably a suspicious government finance analyst would consider this a perversion of the criteria motivated by an attempt to avoid scrutiny in some way. Certification of our cost control system would be hard won. It would also make it difficult for functional management to get insight into their personnel needs since there would be no way for them to accumulate or project future demand for their department's services.

4.4.2.2.4 Three axis matrix. It appears we have three axes of interest in our data, PDT, WBS, and the functional organizations, and we are trying to manage with a two-dimensional matrix. So why do we not deploy an accounting system that can handle three axes or a mapping of budgets three ways instead of the present two? While it does not appear that this would be out of step with the criteria (we would have an organizational structure to which we could map budget), it would likely be hard to certify the system because of its unusual nature. This would also require some new development for the accounting system.

A variation on this theme could assign pseudo department numbers to PDTs on programs, but other systems interfacing via department numbers may be disturbed. This alternative has a lot going for it actually. This approach can be implemented by simply assigning a unique department number to each of the PDTs on a program. As the program comes alive, we transfer personnel from their home functional departments to the PDT departments. Each PDT department manager can function as a CAM, giving them the control we are seeking. The program office WBS Managers provide control from the WBS perspective.

4.4.2.2.5 Projectized organization. There is nothing in the criteria that requires us to organize in a matrix. Our company can be projectized, which would allow us to assign the organizational matrix axis to our program PDT structure. Such a company suffers from a difficulty in making continuous improvements because it has no one methodology — no central focus for its specialized functions. Some companies avoid this problem by splitting off each program into a separate business unit and allowing them all to develop their own identity and process.

4.4.2.2.6 Power to the WBS manager. It is not uncommon for the contract WBS to be under the authority of the finance community in government and contractor organizations. This results in customer finance people coloring technical system composition decisions by virtue of forcing a particular product organization before the engineering community has evolved the most cost-effective product organizational structure based on the need. As a result, the WBS may be at cross-purposes to the product organizational structure about which the contractor would like to establish PDTs.

Customer and contractor finance computing systems are not sufficiently flexible that they can tolerate adjustment of the WBS to align with a technically superior structure; so, the program moves forward with little interest in the WBS by the technical community. The WBS manager, therefore, can become a figurehead position in the program office divorced from direction of the technical effort focused only on cost and schedule concerns. Meanwhile, the PDT leader, focused on the technical aspects cannot control his/her team because he/she has no budget authority. No one person has the authority to manage integrated product development.

The right answer to this problem gives budget and schedule responsibility as well as technical responsibility to PDT leaders. The first step in assuring this capability is to place the responsibility for at least the product component of the WBS in the hands of the PIT. This is not to say that the finance community should be stripped of all voice in this matter. Rather, the decisions on WBS must be cross-functionally derived and not influenced totally by a finance position. Finance should be represented on the PDT and thus have a channel through which to communicate their valid concerns. This joint responsibility between contractor finance and the development team cannot be successful if the customer finance people stonewall and insist on a predetermined WBS.

In the process of making WBS decisions between alternatives, the PDT should weigh both financial and technical aspects of any particular choice but almost always choose the alternative that best enables PDT. This alternative will generally be characterized by minimizing or simplifying the cross-organizational interface intensity between elements under the development management by different PDTs.

With this arrangement in place, we can then proceed to merge cost, schedule, and technical control of PDT into one person. Let us accept that we need to retain the functional organization axis of our matrix organization in our cost accounting matrix. This means that we must use the WBS axis to provide for PDT leadership and control. Instead of making functionally oriented CAMs responsible for budget management, we associate with this job only a responsibility to collect costs for functional personnel planning and staffing decision-making and financial reporting. Since the product WBS is structured in accordance with the product functional and physical structure, we now have alignment between WBS and the basis of assigning PDT responsibilities to the organization of the product. The WBS manager is the obvious person to take on the role of PDT Leader with cost, schedule, and technical leadership.

Under this arrangement, functional organizations cannot be held accountable for cost variances. Since the WBS manager is in complete control, the WBS manager must be accountable for variances. Functional managers and directors should not be upbraided for failure to satisfy program schedules and cost targets. This is a WBS manager responsibility, a program responsibility. Functional managers should be held accountable for providing qualified people, good tools, good procedures, all three of these being mutually consistent, and the continuous improvement of them based on program lessons learned. This division of labor results in a very lean functional management staff.

4.4.3 PDT-stimulated personnel staffing problems

Some people hold that PDT creates staffing problems for the functional organization and this is true if in moving to PDT the functional organization is stripped of any knowledge of future budget availability. The functional organization needs this information in order to be able to ensure that the right number of people with the right kind of training and experience are available when the time arrives to use them on a program.

Some of the alternatives we considered above, frankly, deprive the functional organization of the information it needs to satisfy future program needs. The last one considered and recommended does preserve functional management access to this knowledge.

PDT can also result in PDT leaders contracting with the wrong functional organization for a particular kind of work. Some small companies may be able to handle the resultant volatility and it can even be a source of good restructuring ideas in a rough and tumble, but potentially effective fashion. Large companies will generally have difficulty with this and should have a clear definition of functional department charters with energetic enforcement of those charters by functional management when program management attempts to deviate.

The reason that these boundaries have to be jealously guarded is that you wish to hold the functional organization responsible for continuous process improvement concurrently with high standards of performance. A functional manager cannot be held accountable for maintaining a company's proficiency in a particular specialty if a different department is being contracted to do that work on programs. Suppression of deviations from the functional charter responsibilities is not necessarily, however, the best response in all case where this problem arises.

It is not uncommon that these occasions will be driven by a fundamental flaw in the way the company has assigned charters for specialties. So, when an incident occurs, functional and program management should first consider if there is some value in considering an alternative organization charter responsibility map. If there is not, then the functional perspective should almost always win out. At the same time, during periods when one or more disciplines are understaffed with respect to the demand for work, they may find it useful to establish temporary agreements with other departments to provide personnel to do their work. This same arrangement can be useful at other times as a means to create effective system engineers or simply to increase sensitivity to the needs and concerns of other disciplines while creating personnel with broader qualifications to improve flexibility in satisfying changing program needs.

4.4.4 Personnel evaluation problems

Now, we know that some people, the late Dr. Demming chief among them, are convinced that the devil himself/herself designed the personnel evaluation system used in much of industry. This system results in functional management ranking all department personnel based on the functional manager's perspective of relative worth. This ranking is then used as a basis for salary increases, promotion lists, and, to some extent, work assignments that can influence future evaluations by virtue of experience gained through assignments. It is not uncommon that the evaluation criteria are flawed and applied in an irrational or uninformed fashion besides. Commonly functional managers simply wish to get through this exercise as quickly and as painlessly as possible.

If you remove people from their functional organizations and physically co-locate them with PDTs, you make it difficult for functional managers to observe their performance. If we believe the current PDT/TQM, literature, we accept that performance evaluation should be done by team members and not by functional management anyway. So, how do we provide for personnel evaluation?

If we preserve the functional axis, we must accept that functional management must be responsible for evaluation because they have the responsibility to provide programs with qualified people. However, functional management need not be the only source of input for evaluation data. If an employee were assigned to one and only one PDT for a whole evaluation period (typically 6 months or a year), then the evaluation could conceivably be done by the PDT members in some fashion as Dr. Demming would suggest.

What happens when a person works fractionally on two or more teams over the evaluation period? It will not be uncommon for some specialists in traditionally low budget specialties to work like this but most will fit the pattern. We simply need some mechanism for the accumulation, merging, and distribution to functional department chiefs of team-derived evaluation data. Also, it would likely be useful to increase the frequency of evaluation events to perhaps quarterly with this data merged in some way into annual figures for pay and promotion determination.

This is a valid concern for entry into a PDT work environment, but we should have little difficulty working up a computerized approach to acquiring the data from people assigned to teams, collecting and assessing that data, and providing it to functional managers in a fashion that encourages rational decision-making in rewarding personnel for performance. It will require a change from present methods that generally are arbitrary, counterproductive, and otherwise just terrible. Good riddance!

4.5 Model matrix for this book

Figure 4-2 illustrates the overall organizational structure we will assume. One of several program panes is expanded with program reporting in the vertical dimension and functional reporting in the horizontal dimension. The Program Managers report to the company or division executive as do the top people in the functional organization on the left margin.

Each program has two or more integrated product development teams aligned with the product structure that is coordinated with the WBS. Each team is staffed by personnel derived by the program from the functional departments. In many companies, the functional Chiefs/Managers would control the tasking of the people staffing the intersections of the matrix. In

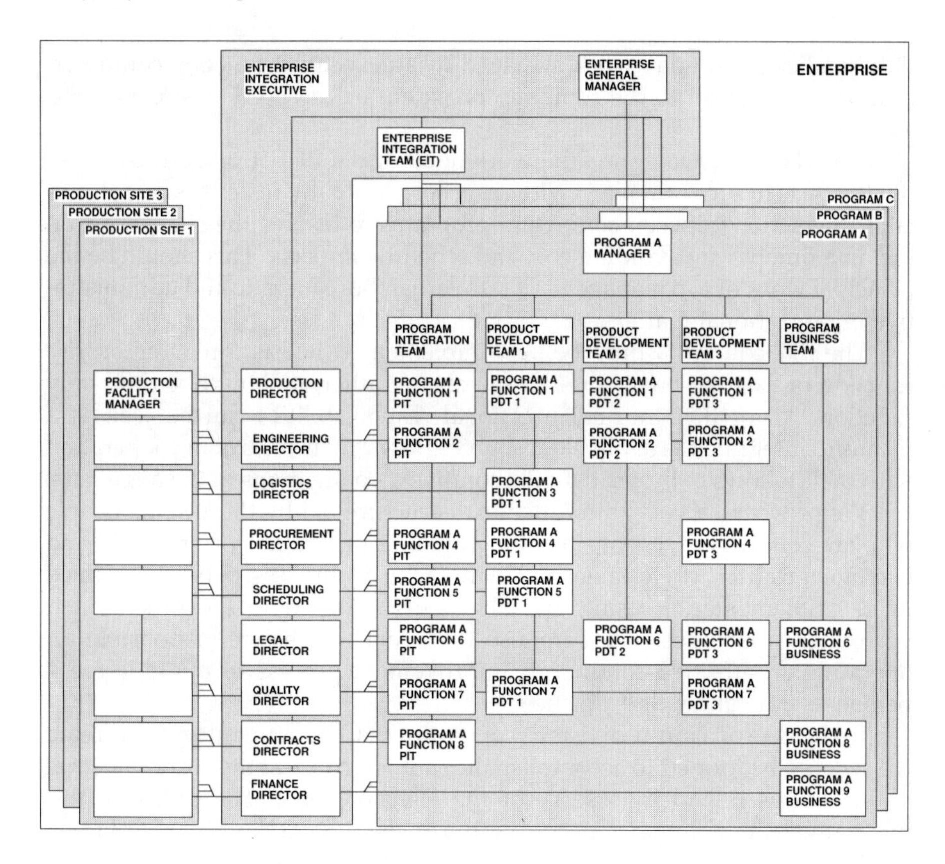

Figure 4-2 Model matrix organization.

our model, we will insist that the functional department Chiefs specifically not provide program task assignments or work direction. That must come through the program channel traceable to a PDT Leader also serving as a WBS Manager.

Each program pane also includes a PIT responsible for system level integration between the PDTs and a business team responsible for program administration, finance, and contracts.

4.6 *Enterprise integration team (EIT)*

If our company commonly has multiple programs running and a common set of resources, such as its manufacturing plant, company management should consider themselves an integration team the author calls an EIT. This team must balance the available resources to serve the greatest good for all programs in the aggregate. This team can be formed as a working level group that handles day-to-day work managed by an executive function composed of those who report to the company president or CEO or Division General Manager.

The EIT acts to integrate the work of all of the PIT. It proactively seeks out conflicts in program plans with respect to company resources and stimulates interaction between conflicting programs to resolve the source of conflict in a timely way avoiding cost and schedule impacts. This should be the principal work of top management rather than bureaucratic and administrative organizational matters.

The EIT concept would be best implemented not as a new enterprise function of existing board or staff members; rather, it should be their only function. All of their normal functional responsibilities can be seen as a means to integrate the available resources under their control in cooperation with their fellows to best satisfy the company's obligations in the aggregate.

We prescribe a very thin functional structure led by these same people who are collectively responsible for providing programs with a sound and common product development and manufacturing process, personnel trained in its implementation using available tools matched to the process and personnel capabilities. They are also the agents of change responsible for operating an effective continuous improvement process to upgrade tools, personnel education, and practices.

In the case of a multi-division corporation, a Corporate Integration Team (CIT) could be formed to accomplish the same purpose at the corporate level as the EIT is intended to serve at the Division level. The utility of this integrated team notion begins to run thin at some point, however. At higher levels of teaming you are dealing with higher levels of abstraction and generality. There is a danger that in seeking a common practice at a high level (say corporate) you may force lower level organizations (divisions, for example) into inefficient operation. While this book encourages a standard process, it does so for a single business entity based on that enterprise's situation in the market place.

So, the challenge for large corporations is to determine the optimum granularity of their division structure with respect to the product and customer base they seek to serve. As that optimum granularity is achieved, each business unit should be encouraged to optimize its identity for their plant and the market segment for which they are responsible. An efficient way to apply the CIT notion would be to make the CIT principally responsible for adjusting the granularity of the Division responsibilities determining when is the right time to consolidate or further decompose business units based on the current and anticipated business situation. This team should be very wary of trying to inflict common practices on its divisions, however. Divisions are formed because of differences and these differences must be respected where they can clearly be traced to a sound rationale.

chapter five

Integrated program planning

5.1 The ultimate requirement and program beginnings

The customer's ultimate requirement is their need, a simple sentence or paragraph that succinctly describes the needed product in terms of its effect on other systems, the environment, or both. This need may be stated by a DoD customer as a new threat that must be defeated, an ordnance transportation and/or delivery demand, or an undersea war fighting capability, for example. It may also be phrased by a commercial concern about their hoped-for customers. On the surface it would appear that the fundamental difference here is that the commercial company must understand its customer base and their needs well enough to know what will sell, whereas the military contractor is told what is needed by their customer in an RFP. Every healthy DoD contractor is in the same boat with the commercial companies actually and must be constantly examining marketing possibilities, developing new ideas that their customer base may find useful, and studying their customer's current situation and potential needs. Also, the DoD contractor increasingly must work to understand all of the stakeholders for a given system, possibly even extending to the population at large.

So, the phrase "customer need" need not scare away those interested in applying this process to commercial practices. In commercial business it just may be more difficult for the company to find out what the customer's needs are. Success may require good market research, good intuition, lucky guesses, rapid development, and a very fine pencil for cost and schedule control. Whether the system is of a military or commercial nature, no one need respond to the customer needs, however they become aware of them, with an ineffective, ad hoc, hit-and-miss process. The planning process included in this chapter can be applied to government or commercial development programs. The first step in this organized process is to understand clearly the customer's need.

In the structured, top-down development model encouraged in this book, this need is expanded into a system operational requirements document, system requirements document, or system specification through requirements analysis, system modeling and simulation, mission and operations analysis, environmental and environmental impact analyses, and logistics and basing analyses early in program phasing. In addition to the system requirements identified in these early phases, the system architecture is also defined based on the functionality needed to satisfy the need and it is used to fashion a product work breakdown structure (WBS), a hierarchical arrangement of product material and services cost elements. A Department of Defense (DoD) program will use the appropriate MIL-STD-881 appendix as a guide for the WBS but the evolving product architecture based on allocation of needed functionality should be used as the principal input for the WBS. Remember our concerns in this area expressed in Chapter 3. The customer's needed functionality must drive the system architecture, and therefore the WBS, and not the finance community and their inflexible cost-tracking computer implementations.

The program planning work must begin in the proposal period, or early marketing efforts in a commercial situation, and continue during program execution. Where a proposal is required, the management volume or section should reflect much of the content of the program plan. In some cases, the contractor will be required to provide a formal program plan, system engineering management plan (SEMP), or integrated master plan (IMP) with the proposal in draft, preliminary, or final form. In the commercial situation, a new product is a good reason to frame a new development plan or update an old one for new conditions, however simply it may be stated. Granted, there are some commercial product lines richly dependent on serendipity exposed through basic scientific research, an uninhibited creative mind, or good intuition where a degree of non-structure may be conducive to success. But, even in these situations, once the product possibility reaches the conscious mind, its development and distribution can fit into the pattern described here with some tailoring.

At the time this book was written the world was unwinding from several decades of fierce competition between the East and West involving an intense system development rivalry to develop the unstoppable weapon system and to counter the other side's systems. This competition energized a search for technology that sometimes resulted in new systems requiring a clear sheet of paper approach with several facets pushing the state-of-the-art simultaneously. The systems approach matured from this clean sheet of paper environment, but it can be applied effectively on programs entailing mild to massive re-engineering or modification of existing systems to accommodate new or changed conditions.

In these cases the system engineering products described in this book may or may not have been created when the system was new or they may have long since been lost or disposed of. So, even if a world class system engineering job had been done on the original system, the results of that work may not be available for use in the changing of that system. In these situations, what can be made clear is the boundaries of the system in terms of its architecture. We can then determine what new functionality is needed and which of the elements of the system architecture must be changed, deleted, or replaced to achieve the new goals set for the changed system. Once this is known, we can apply essentially the same system engineering process to the development of the affected items as we would apply to a new system. It may be a case of working from the middle out rather than from the top down, but it can be done in an orderly process following the tenets of this book.

At the time this book was written U.S. Air Force customers had for years been required to follow MIL-STD-499A in the implementation of a system engineering program. This standard required contractors to develop a SEMP telling how they intended to perform the technical and management activities required by the contract. Some contractors wrote these plans as extensions of the proposal in the same sales language used there rather than as a plan for their own guidance in program execution. This pattern of behavior has been a mistake. Whether a customer requires a management plan or not, your company needs one and it should be written by your company for your company based on a marriage of your customer's needs and your capabilities. This is a fairly complex integration process, linking customer needs and contractor capabilities, composed of many specialized component parts.

In this book we will encourage the use of an expansion of a U.S. Air Force initiative referred to as integrated management system. It is a variation of the planning process described in MIL-STD-499B (rejected by Department of Defense at the time this book was written) that encourages the development of an Integrated Master Plan rather than a SEMP. The reader interested in commercial markets should not become disinterested because we are going to use a U.S. Air Force model. The fact is that the Air Force has really seized on an excellent planning and management model applicable to most any development situation.

To complete the picture from the contractor's perspective, we will glue onto the Air Force model an internal contractor generic planning capability and continuous process improvement module. We will also merge the planning requirements of the integrated management system.

In applying the resultant planning system, the contractor or commercial company first must understand themselves, their capabilities, and their best practices leading to good customer product value. Next, for each program, they must accomplish a transform between this knowledge of their capability and an expression of the program plan in terms of the customer's need. Then this plan must be well executed. Finally, you must take advantage of lessons learned from each program execution to continuously improve your generic self. This provides an integrated environment within which to attain and maintain a world class capability in your product line. And system integration plays a central role in realizing this goal because we must accomplish our work through the integration of the work products of many specialized people both in the program planning and execution phases.

5.2 Program plan tree

We would all doubtless accept that a specification tree is necessary on a large program to introduce order into the product requirements development effort. What is not as universally accepted is a similar tree for program plans that capture the requirements for the development and production process. All too often programs are implemented allowing planning documentation to be autonomously prepared by the several functional departments (engineering, manufacturing, finance, etc.) contributing work to a program. These plans may be generic company plans or procedure manuals applied to the program or specifically written for the program. Many companies become caught in the trap of trying to be totally responsive to every customer's initially stated process requirements to the extreme that they redesign themselves for each customer in terms of these plans. A future-looking company will apply a continuous process improvement concept to their generic procedures in combination with a rigorous customer procedures tailoring effort on each program to take advantage of the practice-practice-practice notion that world class athletes use on the road to greatness.

Autonomous, functional department planning, disconnected from program requirements, is the target in this chapter. Program process plans should be architected just as the product systems requirements and components should be. Plans should exhibit traceability from the top level plan down through the lower tier plans. A structured, top-down planning process will ensure mutual consistency of all of the program plans with a minimum of surprises during program execution. In the process of preparing program plans, it should not be necessary to come up with a new program design for each proposal and program. It should not be necessary to re-design your company for each customer. We need to find out how to apply the practice-practice-practice technique to our work through careful program planning for specific programs and apply continuous process improvement to our methods with a long term view. Company personnel should be applying the same proven process, incrementally improved in time, to each program and in the process become expert in their specialized disciplines.

Very little of the business that a company tries to gain through proposal or marketing efforts involves radically new initiatives. Most of our energy is applied to prospects that are close to our historical product line. Therefore, most of a company's procedures and plans should apply in any new program. This is especially true if your company already does apply an energetic continuous process improvement program.

Given that we are organized in a matrix structure, our functional departments should have procedures covering how they perform their function on programs as members of cross-functional teams. The program must knit the functional methods into a coherent process appropriate for the particular product system under development as appropriate to the development phase. In this process, programs should not be allowed to substitute alternative processes creatively without acceptance by functional management because the functional departments should be deploying the very best methods they have developed over time based on continuous improvements fed by lessons learned from prior program experiences.

There are, however, two sound reasons for permitting programs to deviate from the current best practices. First, it is through program implementation that improvements can be developed and tested. A particular program may be asked to experiment with a particular technique in defining interfaces, for example. Perhaps the company's history is to use schematic block diagrams and the proposition is that program XYZ will use N-square diagrams instead. The second reason is that the customer may have a valid need for a job to be done differently than our current best practices cover. Perhaps the company uses a particular computer program and related procedures to capture logistics support analysis data. The customer may have a big investment in capturing the data in a different computer data structure and have a perfectly valid reason why they need to collect different data than your system will address. You will simply have to adjust your practices to the customer's in this case. In the process of doing so, you may find improvements that can be woven into your preferred practice. But, the suggestion is that this should be the exception and the rule should be to follow internal procedures while incrementally improving them.

Given that we all accept that programs should be conducted in accordance with prepared plans for each activity, what plans are needed? At the top of this set of program plans rests the program plan. The program plan can be very simple giving the overall schedule in very broad terms, the customer's need, ground rules and policy, top level program organization and responsibilities, and reference to other documentation. Figure 5-1 is based on the integrated managed system but includes a SEMP and subordinate plans to satisfy the IMP requirement for narrative material covering certain planning areas. The Figure 5-1 plan tree will satisfy MIL-STD-499A requirements as well as the integrated management approach. The indicated plans provide requirements for the process that will result in the product system. They should be mutually consistent and this can be demonstrated by establishing traceability between the plans in the patterns suggested by the plan tree.

These plans, in combination with program schedules, tell the humans populating the program what to do, when to do it, and who should do it. Where do these plans come from? What are the right plans to write? Who should prepare them? As mentioned above, we can allow our functional departments to autonomously develop program plans and then try to fit them together during program execution. This is a bottom up or grass roots approach to planning. Alternatively, we could approach program implementation planning as we should approach product system development — systematically from the top down. The remainder of the chapter focuses on this approach.

The fundamental difference between past planning processes and integrated management system is that the integrated management system defines the work to be accomplished very rigorously linked to the program statement of work and WBS by a common work identification coding system. The functional plans become detailed narrative descriptions of how the work defined in the IMP will be accomplished. Also, the integrated management system calls for an integrated master schedule (IMS) rather than the system engineering master schedule (SEMS) called for in MIL-STD-499B. These are

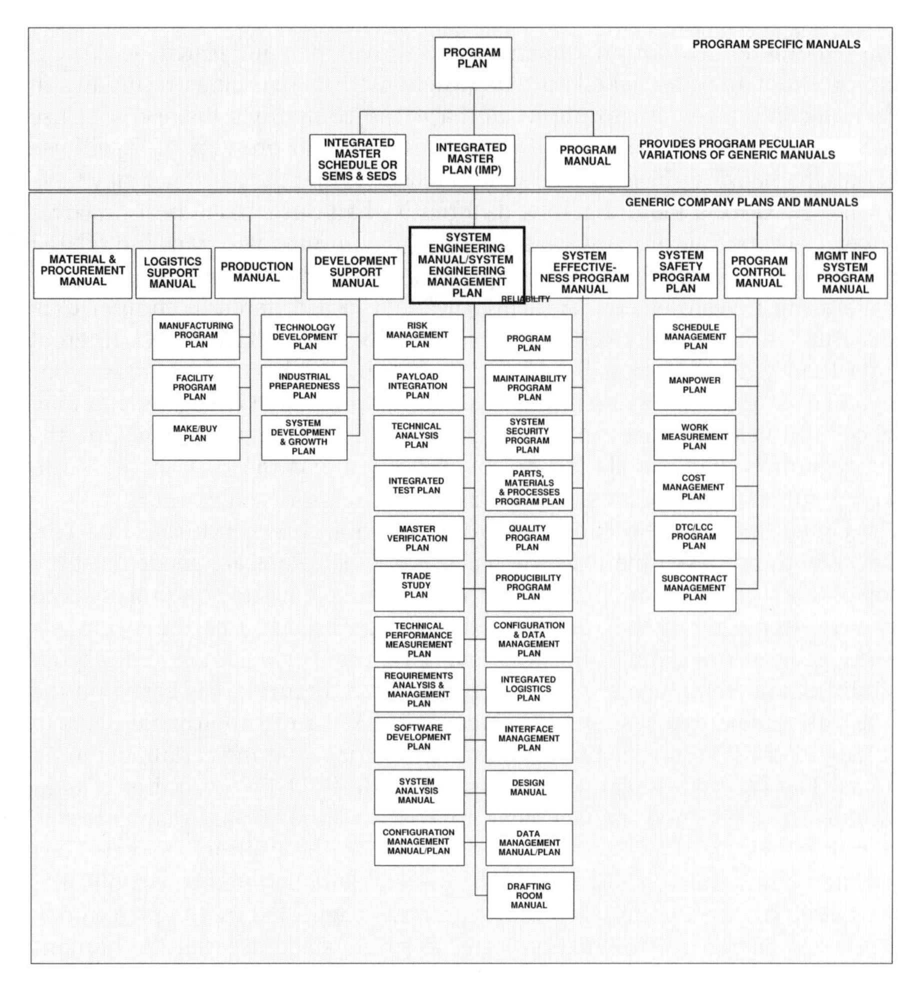

Figure 5-1 Program plan tree example.

essentially the same except in name. The advance made by the integrated management system is to absolutely link all work defined in the IMP with the schedules in the IMS using a common work identification system. The remainder of the chapter focuses on the integrated management system approach.

If you are forced by a DoD customer's requirements to plan a program using the pure MIL-STD-499A approach excluding the integrated management system, the IMP is simply removed from Figure 5-1 and the tree collapses to the program plan and the IMS becomes the SEMS. You may conclude from this that the IMP is an extraneous layer of planning, but you are encouraged to finish the chapter before making that conclusion a permanent part of your belief system.

In the integrated management system approach, you would not normally prepare all of the plans subordinate to the IMP. The content of all of those plans could be incorporated as part of the IMP in narrative form. They are separated here for two reasons. First, a company may have to respond strictly to the MIL-STD-499A requirements, so the book should offer people in that situation this alternative. Second, the principal approach offered in this chapter recognizes a planning approach that makes best use of the limited time available during proposals (by using the generic planning data as narratives referenced in the IMP); focuses proposal work on how these data can be blended into a program-specific IMP; and encourages company use of standard procedures, incrementally improved over time, that provide customers best value. You cannot develop the latter unless you are allowed to practice-practice-practice. If you re-design your company for each new contract, you will never realize a single identity and your work force will not be able to benefit from repetition.

Regardless whether your common customer base can be expected to call MIL-STD-499A in your RFPs, you are encouraged to develop the planning approach offered in this book. You should, of course, sound out your current and potential customers on this planning approach and get their feedback on acceptability. Generally, you will find them to be delighted with the thoroughness of your approach.

Please understand that the intent is that all of the generic plans subordinate to the IMP in Figure 5-1 are intended to be available as a function of ongoing functional department planning that defines generically how the various tasks in department charters are to be accomplished. On a given proposal and subsequent program work, that basis should already be in place. The proposal work should focus on fashioning a program plan, IMP, and IMS appropriate to the customer's needs drawing on your standard planning data as narratives referenced in the IMP. If you have ten programs in house, each of the ten programs would have its own program plan and IMP but each IMP would reference the same set of functional manuals that tell the people on the programs how to do their tasks.

It is true that some customers will be uncomfortable with your generic planning in that you can change it without their approval after a contract has been awarded. Some customers will want review authority on your internal plans and that can be a nightmare when you have multiple customers with very different interests. In such cases you may have to run a copy of a generic plan and re-identify it for specific use on a program. This program-specific copy could thereafter be changed with customer review without causing chaos in your internal documentation. This is a slippery slope that will force you to deviate to some extent from the practice-practice-practice notion. If all of your customers take this tack then you may not have achieved very much with a "standard" process. You must understand your customer base and ensure that your standard process will not offend them. You cannot construct your generic process in a vacuum.

The party line in your proposals and in conversations with your customers should be that you are motivated to continuously develop generic procedures that will provide all of your customers with the best possible value. In cases where the needs to satisfy two or more customers are in conflict, you need to consult with those customers as part of the process improvement activity. You should also offer your customers access to your generic internal planning data on the basis that it is proprietary. Let them act as one of the pathways through which you can detect incompatibility using management by exception. One excellent way to do this is to provide internal read-only access to all planning data via computer network throughout your facility and extend this access to your customer base either at their facilities or only at your own.

In any case, the specific work required for each customer's program would be clearly defined in their program-unique and customer-approved program plan, WBS, SOW, IMP, and IMS. These documents would reference the generic planning data that only tells how to do these tasks. The principal obstacle in implementing this approach is, of course, that everyone has not done a fine job of documenting their practices. In the author's opinion, this is not a valid basis for rejection of the offered planning method, nor does it represent an impossible barrier. It means that you must begin developing these generic practices now and keep improving them over time. Your competition may yet give you time to improve your performance for they are very likely in every bit as much difficulty as you are in this respect.

In the case where a company is organized in a projectized fashion, or only deals with a single product line, you still have to acquire your personnel from the same source as everyone else, from the human race. We are all knowledge limited and are forced to specialize. So, even if you have no matrix and only a project structure in a company with one or more projects, you will have to accomplish program work using specialists whose work patterns should be standardized, one program to another. This standardization can be captured in the kinds of documents shown in Figure 5-1 under the IMP.

5.3 Know thyself through generic program planning data

In the ideal situation, our company will have been involved in a continuous process improvement program for some time and we will have developed a generic set of good practices coordinated with our tools, personnel knowledge, and skills base. This data should include: (1) a functional department charter listing all of the tasks for which each functional department is responsible for maintaining and improving company technology and capability, (2) a generic process flow diagram that hooks these tasks into relative time, (3) a task planning sheet for each charter task (to be explained later), and (4) an integrated company System Engineering Manual and generic System Engineering Management Plan (SEM/SEMP).

Attached as part of this book you will find a SEM/SEMP model which includes in appended data a generic process flow diagram (some parts of which appear in the body of this book) and an accompanying coordinated company charter task list. The generic process flow diagram provides a framework into which all of the charter task descriptions fit as pieces into a puzzle. In this model, the process diagram is a simple process flow diagram. It could be expressed as a PERT, CPM, or other network diagram; V diagram; generic schedule diagram; or Gantt chart. The time axis could be in some nominal time measure or the complete period of execution equated to 100%. We can picture this diagram as a rubber sheet being stretched or compressed to satisfy a particular customer's needs with some paths being deleted, others possibly added, as a function of the contract.

Figure 5-2 offers an example of a generic planning sheet for one of the tasks illustrated on the flow diagram and listed in the task list. It happens to be a reliability task. Much of this information will have to be changed as it is applied to a specific program, in particular the intensity of task application, but it is helpful to have this information as a beginning point. If you have these sheets available as a generic input to the program planning process, you can save time and introduce your company's standards for process quality into the program planning process commonly accomplished during the proposal development period. If you have no enduring generic planning data now, these forms could be filled in initially as part of your next proposal task estimate and later iteratively improved upon in each subsequent proposal process using the incremental continuous process improvement notion.

Some of the charter tasks expressed on the generic process diagram will entail personnel from only one department working alone to complete them. Many tasks will require cooperative effort on the part of personnel from two or more departments as a function of how you are functionally organized and the business in which you are involved. Therefore, the atomic structure of generic task definition includes a task number and a department (or specialty) number. The generic task described in Figure 5-2 would be referred to as 1051822312-834. This identification number appears quite long due to the number of process layers included, but it is consistent with the generic task definition data in the appended SEM/SEMP.

With generic planning data in hand, we can apply this data in a tinker toy or erector set fashion to particular customer needs expressed in their request for proposal. What we need is an efficient transform process between generic planning data and specific program planning data. If you look at the planning data prepared in many companies, it includes a mix of functional department and product-oriented data. Our transform process must map the generic planning data into a program and product context understandable to our customers and valuable to us as a basis for managing the program through product-oriented integrated product development teams.

We have assumed throughout this discussion that the company is organized into a matrix structure. You may not be organized in a matrix, but you will still need to focus on the many specialties the technology base associated with your product line demands. The advantage of the matrix is that the functional organization can provide process continuity and a continuous improvement focus while the program organization runs the program. The

FUNCT. TASK ID	1051822312
DEPARTMENT	834, RAM Engineering
TASK NAME	Reliability Allocations (Task 202)
PROCEDURE REF	Company Procedure 42-8; MIL-STD-756B (tailored); *Reliability Engineering*, ARINC, Prentice Hall, Chapter 6.
OBJECTIVE	Define a quantitative reliability figure for each system item
DESCRIPTION	The analyst studies the planned design concepts provided by the PDT designer(s) and allocates the system or lower tier failure rate to subelements from top to bottom in each architectural layer and branch based on available historical data or predictions determined by the item composition.
SIGNIFICANT ACCOMPLISHMENT	Reliability figures assigned to all system items to a depth defined for the program.
COMPLETION CRITERIA	The task is complete when every configuration item, procured component, and other items as defined on the program have been assigned a reliability figure and the complete set of data has been checked for internal consistency, customer requirements compliance, and been approved by the PIT.
EVENTS	Model structure complete at SDR, allocations loaded by PDR, and predictions loaded by CDR.
RESOURCES	
TOOLS	Five IBM compatible computers each running RAMEASY, Microsoft Word 5.0, and Microsoft Excel for first two quarters. Thereafter, two machines adequate.
PERSONNEL	At peak work load, 6 experienced reliability analysts skilled in failure rate prediction and reliability math and familiar with equipment used in missile systems.
FACILITIES	Each analyst requires a standard company engineer's workstation. Personnel will be assigned to PDT with 2 assigned to PIT and four to product teams (1, 3, 4, and 5), one per team during the first 2 quarters of the program. Thereafter, two analysts on PIT only.
INPUTS	1. Design concept information from item designer 2. System architecture definition
OUTPUTS	1. Reliability Model complete with a quantitative reliability number for each configuration item and procured item. Data users include: PDT as a source of reliability requirements; availability analyst as a source of reliability data for computing availability

BUDGET EST PER QUARTER	1	2	3	4	5	6	7	8	9
HEADS	2	5	5	5	2	2	2	2	1
MAN-HOURS	1008	2520	2520	2520	1008	1008	1008	1008	504

Figure 5-2 Sample functional task definition form.

danger occurs when the functional organization is allowed to enter the program work supervision path.

5.4 Integrated management system overview

During the late 1980s and early 1990s the U.S. Air Force, working with several contractors on several major programs including the F-22 Fighter Aircraft Program, evolved an extremely important acquisition management methodology called integrated management system. One of the beautiful aspects of this Air Force integrated planning initiative is that it strips away a buildup of past confusion and shines a brilliant light on the important things in the program planning process. The fundamental notion is that we should first understand the requirements for the product system and then apply a structured approach to design a program that will produce a system compliant with those product requirements. The program design is captured in a system specification, work breakdown structure dictionary, statement of work, a list of deliverable data, a set of program plans, and program scheduling data. The content of the plans is requirements for performance of the process that will produce the product defined by a customer system specification. There should be no work planned or performed on a program that does not contribute in a perfectly clear way to satisfying the customer's product needs expressed in their system specification. This shockingly simple concept is, sadly, not always realized in practice.

This methodology recognizes six fundamental documents that collectively contain the requirements for the product system and the program through which the product will be created. Figure 5-3 illustrates the traceability relationship between these documents from the ultimate requirement, the need. The system specification should be driven by the customer's need and it should recognize a particular top level architecture that is expanded into a WBS dictionary for the purpose of structuring the program for management purposes. A statement of work (SOW) defines the work that must be accomplished for the product system and each WBS element of the system. The required work must then be planned in an IMP that fits the work elements into a framework of major program events. Finally, the planned work hooked to major program events is scheduled in time in the IMS. The system specification, SOW, and IMP all represent the top ends of trees of specifications, statements of work (including supplier SOWs), and plans, respectively.

Some large system customers, like DoD, use a device called a contract data requirements list (CDRL) to define their requirements for formal delivery of data about the product and the process for creating it. This document lists each item of data that must be delivered and tells what format it must be in (referencing a data item description, or DID), when it must be delivered, to whom it must be delivered, and how many copies are required. These items are commonly tied to paragraphs in the statement of work and WBS

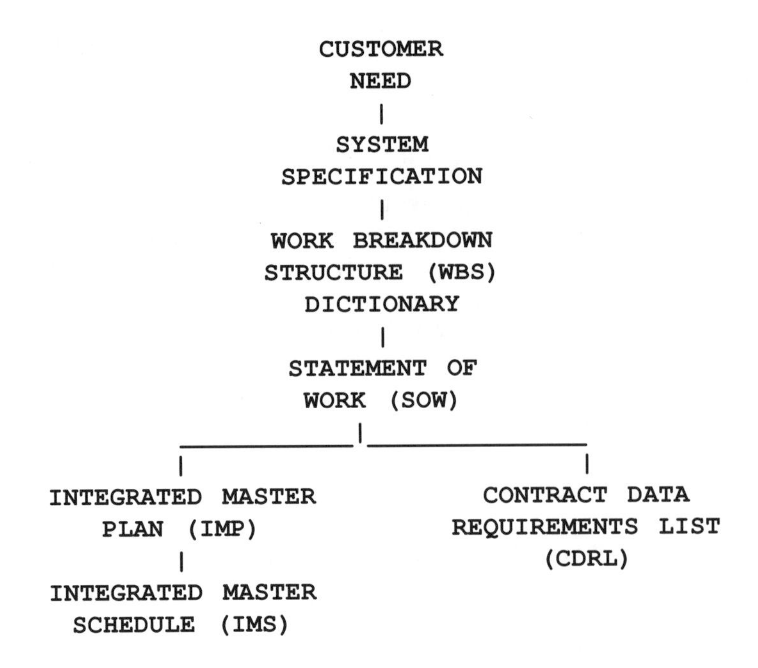

Figure 5-3 Program planning document stream.

numbers for management purposes. Some of these information products will be part of the delivered product, such as technical orders that customer personnel will use to understand how to operate and maintain the product system. Other data items, such as cost schedule control system reports, meeting minutes, and schedule updates, will be focused on reporting progress in the product system development and manufacturing effort.

5.5 Generating the six primary documents

5.5.1 The system specification

Depending on the program phase we are discussing, the system specification may be available to us or may not yet exist. Preparing the system specification may be one of the program tasks if the contract involves a very early program phase. More likely, the contractor will receive with a request for proposal some kind of requirements document: such as a system requirements document, operational requirements document, or draft system specification from the customer. This input will have to be completed or transformed into a system specification to the mutual satisfaction of the customer and contractor. Commonly, this transform happens during the proposal process with a system requirements review (SRR) scheduled very early in the program execution. Refer to the book *System Requirements Analysis* by the same author for a description of a process that will result in a quality system specification.

It often happens that the customer's original need has become lost by the time they are ready to let a contract for the development of their system. Sometimes a need statement is never phrased. But even when it is prepared, it is not uncommon for it to have been so thoroughly digested in the process of developing the system requirements that it simply passes from view. Some people would maintain that once you have the system requirements defined, you do not need one anyway. The author disagrees. The need is the ultimate requirement and, if it truly represents the customer's need, it offers useful guidance throughout the development process.

If you cannot find a need statement in materials provided by the customer, you should ask them for it. If they cannot or will not produce one, write one and get customer acceptance. This may be very difficult, especially where the customer has many faces including one or more users and a procurement agency. But because it is difficult, you should not conclude that it is not needed. Quite the contrary, the harder it is to phrase and gain acceptance of the customer's need, the greater the need to press on to a conclusion. A lot of good work will result that will eliminate many false starts during later program execution. This same policy should be pursued in the definition of system requirements — the harder it is to gain acceptance, the harder you should work to understand the customer's needs and achieve agreement.

All of the requirements in all of the lower tier product requirements documents should be traceable from and to this document. All of the requirements in the system specification should be traceable to the customer's need through a logical process of functional decomposition, allocation, and requirements analysis.

5.5.2 The WBS dictionary

The WBS has, in the past in the DoD market, all too often been crafted by finance people in industry and government based on one of several MIL-STD-881 appendices reflecting different kinds of systems. Even though MIL-STD-881 encourages that its appendices be used as a guide only, the mindless way it has been applied has often had a chilling effect on the application of a sound system functional decomposition approach to the development of product systems. Despite the evolution of a conflicting functional system architecture on such a program, the inflexible computer cost management tools used by the customer and contractor finance communities have often inhibited alignment of the WBS to the needed functionally derived architecture. Under the Air Force initiative, the government has not shown complete flexibility on WBS adjustment, but the climate is more favorable than in years past.

The WBS dictionary provides a hierarchical organization of product material and services needed to satisfy the customer's need. The WBS must span the complete system allowing everything in the system to be placed in some WBS category. The content of the WBS must reflect what is needed to

Missile System
- Air Vehicle
  - Propulsion (Stages 1...n, as required)
  - Payload
  - Airframe
  - Reentry System
  - Guidance & Control Equipment
  - Ordnance Initiation Set
  - Airborne Test Equipment
  - Airborne Training Equipment
  - Auxiliary Equipment
  - Integration, Assembly, Test, and Checkout
- Command and Launch Equipment
- System Engineering/Management
- Systems Test and Evaluation
- Training
- Data
- Peculiar Support Equipment
- Operational/Site Activation
- Industrial Facilities
- Initial Spares and Repair Parts

Figure 5-4 Example of a MIL-STD-881 work breakdown structure.

satisfy product system functionality and should not be chosen rigidly and arbitrarily based on some financial model. Figure 5-4 lists the partial content of Appendix C of MIL-STD-881B for missile systems. Only the Air Vehicle, a 2nd level WBS element, is expanded to the 3rd level.

MIL-STD-881 does not prescribe a numerical coding system for the structure, but one is always assigned. For example, in Figure 5-4 the Air Vehicle might be assigned WBS 1000 and the other 2nd level items 2000, 3000, and so on. WBS 0000 would be assigned to the whole system in this case. If the system included an Air Vehicle and other things, like a launch site and a final assembly factory, WBS 0000 would be assigned to the complete system and each of these elements, sometimes called segments, would be assigned thousand level WBS codes. Some programs are so complex that prefixes are assigned to permit cost accumulations in different useful patterns. The WBS 012-1000 might be assigned for non-recurring development of the Air Vehicle while 014-1000 is assigned to recurring manufacturing of the Air Vehicle. Costs can be accumulated in WBS 1000 across all WBS prefixes for the complete Air Vehicle cost and within prefix 012 for all development cost.

Lower tier WBS identification is accomplished by using different numbers in the hundreds and/or tens place of the four digit WBS code. For example, WBS 1100 might be assigned to the first stage propulsion system, 1200 to the second stage propulsion system, and 1210 to the rocket engine of the second stage propulsion system. We have chosen a four digit code so far but there is nothing to prevent the selection of a five or six digit code. In a very complex system, a decimal expansion could be added, such as 1210.05

for an engine turbopump. In this fashion, the WBS can identify everything in the product system in a very organized fashion to any level of indenture desired. It is not necessary to apply a unique WBS number to each product item throughout the system hierarchy, however. The WBS is a management tool and should give managers insight into the system structure. The WBS is actually (or should be) an overlay on the functionally derived system architecture which should be expanded to the component item level (valves, black boxes, mechanical assemblies, etc.). Not all of this detail is needed for management of the system development.

5.5.3 The statement of work

The next step in the programmatic requirements development process is to determine what work must be performed to develop, design, manufacture, test, and deploy every product element depicted in the WBS. Every bit of work we perform on a development contract should be included in the SOW at some level of detail and should be traceable to the product requirements in the system specification. The SOW, therefore, will tell what work must be accomplished for each product WBS element at some level of indenture. Only then can the program work be said to flow from the system requirements.

In the past, the SOW has often been prepared by someone in the customer's program office copying customer-created boilerplate SOW material from a similar past program into the new program SOW. The integrated system management approach calls for the contractor to write the SOW based on a WBS derived from the same work that produced the definition of the system contained in system specification. The contractor must decide what work must be done within the context of his plant, personnel base, and product history in order to create a product system that satisfies the provided requirements and organize the work around the breakdown in the WBS.

Two competing contractors may very well offer the customer two very different SOWs with their proposals because they have different plants, product histories and experiences, and personnel mixes. Neither may necessarily be better than the other, only different for these reasons.

If we have properly identified program work, it should be possible to establish traceability between the paragraphs of the SOW and the paragraphs of the system specification. Where DID CMAN 80008A is called by a DoD customer to define the system specification format, the product-oriented SOW paragraphs can be traced to paragraphs under system specification paragraph 3.7, which captures requirements for major items in the system architecture or WBS. Paragraphs under 3.7 should have been initially conceived from an orderly functional decomposition of the customer need and the WBS developed from that same analytical process.

We can picture all of the supplier statements of work strung out from the system SOW in a tree structure branching from the SOW element of Figure 5-3. Work traceability should exist between the process requirements we accept in the applicable documents referenced in the statement of work with our customer and the process requirements we lay upon our suppliers in procurement SOWs. Otherwise, some elements of the complete product delivered by us (containing supplier elements) may not be compliant with our customer's requirements. The Air Force planning initiative brings into clear focus the need for traceability not only through the specification tree but through the SOW tree as well and between a SOW and its companion specification. This means that the complete product and process definition for a system can be unfolded from the customer's need in a structured top-down development effort.

Commonly, the system level SOW is written to cover all of the work in several WBS indentures. We prepare supplier statements of work for major suppliers but the system SOW commonly provides the only work definition coverage for the prime contractor. The product development team (PDT), or concurrent engineering, paradigm suggests an interesting alternative to this arrangement. We could prepare the system level SOW to cover only the system level work under the responsibility of the PIT and write internal SOWs for each major system element identified in the WBS and to which a PDT will be assigned. These internal SOWs would each define the work that must be accomplished by one of the teams. The product element would be covered by a specification defining the product requirements and the team SOW would define the process or work requirements that must be accomplished to satisfy the corresponding product requirements.

This arrangement results in product requirements and work definition documents aligned perfectly with the product system elements and the development responsibility definition and should result in great precision in management of the program. In order to be successful in this approach, you have to have thoroughly studied the customer's need and requirements and decomposed their need into a stable architecture that can be assigned to product teams. Instability in this whole structure can result in a tremendous amount of parasitic work.

The SOW, as the name implies, identifies work that must be accomplished at a high level to provide the items identified in the WBS. Now, how do we identify or describe the work that must be done in detail? A three-step process is suggested.

The first step in transforming generic planning data into the specific program plan for a given program is to map the generic tasks to the WBS. Figure 5-5 illustrates this process. It can be accomplished in a top-down or bottom-up fashion and be accomplished at any level of management desired either by the program team or functional management. Two alternatives for these three planning process components are listed below under headings in the form Development Direction/Planning Level/Planner and the reader can imagine other possible combinations:

a. Bottom-Up/Department Chief Level/Chief — Each functional department Chief, or their representative, lists under each WBS their department tasks that must be performed.

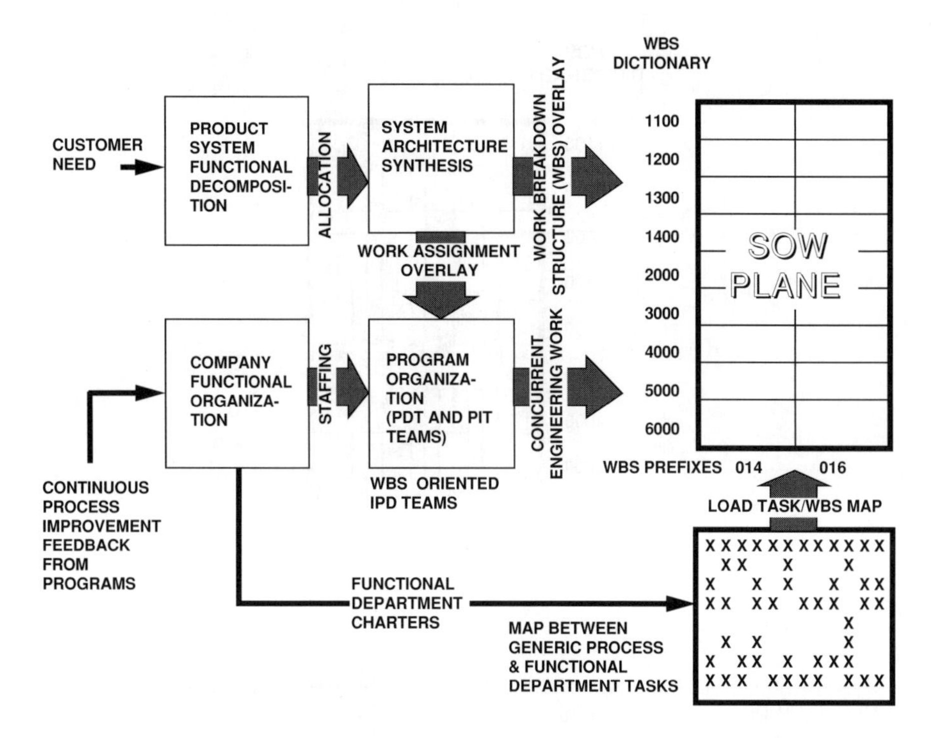

Figure 5-5 Initial SOW task loading.

b. Top-Down/Upper Management Level/Program Planning Team — A program planning team, composed of people from functional departments, accomplishes the map between the functional department charters and the WBS. The planners shop the functional charters for tasks that satisfy the customer's needs expressed by the WBS dictionary.

Step 2 is to organize the functional task inputs into a program context within the WBS-driven SOW hierarchy. In the author's preferred scenario, no matter how we orchestrated the first step above, the program should form an integrated planning team composed of people from the functional departments assigned to work on the proposal or program initiation work. Ideally, these same people will later be assigned to work on the actual program from those departments.

The planning work for each WBS should be led by the person who will be responsible during program execution for that WBS, if possible. In the case of product-oriented WBS elements, at some level of indenture this lead person should become a PDT leader. The planning team members for a given WBS must study the functional charter task sheets (see the sample in Figure 5-2) assigned to their WBS and organize, integrate, or synthesize them into a specific set of major program tasks that must be accomplished to satisfy the work element goal or purpose. The aggregate results of this work forms the SOW plane of the IMP Work Space shown in Figure 5-6. We might conclude,

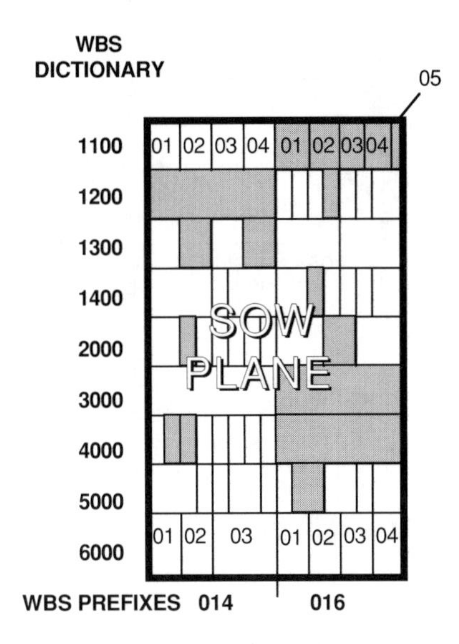

Figure 5-6 The SOW task plane.

as a result of this planning exercise, that all of the work that must be done for WBS 014-1200 (prefixes not necessarily used in a specific situation) can be achieved within the context of major program SOW tasks 014-1200-01 through 014-1200-07, 016-1200-01, and 016-1200-02, for example.

Some of the detailed functional task sheet inputs may be gold plated, even unnecessary empire-building attempts. Some may be understated or even missing. They must be assembled into the fabric of the program by people experienced in the company's product line, organizational structure, and functional department charters and the needs of the customer. This is an application of ALL CROSS integration seeking out a condition of minimized completeness.

We have to assure ourselves that we have covered all of the necessary work and have not introduced any unnecessary work. For each planning sheet offered, we have to ask, "What would happen if we did not do this work or did it at a lower level of intensity?" For each WBS element we have to assemble the complete stack of planning sheets (physically or in our computer) and ask, "Have we covered everything that needs to be done to satisfy this work element?" There is, of course, no substitute for successful program planning experience in this integration process no matter what kind of whiz-bang computer system you are fortunate enough to have.

Prior to kicking off this SOW task planning exercise, the WBS should be well developed and made available to everyone involved in the work planning exercise. It is a great source of confusion when the WBS is being changed twice a day in the middle of the work planning activity. Of course, the WBS may have to change based on new knowledge developed in this process, but the more stability that can be introduced into the WBS prior to beginning the more detailed SOW planning work the better.

Very little of the planning information shown in Figure 5-2 need be included in the SOW, but it is needed for the overall planning process including budget determination (cost estimating), detailed program planning, and program facilitization, staffing, and equipping. Planning sheets like that shown in Figure 5-2 received from the detailed task planners form the atomic structure of the detailed planning activity that will result in the IMP and IMS. These inputs must first be organized within the WBS, as discussed above, and woven into an understandable presentation in the SOW. Figure 5-7 includes a sample SOW fragment framed in the integrated system management style organized not by paragraph numbers but by an expanding work ID number formed from the WBS prefix (if used), WBS number, and SOW Task number.

These SOW tasks are fairly generic in nature and are appropriate for a wide range of programs. Note that in the two alternative mapping processes we discussed above you could have used a top down list like this created by the program prior to the functional department loading of the SOW tasks. The alternative is that you allow the functional departments to load their tasks into the WBS numbers and integrate the result to gain insight into the SOW tasks by combining and unifying. The author prefers the former method, but no matter how you gained insight into them, they are the program tasks to which we have mapped all of the functional charter tasks.

Twenty different departments may have mapped a charter task to SOW task 014-1200-01. A system engineering department could have indicated they will coordinate the requirements development process for the item/team. Fifteen different specialty engineering departments (reliability, maintainability, mass properties, design-to-cost, etc.) may have indicated they must develop a particular requirement for the item. Functional charter task 105182312-834, displayed in Figure 5-2, could be one of these. A manufacturing engineering organization and logistics department will both cooperate in development of requirements that are consistent with manufacturing and system operation needs. A test and evaluation department will have to define verification requirements. And finally, an operations analysis department may have to perform a special simulation to define the most cost effective guidance accuracy allocation for the item.

In the process of combining all of these functional department charter tasks into SOW task 014-1200-01, we have also defined the specialists who must be brought together in a team environment to concurrently develop the requirements for the item. The WBS leader/PDT leader who performed the task integration process will have derived knowledge of his/her team composition as a function of the planing experience. Table 5-1 illustrates a fragment of a SOW task responsibility matrix that we could now begin to build. The matrix will be useful in controlling program implementation.

Each SOW task is identified by number and name (using a subset of those identified in Figure 5-7) followed by the task status (STAT) column, PDT responsibilities definition (left blank because we discuss it next), and needed specialized functional department participation. Given that we have

014	Non-recurring work.
014-1200	Non-recurring upper stage development work.
014-1200-01	Define upper stage requirements. Conduct analyses to define appropriate requirements for the upper stage based on system requirements and the evolving system concept.
014-1200-02	Review and approve upper stage requirements. Subject the requirements to formal review and approval prior to design authorization.
014-1200-03	Design upper stage. Develop a preliminary and detailed design fully compliant with item requirements.
014-1200-04	Provide design test and analysis support. Perform analyses and tests for the purpose of validating the adequacy of the design concept.
014-1200-05	Review and approve design. Subject the design in preliminary and detailed stages of maturity to a formal review by qualified specialists and approve same with possible re-direction.
014-1200-06	Verify design by analysis. Analyze upper stage design for compliance with requirements.
014-1200-07	Verify design by test. Conduct testing to verify that the design complies with item requirements.
014-1200-08	Identify material sources. Define the source of each item of material needed to manufacture the item (make/buy).
014-1200-09	Develop manufacturing process. Define facilitization, tooling and test equipment, and personnel needs.
014-1200-10	Develop inspection process. Define inspection steps necessary in coordination with manufacturing process definition.
016-1200	Recurring upper stage work.
016-1200-01	Acquire and distribute material. Execute material acquisition plans, stock received material, and make available to the production process.
016-1200-02	Manufacture, assemble, and test upper stage. Perform all necessary steps to assemble materials defined in engineering drawings in accordance with manufacturing planning and inspection data. Test upper stage in accordance with approved written instructions for the purpose of assuring item readiness for acceptance by the customer.
016-1200-03	Deliver upper stage. Package and ship item to launch site.

Figure 5-7 Sample SOW fragment.

progressed in the program to the point that requirements are complete and approved, the "C" in the STAT column would convey that condition. The "A" in the design row would indicate the team may perform design work and supporting analysis and test work.

5.5.4 *Integrated master plan and schedule*

Now that we know how the product system and development process will be organized (WBS) and what work must be done (SOW content expanded

Table 5-1 Task Responsibility Matrix Fragment

SOW Task	Task name	Stat	Team			Functional dept						
			1	2	3	A	B	C	D	E	F	G
014-1200-01	Define upper stage requirements	C	X					X	X		X	
014-1200-02	Review and approve upper stage requirements	C	X				X					
014-1200-03	Design upper stage	A	X				X		X	X	X	X
014-1200-04	Provide design test and analysis support	A	X									
014-1200-05	Review and approve design		X				X					
014-1200-06	Verify design by analysis		X					X		X		
014-1200-07	Verify design by test		X				X					
014-1200-08	Identify material sources		X						X			
014-1200-09	Develop manufacturing process		X								X	
014-1200-10	Develop inspection process		X									X

by the detailed planning sheets mapped to the SOW tasks), it is necessary to determine who will be responsible (not in the functional organization sense but within the program PDT structure), how it will be done, and when it will be done. We could establish a tree of plans to define these things like that illustrated in Figure 5-1, topped by a program plan flowing down to a SEMP and sub-plans such as: manufacturing plan, procurement plan, quality assurance plan, etc.

The Air Force integrated management system initiative calls for an IMP and that is the pattern we will follow here. This does not mean that a program team should overlook all the other plans shown on Figure 5-1. Many of those are still appropriate as IMP sub-plans or integral IMP narrative content. The IMP can replace the program plan and SEMP but may not provide the detail needed by manufacturing, quality, and many other functions. The integrated management system initiative recognizes that all IMP content does not fit neatly into the highly organized structure explained here. It also calls for narrative sections where appropriate. The content of the functional lower tier plans, if prepared, can provide the narrative data encouraged in the IMP.

The author proposes tying the MIL-STD-499A prescription together with the integrated management system initiative by recognizing that the SEMP is essentially the system engineering narrative of the IMP. Other functional plans can have the same relationship to other IMP narratives. The author's idea of the ideal situation is that the contractors should have a set of functional department plans or manuals that describe how their chapter tasks are to be accomplished. One of these plans would be a generic SEM/SEMP that would tell how the company accomplishes the system engineering process on all contracts. If the company deals with the military as a principal part of its business, then its generic SEMP should map to MIL-STD-499A content with possible tailoring clearly defined in any contract calling for a SEMP. The appended SEM/SEMP is intended to fill precisely this need.

For a given contract/program, the company would write a WBS dictionary/SOW/IMP/IMS document set and in the IMP, where the customer requires narrative materials, refer to the lower tier functional plans for detailed coverage. This arrangement allows company personnel to follow the same practices on all programs developing and maintaining skills in a single process. Continuous process improvement based on program lessons learned can then be applied to a sound base to move toward the very best possible capability constrained by available resources.

The IMP expands on the SOW telling how and by whom the work will be accomplished and provides a means to determine successful work completion. For each SOW task, at some level of WBS indenture, the IMP must contain planning data that identifies a series of major events through which the development of that product WBS will be accomplished. These events are selected with effective program management in mind. We do not want so many events that we are spending all of our time reporting. Nor do we want so few that we know not what is going on. This is like selecting test points in a system design that allow rapid identification, isolation, and correction of problems. Too many test points cost money unnecessarily and are tedious to use in isolating faults while too few test points lead to ambiguity in fault isolation. These same extremes apply to program health monitoring systems for management.

5.5.4.1 Program events definition

The principal event discriminator will be defined by a customer or by your marketing department's determination of a potential date that commercial competitors might be ready to market a competing product. In DoD this is called an initial operating capability (IOC) marked by a specific date in the future. This date, or its commercial equivalent, defines a point in time when the program must have completed testing and have produced enough product articles to equip the customer for what they have defined as IOC. This might correspond to two aircraft squadrons, one armored battalion equipped, or production of the first day's assembly line run of new cars.

Your scheduling experts will then have to fit all program activity in between the program beginning time and IOC with some manufacturing and logistics events extending beyond IOC. A generic process diagram, drawn on a rubber sheet in our imagination, can be useful as noted in Figure 5-8. The major events on the generic diagram can be pushed and pulled into alignment with the realities uncovered by the scheduling experts.

Depending on the customer, they will insist on a particular stream of major reviews, possibly those included in Table 5-2. Other major events, or milestones, unique to your product line can be added based on the kind of product, maturity of the product development process, and degree of intensity with which customer or company management wishes to manage the program. The list may include only a single program phase or recognize the full sweep of the development, deployment, and operational steps in the development portion of the system life cycle.

The list of program events in time must now be applied to the IMP space as shown on Figure 5-9. The rubber sheet generic process diagram now can be stretched and compressed to conform with the program-specific events and be used as an aid in transforming the rough SOW task planning results into the fine structure provided by IMP task planning.

All of the SOW tasks now need to be laid into this time axis. The team responsible for each SOW task should expand the planning detail for their activities by linking up each SOW task with the events. This breaks up each SOW task into a number of segments like the sample illustrated in Figure 5-9. Some of the periods of time between events will correspond to voids for some SOW tasks as indicated by shading in Figure 5-9, meaning no work is planned for those periods and tasks.

5.5.4.2 Final work definition steps

The work components formed by mapping generic functional tasks into the WBS dictionary, partitioning and combining the work mapped to specific WBS numbers into major SOW tasks, and the time delimitation of those SOW tasks into events provide us with manageable increments of work that are

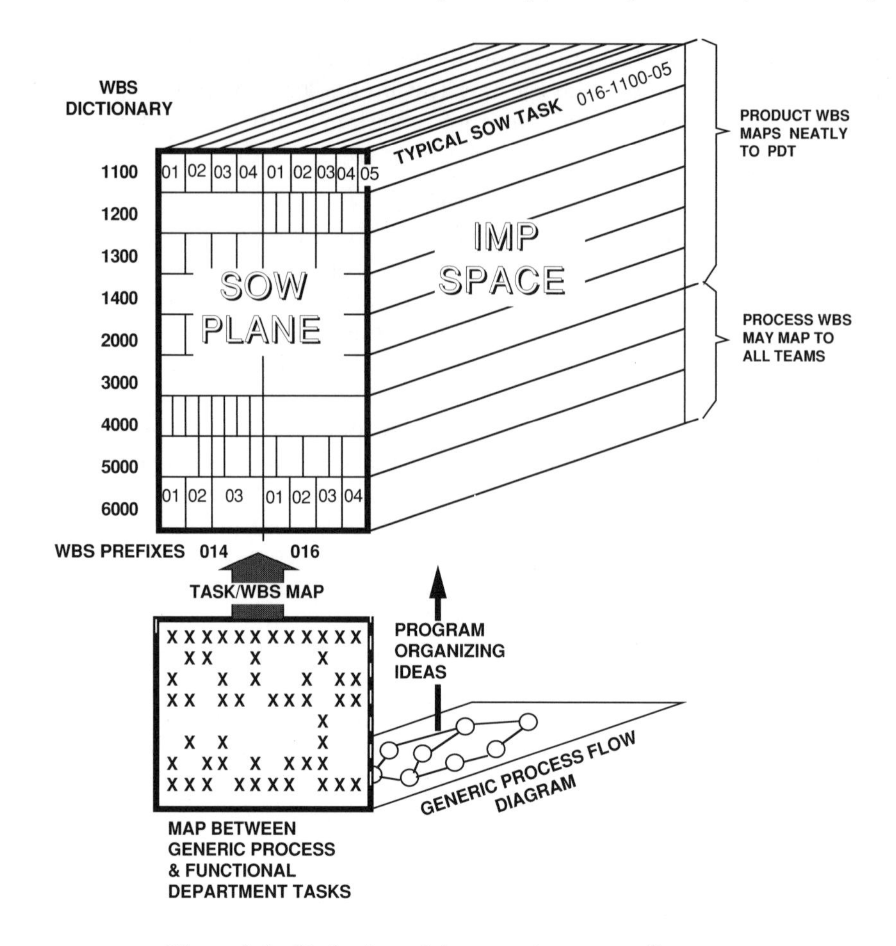

Figure 5-8 Projection of the generic process diagram.

associated with specific WBS numbers, most of which will map to PDTs in an unambiguous fashion if we have organized the teams properly to reflect the product structure. To enhance the management potential for these tasks, the integrated system management initiative calls upon the planner to identify one or more significant accomplishments for each of these IMP tasks and for each accomplishment to identify a measurable accomplishment criteria through which we can objectively determine when the accomplishment can be claimed.

Figure 5-10 illustrates these final IMP task definition parameters. An IMP task is a work increment between two events within a given SOW task. It is composed of one or more sub-tasks each of which is characterized by a significant accomplishment statement. Each significant accomplishment must have identified for it one or more unambiguous accomplishment criteria.

Table 5-3 illustrates one simple string in this planning process using a dash delimited numbering system. Paragraph 3.7.1.2 of the system specification calls for a space launch vehicle upper stage to accomplish some previously defined functionality. WBS 014-1200 was selected to identify upper

Table 5-2 Major Program Event List

ID	Acro	Event name	Event description
01	SRR	System Requirements Review	Joint understanding of requirements reached.
02	SDR	System Design Review	System requirements validated with risks identified and mitigated to an acceptable level.
03	IRR	Item Requirements Review	Approval of item requirements as a prerequisite to item design work.
04	PDR	Item Preliminary Design Review	Acceptance of readiness to undertake detailed design.
05	PDR	System Preliminary Design Review	All item PDR complete and acceptance of detailed design entry.
06	CDR	Item Critical Design Review	Design acceptance and authority to initiate manufacturing of test articles or low rate production.
07	CDR	System Critical Design Review	All item CDRs complete.
08	FCA	Functional Configuration Audit	Proof that design will satisfy requirements.
09	PCA	Physical Configuration Audit	Proof that product will comply with design, quality, and manufacturing planning.
10	IOC	Initial Operating Capability	Enough resources produced and delivered to provide a complete operating capability at some service level.

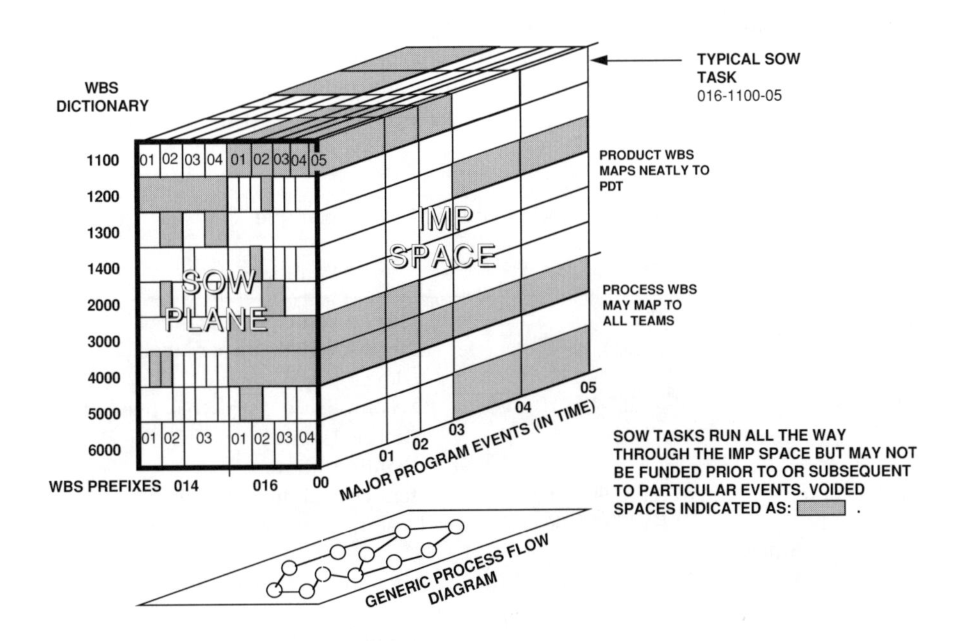

Figure 5-9 IMP space partitioning by event.

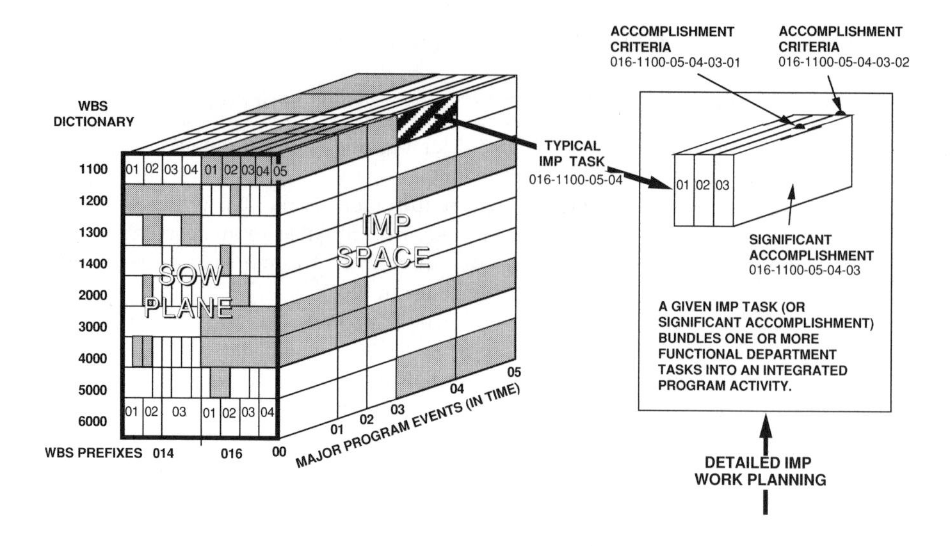

Figure 5-10 IMP task definition.

stage development work. The SOW has expanded on WBS 014-1200 to include several work tasks needed to satisfy the WBS 014-1200 requirement. Based on an analysis of the evolving IMS (or SEMS), we selected completion of design as a critical event through which to manage the program such that we could review design progress before we committed to manufacture of the flight test vehicles. Traditionally in the DoD environment, this event is called the Critical Design Review (CDR). We will accept that the launch vehicle design is complete if 95% of all planned engineering drawings have been formally released at that time. Scheduling, with the concurrence of program management, determines that 20 January 1994 would be the ideal time at which to hold the CDR.

Table 5-3 Single Planning String Example

Document	Reference	Planning content
System spec	3.7.1.2	Launch vehicle upper stage
WBS	014-1200	Launch vehicle upper stage development
SOW task	014-1200-01	Accomplish design work needed to define the upper stage configuration.
IMP event	014-1200-01-06	Item critical design review
Significant accomplishment	014-1200-01-06-01	Design complete
Accomplishment criteria	014-1200-01-06-01-01	95% of drawings released
IMS Event/date	014-1200-01-06	20 January 1994

This is only one planning string. It may require thousands of these branching strings to complete the whole planning database for a large program. It quickly becomes obvious that computer technology could be helpful just to capture and organize all of this information. At the time this book was being written, several of the computer tool companies with requirements database tools were beginning to see a potential market for the application of these tools to programmatic as well as product requirements development.

All of these documents have a similar structure of a paragraph, WBS, SOW, or IMP task number; title; and text. These tools could be used to capture the specifications, WBS dictionary, SOW, and plans and include traceability across the current valleys between them as well as generate the documents, or parts thereof, on demand.

When you have linked up all of the work in strings like the one illustrated in Table 5-3, using manual or computerized methods, you should have a mutually consistent network of management data that provides implementation guidance for people doing the work and management test points for effective program health monitoring.

5.5.4.3 *Final IMS development*

As we are developing the IMP content we do not obligate ourselves to define scheduling constraints, only to identify the work that must be done. Once the work is defined, scheduling must arrange the tasks corresponding to the significant accomplishments in time, constrained by the events previously located in time. In the process, we will find that some things cannot be accomplished as planned in the available time, requiring adjustments in event timing and span times between events. This may require several iterations before all of the inconsistencies shake out and the complete set of planning data is ready for use.

5.5.4.4 *Planning process summary*

Figure 5-11 combines all of the steps described over the past few pages into a single diagram to better illustrate the intended continuity between the steps outlined. Our approach has been to associate generic functional department work descriptions with WBS numbers defined for the particular product system derived from an understanding of the needed system functionality.

These functional tasks were then combined by the responsible teams into coherent SOW tasks oriented about the product WBS structure. We next determined some major milestones to use in managing program progress and projected these events into a space created by expanding the SOW plane into the time dimension. Finally, we further organized the work content of each IMP task defined by a SOW task and a terminal event into a series of specific significant accomplishments each with pre-defined accomplishment criteria.

5.5.5 *Contract data requirements list (CDRL)*

The CDRL identifies every item of data the customer expects to have formally delivered. This should be driven by the work that will be accomplished

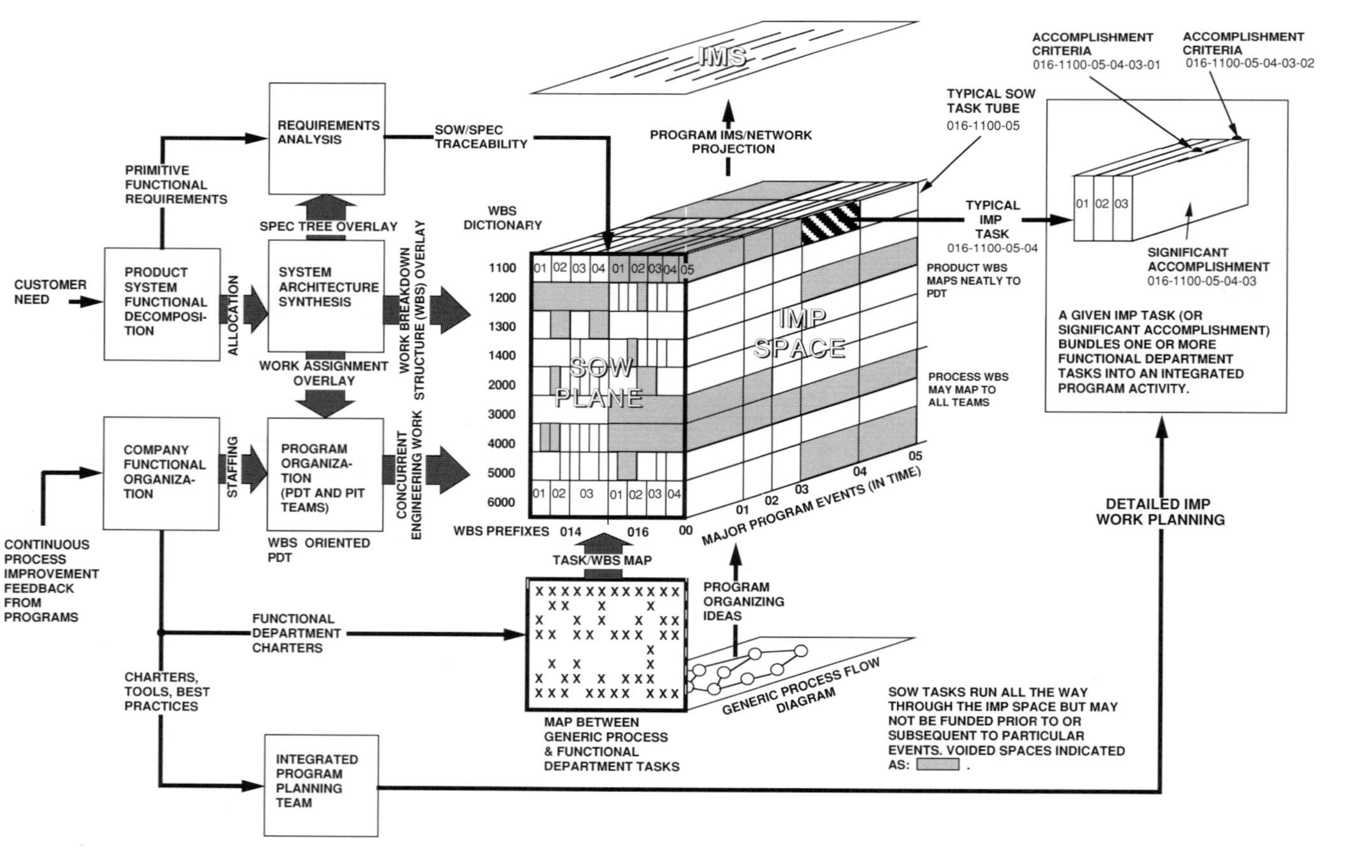

Figure 5-11 Overall program planning scenario.

as defined by the SOW. Every task listed in the SOW that will produce a data item of value to the customer in managing the program or understanding the technical or administrative flow of events should be referenced there in terms of its CDRL identification and all of these items collected to form the CDRL. The CDRL items should be determined from the SOW tasks and, once identified, should be fed back into the SOW for reference.

Customers sometimes mindlessly boilerplate the CDRL as they do the SOW, rather than deciding in an organized fashion what, of the information that the contractor must produce to develop the product and manage the development, they need for the same purposes. Customers realize that formally delivered data costs contractors more than data generated only for internal consumption so there is a naturally limiting cost boundary at work. The customer should be able to acquire data developed on the program anyway whether the item is on the CDRL or not. This data follows the contractor's format and content definition rather than the customer's, as in a CDRL, and requires a special request on the customer's part to obtain it.

The contractor should maintain an organized list of all program data and make it accessible to all contractor team members in a program library. A DoD customer calls this list (exclusive of CDRL items) a data accession list (DAL). This list names all of the data the customer can acquire by specifically requesting it and paying a separate fee under the terms of the contract. The contractor benefits by having this same data available for internal use rather than it becoming lost in desk drawers and waste baskets.

The CDRL must include schedule and financial reporting data that allows the customer to determine the health of the program. In addition, the customer should wish to receive technical performance measurement (TPM) data that offers insight into how the technical development is progressing in terms of meeting a small number of key system requirements. Together these three items inform the customer about evolving program cost, program schedule, and product performance.

The customer will normally wish to make the system specification, WBS, SOW, CDRL, and IMP contractual documents requiring a formal process (contract or engineering change proposal) to change. The IMS, however, should be a CDRL item because it will have to change as a function of showing progress, if for no other reason. All other CDRL items should be selected with care from the work identified in the SOW to provide the customer insight, at reasonable cost, into the evolving product requirements and design synthesis and into the product development, testing, and production processes.

Ideally, the DAL and CDRL items should be in electronic media stored on a computer network server and accessible from all work stations in the contractor's facility. The contractor can then also quickly respond to a customer request to electronically deliver any item listed in the DAL and in the program library. Very little data is produced today by means not involving a computer. Therefore, most items that a customer would be interested in receiving as a CDRL exist in electronic media within the contractor's business.

Yet many customers continue to insist on delivery of tons of paper documents that will fill ever-expanding storage space. Alert customers will begin to require delivery of CDRL items in electronic media. This will change the nature of CDRL delivery. Instead of periodic delivery of paper copies of updated documents, the contractor will be required to refresh a customer database at a particular interval or at particular milestones.

The DAL should exist as an electronic data delivery conduit from the program on-line library to a customer. A customer could be charged a periodic rate plus a fee for each call to gain access to this library section. There are some fine precedents for this kind of service in the form of on-line databases open to public access like CompuServe. This kind of automated process would result in important savings for large customers like the government. This arrangement will require special provisions in the contract and good discipline on the part of customer personnel to avoid over-running cost targets.

The company that is able to integrate all of this information product into a central program library has a distinct advantage in concurrent engineering because concurrent engineering is, or at least requires, effective communication of ideas. The people working on the project will also never have to worry about using out-of-date paper copies of documents. And, incidentally, the facilities people in both customer and contractor ranks will even benefit from fewer file cabinets.

5.6 Work responsibility

The IMP must relate program work to work execution responsibility. First we must decide how we will organize to execute the program. In Chapter 4 we discussed the organizational structure preferred by the author, namely matrix management characterized by: (1) programmatic integrated product development teams and (2) functional departments that provide qualified people as well as proven tools and procedures to programs. Of course, other forms of organizational structures can successfully execute a program. For small companies with little product line differentiation, forms other than matrix may even be an advantage. This book attempts to focus on a situation where a matrix is advantageous for the reasons suggested in Chapter 4.

If we are to follow this pattern, our program work must not be assigned to functional department directors, managers, and chiefs, rather to responsible PDT. In the process, if we are not careful, we may develop a team structure that is completely unworkable in execution. The team structure must be aligned with the WBS (product architecture) in order to preclude conflicts in assignment of budgets and tasking to teams.

The product teams must be aligned with the product structure reflected in the WBS because the budget will be aligned with the WBS in order to satisfy the cost/schedule control system (C/SCS) criteria required by a DoD customer or an equivalent entity by a non-DoD customer. In 1991 the criteria for DoD contracts were re-located from DoD Instruction 7000.2 to newly released DoD Instruction 5000.2 but remained unchanged in content. As discussed in Chapter 4, the C/SCS criteria require us to manage budget through intersections in a matrix of WBS and functional organizations.

We wish the teams to align with the WBS so the budget for all team tasks can be simply assigned. The larger the number of crossovers that exist between WBS, teams, and tasks, the more complex the program will be to manage. At the same time, the WBS must align with the functional architecture allocated from needed system functionality. This means that the WBS must track the evolving functionally derived architecture reflected in the system specification. This combination will result in the fewest possible crossovers between organizational interfaces and product interfaces, referred to as cross-organizational interfaces, which lead to program problems.

This combination also results in the simplest possible integration task. System integration is a difficult task no matter how expansive. We can do a better job at it if there is less of it to do. Coordination of the organization of the work, product, and performing organization leads in this direction. Figure 5-12 illustrates this point by showing a perfect correlation between a product N-square diagram and an organizational N-square diagram. An N-square diagram displays the relationships between N items by noting these relationships in the square matrix intersections. The N items are listed down the diagonal.

On the product N-square diagram we see a requirement for an interface between two specific items identified on the diagonal. On the organization N-square diagram we see a need for the two teams corresponding to the product items joined by this interface to communicate about this interface. If our PDTs are organized about the same structure that the product system uses, the team communication patterns will match perfectly the product item cross organizational interfaces. It is precisely these cross organizational interfaces that traditionally lead to development problems. If there is a complex relationship between product development responsibility and product composition, there will be interface development problems leading to unnecessary cost and schedule impacts.

Either the SOW or the IMP can cross-reference the work to the program organization team structure. Since the SOW lists all tasks, you might conclude that it would be a better place to locate the task responsibility matrix than the IMP. The IMP selectively expands on SOW content, linking work to specific events and accomplishments corresponding to those events for management purposes. The author encourages that this matrix be placed in the IMP because the SOW tends to be a simple task list for people with green eye shades while the IMP is a plan for us humans. As the product and team structures mature, we must check for emergence of misalignments between the WBS and the team structure and work toward nulling them out.

Every task identified in the SOW should have some kind of procedural coverage that tells how that task will be performed. This may be in the form of a customer standard (such as MIL-STD-490A), internal company procedures, or commercial standards (such as the ANSI series). This can easily be defined in a task/procedures matrix or integrated with responsibility assignment into a SOW task/responsibility/procedures matrix such as the fragment

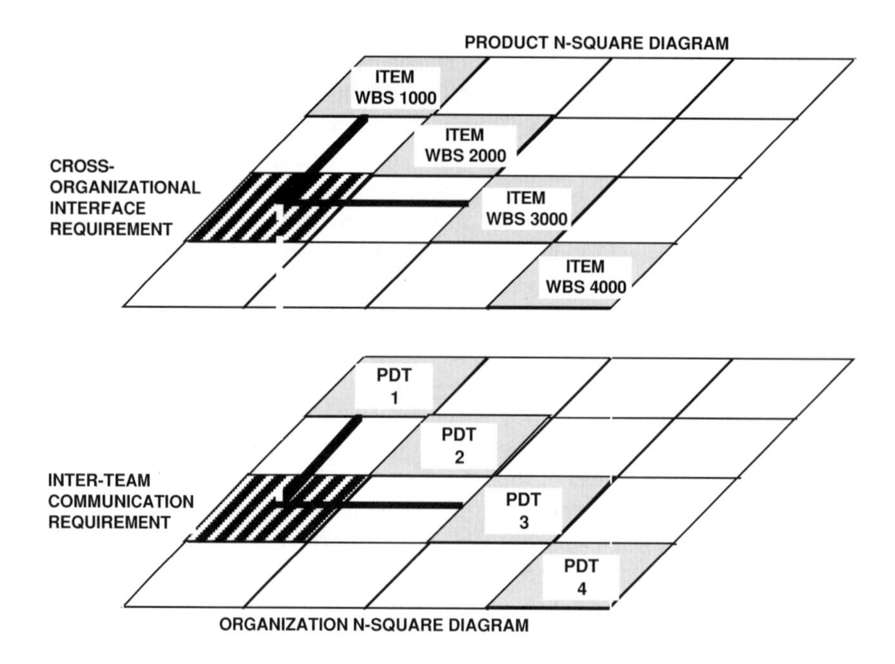

Figure 5-12 Product and teaming N-square alignment.

illustrated in Table 5-4. Ideally, these references will be to generic company planning as previously discussed.

5.7 *Who plans the program?*

We have left one thing unanswered up to this point. As a result of our wonderful planning work, we know what must be done, who shall do it, when it shall be done, and how it shall be done. But, who should have accomplished this planning work in the first place? Should we let the functional departments engage in bottom-up, grass roots planning integrated by the program? Should the program or proposal team do all of the planning?

The author believes that few proposal teams or program staffs at the time this book was written accomplished their program planning activities in a purposeful integrated fashion, including defining traceability throughout the product and process requirements stream as suggested in Figure 5-2. If the systems approach (or concurrent development) method is useful in developing product systems, why should we not apply it to the program planning process? We, the program team, are after all a system. Let's try it.

First, who should participate? Table 5-5 lists the planning documents discussed above and correlates them with some generic principal functional organization responsibilities. Your company may be organized differently, but these organizational entities are probably fairly widely recognized. While these documents should be developed in approximately the order listed (from top to bottom), we should not permit autonomous work on any of

Table 5-4 Task/Responsibility/Procedure Matrix

WBS	Responsible team							
Task	1	2	3	4	5	6	7	Procedural reference
1000.01	X							MIL-STD-1422 (tailored)
1000.02		X						Company practice 128.23E
1000.03			X					MIL-STD-490A
1000.04	X							Company practice 153.5
1000.05						X		Program manual 14.24B

them. There will be insufficient time to sequentially develop these documents during a proposal preparation period of 30 to 60 days. More importantly, their content must be mutually consistent. Figure 5-13 offers a rough schedule showing the relative timing suggested.

The recommended integrated planning approach involves forming a program planning team with membership by representatives from each functional department noted in Table 5-5. Someone identified by the proposal or program manager should lead the team. All members should be physically collocated in close proximity, encouraging easy conversation and a close working relationship. They must have available to them good telephone and computer data communications facilities as well as adequate wall space for posting information for integrated viewing.

The suggested integrated planning approach is not rocket science any more than is the concurrent or integrated product development approach. It simply requires clear definition of responsibilities, cooperation between the parties, excellent interpersonal and communication skills, and a shared appreciation for the discipline of traceability. The system architecture identified in the system specification must be respected in the WBS. All work must be listed in the SOW and linked to the WBS. All events, accomplishments, and criteria in the IMP must link to the tasks defined in the SOW. All tasks appearing on the IMS must correspond with the SOW and IMP tasks and all events on the IMS must correspond to those respected in the IMP.

Throughout the planning process, the selected PDT Managers must contribute to WBS and SOW development and maintain vigilance for crossovers between WBS, team definition, and work responsibilities. They must also work to develop a cost estimate in a proposal situation. Also in a proposal situation, this team will have to coordinate their work with the management volume writing team.

This planning activity probably cannot be accomplished in a straight line fashion. While preparing the IMP we will get insights into changes in the SOW that may ripple through other documents. All team members must have available to them the full content of all of the evolving documents and be familiar with that content through an almost continuous conversation between team members and access to the actual text on their computer screen and a stickup on a wall in close proximity.

Table 5-5 Program Integrated Planning Team Responsibilities

Document	Functional responsibility
System specification	Systems Engineering
Work breakdown structure	Systems Engineering & Finance
Statement of work	Systems Engineering & PDT Team Managers or candidates
Integrated Master plan	Systems Engineering
Integrated Master schedule	Scheduling
Contract Data RQMT List	Data Management

Document	Program planning period
System specification	<---------->
Work breakdown structure	<----->
Statement of Work	<------>
Integrated Master Plan	<---------->
Integrated Master Schedule	<------------->
Contract Data RQMT List	<----->

Figure 5-13 Program planning timeline.

Team members must respect the hierarchy in Figure 5-2 and the content of the documents should be developed in the sequence illustrated there. Each component of each document expands into lower tier document details. This pattern repeats through the hierarchy. As this expansion is developed, the traceability links should be captured. The information developed by the team members assigned to each document must be constantly checked for traceability and consistency by someone responsible for planning integration.

Some readers may think that the material in paragraph 5.5 should have been placed prior to the planning material in paragraphs 5.3 and 5.4. At the same time you have to be familiar with the Air Force initiative before appreciating the opportunities to accomplish the planning work in an integrated fashion. The author agonized over this dilemma for some time until concluding that the reader should first understand the planning relationships outlined in the Air Force initiative, then be exposed to the integrated planning team concept not explicitly included in the initiative. You should now scan the previous material in this chapter with the perspective of the integrated planning team concept in mind.

5.8 A generic SEM/SEMP for you

Some people greeted the workup to release of MIL-STD-499B with screams of, "The sky is falling." You would think that, after many years of contract performance under its predecessor 499A, all DoD contractors would have put in place the fine system engineering activities that they had been describing in the many SEMPs that they had written over the years and submitted with proposals. The SEMPs were never contractual and, sadly, often were never opened by the contractor subsequent to the contract win. So, the adverse reaction to 499B was often based on not then having in place, despite the fine stories told in SEMPs past, an effective systems approach and a concern for how to acquire an effective process in a reasonable time. Many companies dispatched people to the several MIL-STD-499B short courses offered only to have them return with confirmation that they were in very big trouble.

Whether your company must respond to MIL-STD-499A, some other customer standard, or has no constraints on your process, you really should have in place an effective systems approach because it will result in a product with better value for your customer and more business for your company. If your competitors are able to put in place an effective systems approach and you do not, you will have great difficulty matching their product cost, schedule, and quality.

As noted earlier, you will find a copy of a complete generic integrated SEM/SEMP attached to this book. This document can be used as a basis for your company SEM/SEMP. The computer disk supplied with the book contains this document in editable format in Microsoft Word.

It does not follow, of course, that by acquiring a good SEM/SEMP that you will overnight become an excellent systems house, though the process of writing one can be quite an education in the right direction. This document must be matched to a current reality within your company. You will need to monitor your performance to this standard and provide the machinery to force compliance with your own standards. The techniques discussed in this book and elsewhere on continuous process improvement should be applied to your system engineering process to uncover weakness and to understand useful priorities for their correction.

You will also need a way to educate your work force in your system engineering process. This can be done at low cost through brownbag sessions in your plant for your motivated employees which happen to be the very ones you most wish to retain. The SEM/SEMP can be used as the text book for these sessions. You should also cooperate with your local National Council on Systems Engineering (NCOSE) chapter and a local college or university to establish or support a system engineering certificate program that qualifies for your company's tuition reimbursement program.

Given that you have a written procedure for performing system engineering, a way to educate/train your work force in performing that procedure, a means to continuously improve upon that procedure, and a way to encourage compliance with your process, you are on the road to success in performing system engineering effectively. You will also have little to fear from the standards any customer may impose upon you.

5.9 Rapid identity documentation

In Chapter 1 we said that it was possible to maintain a current enterprise identity within an environment of a high rate of change through the use of computers. It is time to defend that claim. The reader familiar with database concepts will have already observed, through the previous material in this chapter, the image of a set of databases that could be used for this purpose. Yes, databases rather than word processed documents. The suggested approach is to apply the same tools used for requirements capture to planning document capture. If you are still using word processors for specification preparation then you will be one step behind the suggested process and a few words of database encouragement are in order.

Several computer applications on the market as requirements tools at the time this book was written were adequate for planning information capture and several of these tool companies were hard at work more perfectly adapting to enterprise modeling tool sets. Table 5-6 lists several of these tools. These companies (with the exception of IDE) have, in the past, been very generous with their time and talent, providing classes the author has taught at universities with tool demonstrations and requirements talks. Identifying them here as potential planning tool sources is a small repayment of that generosity.

The advantages of using a database to capture planning data is precisely the same as the advantages of using a database to capture requirements. Modern tools for this purpose will not only permit the storage of and access to requirements or planning data just as documents do (in paper or word processor forms), but also permit the coordinated capture of other related information that cannot be retained efficiently and economically in paper documents or even in the associated word processor or spreadsheet form. The principal kind of supporting data of interest is called traceability data. But there are some other subtle and powerful benefits that are not so obvious.

You have seen that the statement of work content should be traceable to the system specification, proving that we are doing no work that is not focused on the product and what it is required to accomplish. Well, how do we prove to ourselves in a fixed price world and show our customer that we have accomplished that goal? The discipline of traceability will go a long way toward satisfying this need. In a modern database of the kinds listed in Table 5-6, traceability fields exist that permit us to link up a paragraph in one document with a paragraph of another document with the meaning that one of these pieces of information was derived from the other or is related to the other. In the process of creating and maintaining traceability data in our database, we will find that certain content is not traceable. This should be a signal that we may have included extraneous material in one document or omitted something important in the other. So, this discipline will allow you to follow the program logic built into your planning documentation subsequent to preparing it but will also encourage the preparation of better planning documentation.

As important as it is, simply providing traceability is not good enough for our purposes. Doing the traceability work will actually slow down the planning process for it is added work. So, how can we accomplish more work faster, allowing us to gain the benefit from traceability while completing the planning work more rapidly? If we have captured our generic planning data

Table 5-6 Potential Planning Data Capture Tool List

Product name	Supplier
RDD-100	Ascent Logic Corp.
RTM	Marconi, IDE, USA distributor
Doors	Zycad Corp., USA distributor
Spec Writer	PRC Division of Black and Decker
Document Director	Compliance Automation Inc.

in a database and our tool set is properly configured, we should be able to generate the program planning data by mapping the generic work to the program work and generate the program plans and estimates from the resulting map. Unfortunately, this capability was not yet fully implemented in the commercially available tools the author was familiar with at the time this was written.

The author has experimented with this approach in his database toolbox called Rosetta Stone with encouraging results. Figure 5-14 illustrates a fragment of this toolbox for discussion purposes. The drums represent database structures. Those with a solid body are program oriented and those with a checkerboard body are generic data systems. The arrows show the flow of work from the customer need through the functional analysis (FUN) tool, the allocation of functionality jointly to the architecture (ARC) and system requirements (REQ), the establishment of the WBS as an overlay of the architecture and SOW content from the WBS, with traceability of work elements to system requirements as a cross-check on validity. Each drum is a database and traceability is retained in related databases as paragraph number pairs from the corresponding documents (in database format). Commercial tools commonly will permit parsing of paragraphs into subsets for traceability purposes to account for poorly written documents with complex paragraphs.

Generic tasks (TASK) are mapped to practices content (PRAK) and generic estimating data (GEST) and these data do not change over time except for the effects of continuous process improvement. The generic tasks are mapped to the very general SOW tasks to form the detailed planning steps (PLN). Generic estimating data is also mapped to the program estimate (PEST) based on the planned tasks. Planned program tasks (PLN) are hooked to program events (EVENT) and linked to the program schedule (SKD).

The program planning data can be created through this process quite rapidly with contributions from many sources into a common database. The planning work can be accomplished by a team formed from functional department personnel assigned to a program planning team each mapping their charter department tasks to the program SOW tasks. If the integrated management system is applied, the planning strings can be formed in the process of this mapping activity and provide the organizing codes including significant accomplishments and accomplishment criteria drawing explanatory data from the generic data. The attached SEM/SEMP content is contained in the PRAK database in this system.

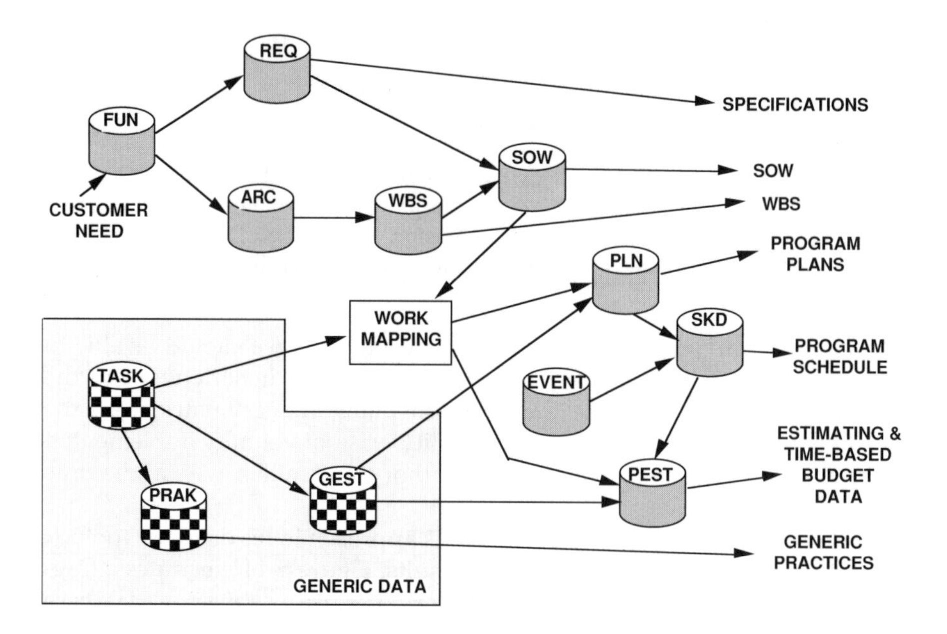

Figure 5-14 Special planning database tool set.

The generic estimating system should be loaded with mean cost data for accomplishing corresponding tasks. As these tasks are mapped to program SOW elements, this cost data should be acted upon by schedule considerations and adjusted, if necessary, for program differences.

It is true that the data in the SOW database, for example, may be generated into a very standard printed SOW document. But, at any time, anyone may also directly access the networked information in the database encouraged by Figure 5-14. The generation of paper from a database is not time consuming but the approval, reproduction, and distribution of the paper originally and for change purposes can be. With paper documents, you also run the risk that the file cabinets full of these plans contain out-of-date information. The only thing that saves you from this class of errors is that no one reads the information after it is printed.

Direct use of planning database content can speed up the planning and the information access processes tremendously if your staff can become familiar with the use of raw database content rather than paper documentation. It also assures that everyone is using current information. Many of us find it difficult to use database content directly because a printed document provides an information integration function as well as simply making the information available. We find it easy to refer from one part of a document to another in paper form. More and more people, however, are becoming comfortable with direct database use and, in time, this will be as easy for many people as you and I might find document use today.

If it is important to have the documents in complete document form rather than using direct database access, the listed tools can generate complete documents that may be loaded into a library hosted on a computer server. These documents will then be available to all via the network. If any one simply cannot use this information efficiently on the screen, they may print one out for their own use.

Most of the tools listed in Table 5-6 offer the direct database user a form with one paragraph or fragment exposed at a time. The Document Director tool listed in Table 5-6 offers the unique feature of appearing to the user like a document even though the content is in database records. Any of these tools, however, can be used to generate a screen or paper output of a document format from the database content for viewing/reading purposes.

So, it is possible to document our current process, change the documentation in accordance with an effective continuous process improvement program, and quickly generate program planning data based on that generic planning data while incorporating traceability information that ensures completeness and avoids omissions and redundancies.

chapter six

Continuous process improvement and process audit provisions

6.1 Is a static identity adequate?

Given that we are successful in once completing our process description, can we then rest and apply it for all time thereafter? Many companies have taken this route in the past. It is not uncommon when a company is flushed with contracts and has the indirect budget to spare to engage in procedure writing. The business cycle eventually returns to lean times and the procedures fall on hard times along with the firm. One would find very few enterprises with a current set of procedures.

The enterprise identity we have described in this book is not intended to be a fixed entity but a continuously changing one. Yes, we will have to maintain our process definition for it to have any lasting value. Now this may be very hard to do. We say we wish to have a documented process but are constantly changing that process description.

6.2 Procedures media

In what form shall we publish and retain our process definition? A computer word processor leaps to mind immediately since the computer is perfectly adapted to continuous improvement. It is very easy to edit these documents stored in electronic files. Alternatively, some computer tool makers, like Ascent Logic, at the time this book was being written, were becoming very interested in applying their tools to enterprise modeling as well as the product system engineering process. It is possible to enter all of your procedures into a database from which practices can be printed. Some of these systems even allow you to simulate operation of your enterprise in accordance with the way you have described it in an effort to find better ways to define your process.

The past standard approach to this would involve printing approved procedures and distributing them to manual holders. There is a lot of unnecessary labor involved in this process including reproduction of paper, distribution of paper, manual update, and periodic audits of manuals to ensure that they are current.

You are encouraged to use an on-line networked medium for publication, distribution, and access for your procedures. In this approach, your procedures are uploaded from the workstation where they are maintained to a network server library of procedures. When anyone needs to gain access to one of these procedures, they only have to call it up on their workstation. Everyone would have access to the very same configuration of the current procedures.

The combination of computerized procedures and a central electronic library will enable an economical approach to maintaining your procedures to take advantage of incremental improvements. Paragraph 5.9 covers this in some detail.

6.3 Continuous improvement process

Figure 6-1 illustrates the suggested process for incrementally improving our process continuously. Lessons learned from programs are studied (block G) for ways to improve the current process description maintained in block L. In block I we study external process descriptions and compare them with our internal procedures. This may lead to insights into improvements or knowledge about our process maturity relative to our competitors.

The results from these two steps are opportunities for improvement actions. In block J we determine which of several alternative improvement approaches we should implement. This may involve personnel training, tools improvements, or procedural improvements. The selected action or combination of actions is accomplished in block K and this work influences the indicated steps.

All of this work should be the responsibility of the functional portion of the matrix providing an infrastructure within which programs can apply the common process adjusted for program peculiarities.

6.4 Process audit

We have gone to a lot of trouble to insist on a common enterprise process. Our rule is that all programs must implement this standard process. How do we ensure that this actually happens? First, proposals should be screened to ensure that they reflect this rule. Secondly, we must audit programs in accordance with some predetermined sampling rule to verify that they are applying the standard process.

A two-dimensional audit process is suggested. It is described below and included within Appendix B of the attached SEM/SEMP. The first dimension verifies that programs are properly implementing the generic process. The second dimension determines how well our practices compare with an accepted standard benchmarking process maturity.

Figure 6-2 shows how this process can operate. We select specific practices to audit from our generic documentation, such as a particular SEM/SEMP process, that maps to particular program process steps. An auditor studies

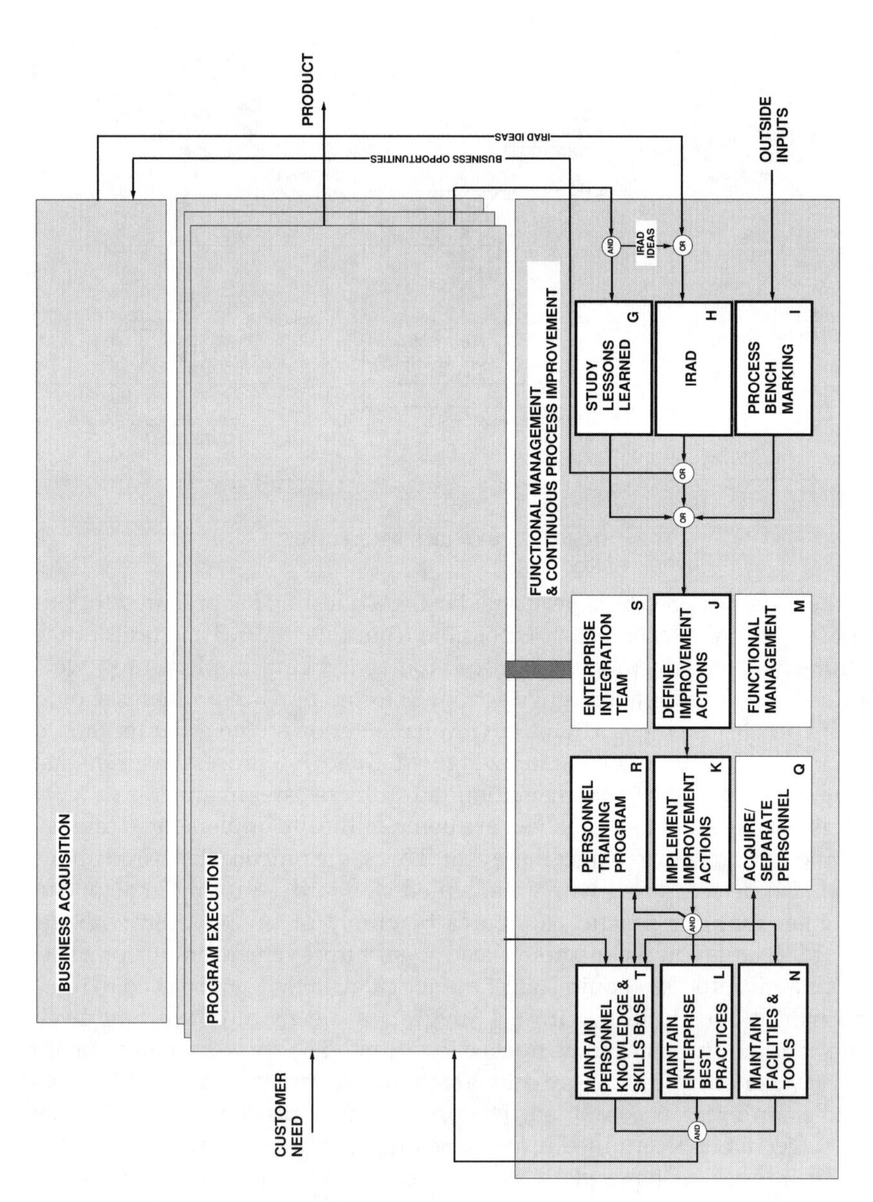

Figure 6-1 Overall continuous improvement process.

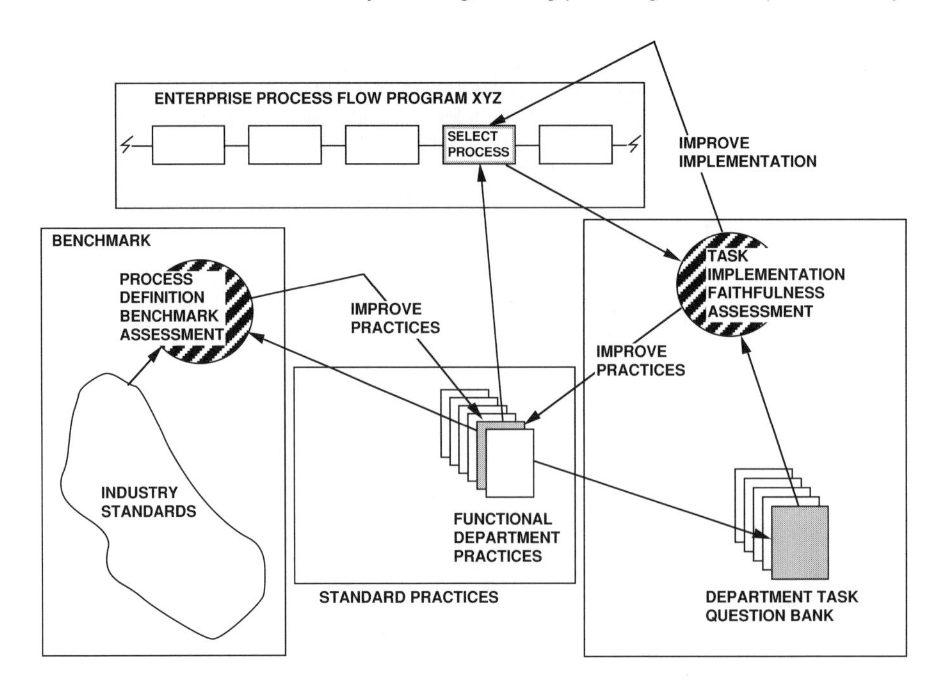

Figure 6-2 Dual track assessment.

how the program process is performed using a question list prepared for this purpose. Several alternatives are possible from this audit. The auditor may conclude the generic process is good and it is being implemented well, resulting in no need for corrective action. Detected practice problems should result in practice changes. Unfaithful implementation on the program should result in changes in implementation. Figure 6-3 shows a process diagram that accomplishes the task implementation faithfulness assessment.

It is possible that bad practices are being faithfully implemented and we need another process to detect this event. This same practice discussed above should also be compared with an accepted standard, resulting in a conclusion either that our practice compares favorably or unfavorably with the standard. If unfavorable, it should lead to an improvement in our practice.

Our audit process should collect numerical data that can be used to track performance over time. Figure 6-4 suggests a means of converting audit results into metrics that can be tracked over time. The answers to implementation assessment questions are converted into a score between 0 and 100, just like an exam grade is computed. The process benchmarking grade must be more subjectively determined in the same range, though it would be possible to define ranges for particular classes of comparisons. Figure 6-4 illustrates several extreme but possible results during audits. We would prefer to receive scores of 100,100 on all audits. Where we fail to achieve a dual high score, a need for changes in our practices, implementation, or both is suggested.

The implementation quality assessment checklists are relatively simple to construct. You begin with the standard practice that covers the activity in

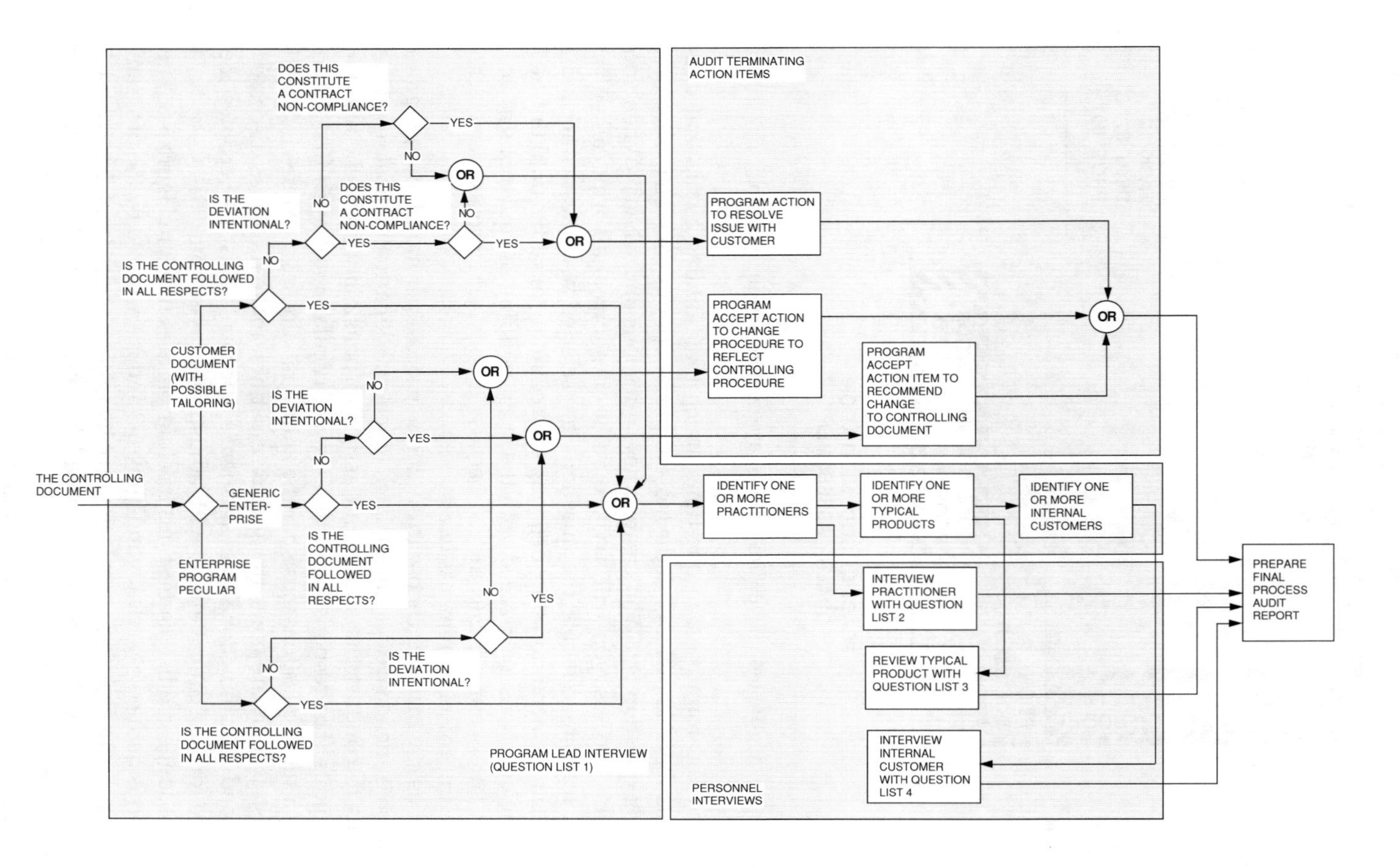

Figure 6-3 Detailed assessment decision network.

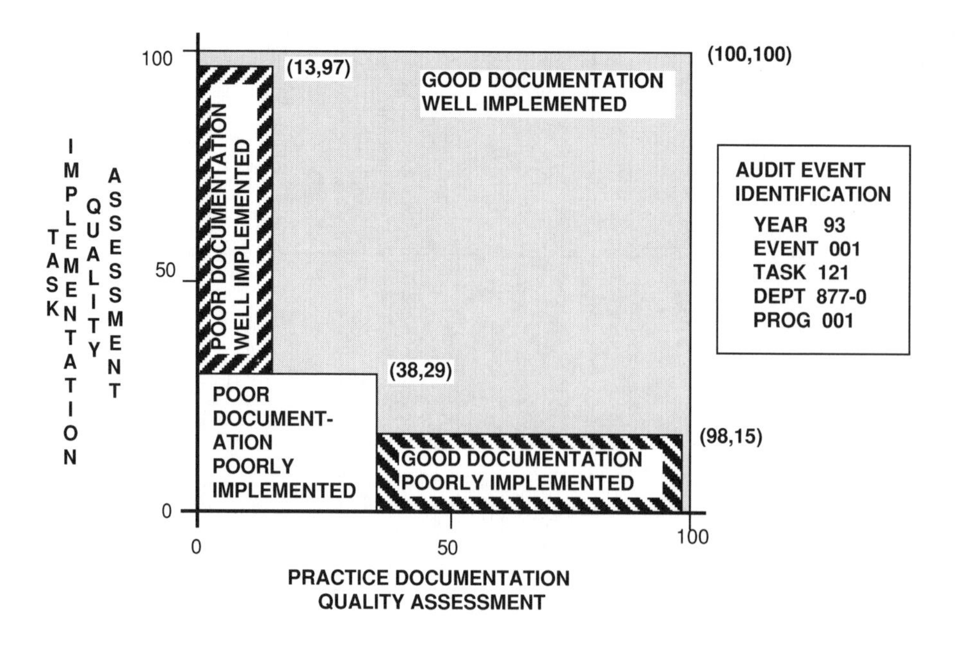

Figure 6-4 Two-dimensional assessment scoring system.

question and form a list of questions, much like building a quiz that tests class members' knowledge of the most important points covered in a lecture. The next question is, "Who creates these questions and when?" When you begin this process, you will have no audit questions at all. You can either form a team to generate the questions for selected activities as an audit process implementation step, delaying the beginning of audits until finished with the questions, or proceed with the audit building the questions as you go. The latter course is encouraged because it begins the process which can then be improved incrementally as you go.

Let us say that you choose to audit one process per month per program per major functional department. Each functional department must select the persons to conduct the audits and assign responsibility and schedule for these audits. The responsible person for a particular audit must first determine if a set of questions exists for the audit. Initially, a set of questions will not exist. If there are none, the auditor must first make the list as discussed above and gain functional management's approval. This list becomes the baseline for subsequent audits with adjustments over time.

The suggested list of questions has four parts as noted in Figure 6-3. The first part is defined generically in Figure 6-3 and is answered by the program lead person for that function. If the problems encountered involve contract non-compliance, then the audit should be curtailed and that compliance issue resolved. If no contract compliance issues are detected, the auditor identifies a practitioner on the program for that activity and applies question list 2 focused on task performance in accordance with the controlling procedure. The practitioner identifies a typical product generated on the program in accordance with the practice and the auditor evaluates the product against question list 3. Finally, the practitioner identifies an internal or external customer for the activity and the auditor applies question list 4 containing perhaps only 3 questions: "Does the product conform to your requirements?", "Are you satisfied with product quality and accuracy?", and "Was it made available to you in a timely way?" Other questions could be added.

Scoring for these questions can be easily done. The questions may all be weighted the same or differently depending on their content and management attitudes of relative importance. The total score is scaled for 100. For any one question during the audit, the auditor must define a grade between 0 and 100 or 0 and 10. Scores are accumulated and influenced by scaling to produce the final score.

The most difficult part of the benchmark quality audit is to select the standard against which you will measure your capability for a particular functional discipline. The evaluation of the internal practice against this standard can only be accomplished subjectively barring an extensive analysis and creation of a checklist that will very likely be more difficult to maintain than the practices themselves. The question can be stated relatively simply, "Does our practice conform to the selected standard?" If so, you should award a score of 100. If it does not, you must assign some numerical grade less as a function of the difficulty in closing the gap through changes in the in-house practice. This does not only involve the written practice but the associated social issues in our organization. The score should be accompanied by an explanation.

Simple computer database systems can be built to capture the results of these audits, provide administrative support, and perform the arithmetic to generate final scores. The author knew of no readily available products on the market to support such audits at the time this book was written.

Given that we have accumulated information about our performance through this audit process, we must have the machinery in place to do something about the results derived from it. That machinery must include methods to change our practices where needed and enforce work performance on programs in accordance with our standards where deviations are not in our best interest.

chapter seven

SEM/SEMP implementation

7.1 First things first

Let us conclude that you are convinced that you must take action to improve the system engineering capability of your enterprise along the lines suggested in this book. You might think the first step on the road to implementing or improving an organized method of developing complex products is to simply write a SEM/SEMP similar to the one attached. Many people fall into that trap. If you take that approach immediately, you will probably be disappointed. Organizations which have not moved off from an old hierarchical, functionally dominated company structure will have great difficulty evolving to the lean functional structure suggested here because of the resistance of its current functional management. So, the first step is to gain the complete conviction, through education, persuasion, or replacement, on the part of top management that this process will not only be documented in a SEM/SEMP but that corresponding changes will take place over a predetermined time frame and you will actually apply this approach on future work.

There are four channels of change that must be concurrently worked: (1) process definition, (2) SEM/SEMP development, (3) organizational structure changes, and (4) the re-education of your work force. Each will be addressed below and your efforts to make changes must also address them together. You will not likely be successful if you attempt to attack these areas one at a time in a serial fashion. Since they are all inter-related, you must approach them concurrently.

7.2 Process definition and improvement

Given these preliminaries, you must clearly understand what your current process is and make comparisons with the process you are trying to achieve. This produces an error signal that you can use throughout all future efforts to make improvements in your process. This is called a closed loop process and is the basis of all goal-seeking machines like guidance systems, aircraft stability control systems, home heating temperature control systems, and the human mind. A closed loop mechanism will provide steering signals to guide your movement toward your goal of attaining a first class systems engineering capability.

Chapter 6 offered such a closed loop system of evaluation of your capability in two dimensions. The evaluation of your use of your internal practices can await development of your new practices. Your first concern should be to visualize a benchmark for your efforts as discussed in Chapter 6. You need some standard of excellence against which to judge your process consequences. MIL-STD-499A, EIA SYSB-1, ISO 9001, and IEEE P1220 all offer such a source as does the attached SEM/SEMP. Pick one of these standards or build one from these and other sources that your common customer base will find acceptable.

Many consultants and books will encourage you to begin your improvement work to satisfy your process goals by trying to define only your current process while consciously excluding any attempt to make improvements. If you serialize this process by building a SEM/SEMP based precisely on your current process, you will likely thereafter find that, in applying the closed loop improvement process covered in Chapter 6, you will have to make too many changes very early in the life of your SEM/SEMP, possibly leading some to feel that it was a waste of time to prepare. You will run the risk of being overcome by changes at a time when you are trying to master implementation of the standard process.

Other consultants and books encourage you to seek your ultimate goal as the basis for your initial process definition. If you wait to implement your SEM/SEMP and process definition until they are perfect, you will run the risk of falling into the Three-P trap, perfection-procrastination-paralysis.

There are some things which simply cannot be intellectualized. You must at some point put your ideas into practice and allow your closed loop improvement process to work its magic. As many serious improvement efforts have died from analysis paralysis or a lack of decisive action to implement as have died from excessive zeal to implement before one is ready.

The author believes that, like so many other things in life, this is an area you should approach with some balance. Your process definition work should unfold in concert with your work to develop a SEM/SEMP, hopefully using at least some of the content of the attached version. If you keep an open mind while trying to understand your current process, you will find some simple, low cost changes that you can make as a part of preparing the SEM/SEMP and the subject of these changes will never appear in your error signal. You can then focus more intensely on applying the content of the plan to your work and less on a disrupting string of needed changes.

Appendix C of the attached SEM/SEMP offers a model that you are encouraged to use in defining your current process. You may even call up the electronic version of these diagrams available on a MAC disk and edit them into a definition of your process. You will require MacDraw II on your Macintosh to do so but they can be easily converted into MacDraw Pro and many other drawing applications. You should start at the top and work inwardly toward the greater detail. For example, you may choose to define one process as product development like task 1 in SEM/SEMP Appendix C. This task can then be decomposed like the diagrams of that appendix with changes reflecting past successes within your company. You can carry this pattern downward to lower levels of detail. Avoid going too far. You probably should stop at the level where a task would be estimated or controlled via budget and schedule constraints on a program.

To create your process diagram, you should form a cross-department team of people with the energetic support of management to define how you presently accomplish development work (amended as noted above for simple changes). Choose a seldom-used meeting room as a team war room or a large wall in a less-traveled part of the plant as a war wall upon which you may post your process diagram and gather your team around for discussions. Install your process diagram on your computer network in a read-only protected server folder/directory. Provide your team with computer projection capability in the war room or upon the war wall. In meetings you can then use the grand sweep of the process diagram to view the big picture and the projected view of a detailed fragment of the process upon which to make real-time changes agreed upon by your team. Team members will also be able to call up the latest version of the process diagram from their normal working assignment location.

You can physically collocate the team as a temporary full-time assignment, make the activity a part-time job with periodic team meetings, or operate as a virtual team using an existing computer graphics, voice, and video networking capability to link the members together. As the process definition unfolds, it should be periodically reviewed by peers and management with adjustments for critical comment. In the process of review it is possible to slide into a condition of having your process designed by a committee. To avoid this, your team needs strong leadership to prevent these review meetings from turning into detailed design meetings and a small number of management personnel as high up in the structure as possible to act as an executive to adjudicate conflicts between top level management personnel.

7.3 SEM/SEMP development

The SEM/SEMP is but one process descriptor your enterprise needs, but it is a central one because it hooks into material, manufacturing, quality, and logistics as well as engineering. You should give some thought to the overall planning tree your enterprise requires. Figure 5-1 offers an example for a program plans tree, but what does the functional generic plans tree look like? And, how are the several program plans trees and the one generic functional tree related in your enterprise? The prescription offered in this book is that the functional plans should guide performance on programs within the work definition provided by program planning documents.

The SEM/SEMP should expand upon your process diagram, telling exactly what occurs within the blocks of your process diagram at some level of indenture. The attached document offers a starting point. You will likely find it necessary to make some changes but the supplied template will remove the impediment of not knowing where to begin. We are all much better at criticism than creativity and that is the advantage of having a template available that no one in your organization need defend.

As you build your SEM/SEMP be very careful to build and maintain a traceability matrix with your common customer standard (such as MIL-STD-499A) and identify and capture any tailoring needed to assure agreement between your final document and the tailored standard. Some of these traceability matrixes are included in the attached template but your changes may influence the content of them.

In order to provide the widest possible customer base access to this data at the lowest media cost, the attached SEM/SEMP is provided on an IBM PC compatible disk using Mircosoft Word for Macintosh as the word processing application. This combination allows you to open the document on an IBM compatible machine and convert to a Word For Windows or DOS document or open it on a MAC with superdrive and place it on a MAC disk for editing. In either case, you should place the document for editing on a hard disk to ensure sufficient disk capacity for making changes. The document includes integrated graphics that may not translate to an IBM word processor perfectly.

7.4 Organizational structure changes

Top management must agree that your company will move to a thin functional structure responsible for good procedures matched to good tools and qualified personnel. There is no simple way to do this but an assurance to those now in functional management of a responsible job may help. If the power in your organization has resided in functional management in the past, then people in those jobs were probably attracted there by power. The thinning of functional management will not be popular with these people. Some of these people will, however, find program jobs as PDT and PIT leaders rewarding. Some may even be qualified and have the right attitude for these jobs. Downsizing concurrent with restructuring can act to encourage acceptance because most will perceive continued employment better, even with reduced pay or position, than termination.

7.5 Re-education of your work force

Many companies and divisions have trimmed their training budget to the bone thinking those dollars not spent will add to the bottom line. The reality is that this has a withering effect on your future capability. You need some means to provide your work force with a continuing education opportunity. Four ways to accomplish this goal are offered: (1) on-the-job-training or a mentor program, (2) in-house, off-budget training classes, (3) a local university certificate program, and (4) seminars and short courses.

7.5.1 On-the-job-training (OJT)

OJT is a very effective training approach because it provides the trainee with a real world situation in which to appreciate the training experience. Train-while-doing is another expressive term for this training form. Some companies carry this process one important step further. OJT commonly is accomplished with little or no structure. People are simply assigned to programs with a specific work assignment and they are expected learn while doing the job with little or no support. A mentor program solves this problem by specifically assigning an experienced person to take a new member under wing. These assignments must change, of course, as program assignments change. Such a program can easily fall apart if not energetically sustained by supervision. This is a valid activity of the thin functional management organization encouraged in this book.

7.5.2 In-house training program

As a functional engineering manager while employed at a large aerospace firm, the author found himself in the fairly standard conflict of responsibility for the training of his department personnel but no funds with which to accomplish that training. The author's solution was to offer lunch time classes on the employee's own time on a voluntary basis.

After putting this solution to classroom training in place, the author was surprised by criticism from the Director of Engineering Administration that he could not continue because personnel were not declaring all hours worked while in class on their own time. This is an example of the kinds of silly constraints that a manager is faced with that can result in a null solution space for this conflict. The author responded that he was going to continue until told to stop by his management and no further action ensued. Good works, it seems, are not enough for the bureaucrats in your enterprise; you must be ready to live or die by your convictions as well.

You have a large reservoir of talent for instructors in these lunch time classes within your current organization. Many of these people would feel uncomfortable in a formal teaching environment, but will do very well in an informal here-is-what-I-know environment. It is true that your personnel may choose not to attend. If you make them aware that these classes will make them more valuable to the company and apply some simple advertising principles, you can increase attendance. Consult with your human resources people about what can and cannot be said about attendance. It should be voluntary which probably means that you cannot use attendance in your performance rating approach. Even if attendance cannot be directly used in performance ratings, you can make the point that those who attend will learn information important to their being successful in the integrated team concept being introduced. Then, of course, your instructor staff has to deliver on that promise with interesting and effective classes.

Given that you overcome the petty internal objections to your program like those mentioned above, you will have to select the subject areas to cover in your program. Five categories of classes are suggested:

a. **Integrated Product Development**. If your enterprise has for years functioned with a strong collocated functional axis, your personnel will require education on how to work together effectively across specialist boundaries. This should include product development team knowledge such as: team building, interpersonal relations, effective communications methods and tools, things that induce teamwork, and effective meetings.
b. **Interdepartment Function Knowledge**. Let your functional managers tell what people in their departments do on programs, what tools they use, and what dependencies they have on other specialists to do their job. This can be a very revealing experience and a necessary one if in the process of restructuring you have changed the responsibilities of functional departments in any way.
c. **Process and Task**. The principal training objective of your whole program should be focused on instruction in how to perform the tasks identified on your generic process diagram. In this area your in-house manuals, like your SEM/SEMP, can be used as text books. If you have gone to the trouble of placing these practices on line in a computer library, your personnel may gain access to them at any time without fear of using obsolete instructions.
d. **Toolbox**. If your enterprise uses complex tools involving computers, you will find it useful to include classes on these tools. As time passes, there are fewer and fewer people who cannot and will not operate a computer even in the most rudimentary way. So, the initial conversion burden from typewriter technology to the computer is nearly spent. There will ever remain the need to train employees on new computer software tools and their application to your process. For example, if you use a powerful requirements toolset or CASE tools, you will have to maintain currency for those applications in your work force. This subject area could be merged with your process and task instruction.
e. **Product and Technology**. The final class of instruction suggested for your lunch time sessions covers the technologies necessary in your product line. This may include rocket engines, guidance systems, auto suspension systems, computer processor units, real-time computer programs, and an endless list of other topics. Members of cross-functional teams require knowledge of each other's specialty to be most effective. We say that the worst problems in the development of complex systems occur at the product interfaces where different parties are responsible for the two terminals. This is true often because of ignorance of the other party's technology leading to a withdrawal from the interface. By educating the two sides to the other's technology, they can be encouraged more successfully to extend themselves

across these interfaces and attempt to understand them from the other's perspective.

Some companies have gone far beyond lunch time classes and offer daytime, evening, and Saturday classes in a near university environment. Motorola University is an enduring example of this approach. You may find after experimenting with the lunch time approach that many employees would prefer to stay after work for these classes. Experimental classes at these times can be used to test attendance. Most enterprises will not have the resources to pursue this solution but can cooperate with a local university and other companies to make the equivalent available to its employees.

7.5.3 Local university system engineering certificate program

Most companies have some kind of tuition reimbursement program where money spent on college classes is refunded to the employee. The details of these programs vary but most will pay the cost of a formal degree or certificate program. The problem in the system engineering field is that in the early 1990s there were few programs available in universities. There are several models of success in this area, however. General Dynamics Space Systems Division and University of California San Diego, Texas Instruments and Texas Tech, Loral Aeronutronic and University of California Irvine, Loral Command & Control Systems in Colorado Springs and Colorado Technical College, and Boeing and the University of Washington are five examples.

The author and David Clemons, as employees of General Dynamics Space Systems Division, approached Dr. John Peak, the Director of Science and Engineering at the University of California Extension, with encouragement for such a program in 1991 and were the first two instructors in the program. That program was still in operation as this book went to press even though San Diego, California had suffered the loss of several large defense-oriented firms including General Dynamics.

At the time we approached the university, we had previously developed our own in-house training program, prepared procedures used both as work performance references and training text material, and defined an overall process within which all the pieces fit together.

A sound formula emerged from this experience that included the following elements:

a. Form a certificate advisory board with membership from local industry and leadership by a university extension program. Lay out a list of courses and for each course an outline and list of objectives.
b. Recruit members of a certificate advisory board from major industries in town as a way of both getting needed input on the need and gaining direct access to the personnel base represented by these companies. Make sure the instructors have access to the planned outlines and objectives to avoid the potential that instructors will teach only what

they know with a great deal of overlap between courses. Members of the board should monitor the first classes offered by an instructor as a means of quality assurance.

c. Recruit instructors from the companies represented on the advisory board through the members' knowledge of the personnel in their companies.
d. Team the advisory board with the local chapter of the National Council on Systems Engineering (NCOSE), if one exists in the area served by the institution, to define the certificate program. You may also find advisory board members in the NCOSE chapter as well.

The University of California Irvine followed a different, but successful, course to a system engineering certificate program. They were approached by Loral Aeronutronic located at the time in nearby Newport Beach, California and asked to provide a system engineering program at their facility. The author was the first instructor in this program attended by 38 students from a relatively small division of a very large company. A complete program was developed by Dr. Mario Vidalon, Director of the UCI Engineering and Information Technologies extension programs, and Mr. Henry Shu of Loral Aeronutronic to fill the Loral need. A program advisory board was formed and the program spun off onto the campus involving essentially the same courses.

The National Council on Systems Engineering Inland Empire Chapter (serving San Bernadino, Riverside, and Ontario, California) subsequently undertook a project under the leadership of Mr. Chuck Brown, retired USAF officer and lecturer at University of Redlands, and Dr. Howard Korman of TRW to help the University of California Riverside form a similar program. This program was still in its formative stages when this book was being writen. So, it is possible for industry and local educational institutions to team up to create local system engineering education resources for the general benefit of the community and the specific benefit of your career system engineers and your company.

The author has found from talking to administrators and professors in institutions about system engineering certificate programs that different institutions offer different avenues of approach. The extension organization may have the independent power as the initiating agent for such a program. On the other hand, the Extension may defer to the Engineering Department Chair to recommend programs. If you approach the wrong office, you may interpret an apparent lack of interest as a rejection rather than an expression of the internal workings of the institution.

7.5.4 Seminars and short courses

Some companies rely totally on seminars and short courses for their training program. These programs do offer a lot of value for a very few new techniques that have not matured into the university education stream, and the limited funding available in most companies should be spent only for this purpose. Persons who do attend these programs should do so on the condition that they agree to teach others the same material in your lunch time classes.

7.6 The database approach

Chapter 5 makes a case for moving from document-based planning to databases to reduce planning time, improve traceability, and encourage program task repetition. In its ultimate implementation, the enterprise will not actually create paper planning documents, rather load program databases from generic databases adjusted for program peculiarities. It may be difficult to move directly to this condition. It can be approached by the initial imperfect application of a tool acquired for requirements work for planning work and continue to improve this capability as the toolbuilding industry improves the functionality of these tools for this purpose.

7.7 Good luck

This book appears at a time when American industry is undergoing tremendous changes driven by fundamental, revolutionary forces. The rate of change continues to accelerate. How can we be sure in this environment that any given decision on our part about our way of doing business is the right one? We cannot be sure. What we can be sure of is that we have to get on this slippery incline of process improvement as early as we can and continue working toward a moving target of perfection, never to reach it but ever improving. The author hopes that the content of this book and his previous two titles will help you move from being a spectator watching the success and failure of others to becoming an active participant in the drive to master the new reality of the market and work place influenced so powerfully by the computer, the dramatic demise of the Soviet Union as one of two super powers, and an unavoidable turn onto the information super highway.

Part II

Contents: Part II — SEM/SEMP

section one

Introduction

1.1 Document purpose

This document is motivated by a company determination to provide customers with products that fully satisfy performance expectations and possess expected life cycle qualities at the best possible price and in keeping with their schedule needs. It is company policy to maximize customer value in the products and services provided by the company. This goal shall be attained by a motivated and trained work force performing in accordance with proven procedures using the best tools available for the job. The content of this document has been specifically selected and tested on programs to provide guidance in satisfying this goal for each program and customer. A continuous process improvement program actively seeks out needed improvements in method based on lessons learned from each program and implements those related to systems engineering through changes to this document. It is the responsibility of everyone in the company to implement the provisions of this document toward satisfying our goal of satisfied customers content with the value they received from the company.

This generic combined Systems Engineering Manual (SEM) and Systems Engineering Management Plan (SEMP) describes the systems engineering process normally employed by the company where the customer does not require adherence to a specific customer standard or where the customer will permit tailoring of their standard for agreement with this generic plan or replacement of their standard on contract with this plan.

This is a generic Systems Engineering functional department manual that provides guidance for program implementation of the systems engineering process by all company program personnel within the context of work defined and described in program documents. It is also a generic SEMP which the company prefers to apply to all programs to prescribe a common systems engineering process. A common process encourages company personnel to become more skilled and experienced at applying the preferred approach through repetition to the benefit of all customers.

1.2 Functional management responsibility

Maintenance and improvement of this plan is the responsibility of the management person responsible for the functional Engineering department. This document is subordinate to the company Engineering Practices Manual (EPM), company manual 02, which in turn is subordinate to the company Standard Practices Manual (SPM), company manual 01.

1.3 Continuous process improvement

It is company policy to apply this generic systems engineering approach on all programs. This SEM/SEMP captures the best systems engineering practices that have been developed for company use at the time it was published. These practices must be applied on all programs in order to encourage company personnel, the company as a whole, and its customers to benefit from process repetition. All functional department heads are responsible for improving their practices based on observed need or opportunity for improvement during program execution, department personnel study of alternative practices, or as a result of benchmarking company practices with respect to widely accepted industry standards. As the functional departments develop and validate new practices, they shall be introduced into the SEM/SEMP replacing old ones or adding new ones. These improvements shall be made incrementally so as not to introduce a major disturbance in the standard practices at any one time or on any one program.

Departures from the practices contained in this standard on particular programs may be made for the purpose of testing a planned practice change. In this event, the Program Directives Manual should call out the alternative practice description and the part of this document that it replaces. Where experience shows that the new approach is preferred, it will be introduced into the generic SEM/SEMP for all subsequent programs.

1.4 Application on programs

This document is intended to be directly implemented on programs rather than used as a guide for writing a program-specific SEMP. The normal implementing initiative on every program is its listing in the Program Directives Manual as the authoritative source for systems engineering practices. However, it may be implemented on a program by any of the following means:

a. Required work is defined on a program in the Program Statement of Work (SOW) with expanded detail in a Program Plan or Integrated Master Plan (IMP) and scheduled either in a program Integrated Master Schedule (IMS) or Systems Engineering Master Schedule (SEMS) and Systems Engineering Detailed Schedule (SEDS) as a function of customer preferred terminology. These program planning documents tell what tasks must be accomplished on a program. Where particular tasks involve systems engineering activities, this generic SEM/SEMP provides guidance on the performance of those tasks. This plan also satisfies the content requirements for the narrative portion of the IMP for system engineering activities, on contracts requiring this kind of planning document, and should be referenced in the program IMP system engineering narrative section.

b. On programs that require application of a customer specified systems engineering standard called out in the SOW, the content of that standard should either be tailored for agreement with this document or the standard replaced in the SOW by this document. Either of these actions will permit the application of these standard practices on the program. If the customer prescribes a particular data item description defining SEMP format and content, that data item description may also have to be tailored. Refer to Appendix A for recommended tailoring for company customer base system engineering standards. Use the tailoring language from Appendix A when responding to customer requests for proposal.

c. Where the customer insists on a program-specific SEMP, the content of this plan may be copied into the customer's format and combined with other customer-required data to provide that plan. In these cases, every effort should be made to retain the maximum commonality with this generic plan.

This document applies to all program phases defined by the Department of Defense (DoD), NASA, and company commercial development policy. Paragraph 3.1 defines the relationships between these several phasing definitions and how the company phasing definition relates to them. The company system development process includes both linear progress toward prescribed program phase milestones and cyclical iteration within phases to improve and refine the definition of appropriate requirements and a preferred solution.

1.5 Technical objectives

The desired outcome of applying the techniques and provisions of this document is to clearly identify a preferred integrated product system design that is consistent with corresponding process designs for material acquisition, production, inspection, test, logistics support, and operations which together meet the customer's need within planned cost and schedule boundaries. The number of avoidable engineering drawing changes and new drawings created subsequent CDR through customer initial operating capability should be less than 5% of the total number of engineering drawings recorded for the product at CDR exclusive of any drawings related to customer-driven engineering changes defined in customer-funded design changes. A determination of the avoidability criteria shall be made by the program integration team (PIT) and this metric reported periodically to program management.

1.6 Technical plan summary

The fundamental systems engineering principal is that specialized engineers must work together on a predetermined set of small problems, decomposed from a larger problem reflective of a customer need, to accomplish the development of complex systems. There are three fundamental reasons for this. First, no one person can possibly master sufficient information to accomplish any significant part of the job by themselves. This is because the available knowledge base appropriate to any given development surpasses that which any one human can master and apply economically in a competitive situation. Secondly, complex problems that evolve into complex systems require many simultaneous activities in the interest of schedule demands, and one person cannot possibly do all of the work even if they could master the knowledge base. Thirdly, the development of a complex product will entail a large number of people who cannot be efficiently managed at the overall program level directly due to span of control problems. The employees of the company must therefore team together to accomplish the work of our customers in ways that recognize the limits of span of control in organizations and encourage the merger of the skills and knowledge of the many specialists required to solve problems common to the company product line.

Given that this is the case, the large problems expressed in customer need statements must be decomposed into sets of related smaller problems that can be successfully solved by small teams of engineers and support personnel qualified in a relatively narrow range of technologies, mutually supported by people from all disciplines in the organization (manufacturing, quality, test, logistics, etc.). The combination of decomposition of the problem posed by the customer need into many smaller problems and the assignment of qualified personnel to two or more responsible teams satisfies the critical distinction characterizing programs that need a systems approach and application of systems engineering and those that do not.

There follows a brief summary of the several sections of this plan.

1.6.1 Introduction

Section 1 includes a general introduction to the document and covers how this document should be applied on specific programs as a function of the customer's mechanism for implementing the systems engineering process.

1.6.2 Standards references

Section 2 lists any applicable documents referenced in the other sections and notes the existence of any company tailoring located in Appendix A. This tailoring is intended to adjust the applicable document into alignment with company practices in order to permit company employees to enjoy the benefits of task repetition and all customers to reap the benefits of an experienced work force functioning at the peak of their individual and collective ability. This tailoring should be used on all programs where the applicable document is called by the customer for use on the program.

This plan follows the general content and structure requirements contained in the Department of Defense data item description DI-S-3618 corresponding to MIL-STD-499A and is consistent with the requirements of MIL-STD-499A as tailored in Appendix A. It should be noted that the referenced data item reflects MIL-STD-499 rather than 499A structure; so, one of the tailoring items for the data item is to cause coincidence between the data item and the standard. The generally sound structure provided by the military standard has been supplemented by the results of some excellent work accomplished in development of MIL-STD-499B and the U.S. Air Force integrated management system during the early 1990s. Other standards were also consulted: ISO 9001, Electronic Industries Association SYSB1, and draft IEEE P1220. All of these inputs were balanced with company experience in development of product over many years.

1.6.3 Development environments

Over the years systems have been recognized; many approaches to system development have been tried. Some of these have achieved a degree of acceptance and support in this company, industry in general, and on the part of customers for complex systems. Section 3 places these several possibilities into the context of 24 specific development environments and gives encouragement to specific environments for particular product types. It also notes the importance of the existence of a good professional work environment, encouraged by program management, within which program personnel will be motivated to contribute to program goals in full measure of their ability. Section 3 also addresses the need for different ratios of order and discipline vs. creativity as a program progresses from conceptual studies through production.

1.6.4 Technical program planning and control (MIL-STD-499A, Part I)

Section 4 describes the company systems engineering management process as it is applied on each contract phase. It defines how work responsibilities are assigned in the context of an integrated product development teaming environment and who is responsible for systems engineering activities within these teams and across the program and product system. Process inputs and major deliverables and results are defined. Section 4 describes how required program work shall be controlled toward program goals and scarce resources allocated and controlled.

Systems engineering activities that support effective engineering and program management are also covered in Section 4 such as: technical performance measurement, requirements margin management, configuration management, data management, interface management, risk management, and decision support systems entailing action items and design reviews. This section also corresponds to MIL-STD-499A Part I.

1.6.5 Systems engineering process (MIL-STD-499A, Part II)

Section 5 describes how a functional decomposition of the project need will be carried out and to what end. Allocation of needed functionality is described as are the methods for transforming functional allocations into appropriate performance requirements for the items to which the functions are allocated. Where a system is being formed largely through modification or adaptation of an existing system or a large part of its resources, provisions are included for grandfathering those resources and applying structured decomposition only to those areas where new development is required and re-engineering or disposal of the residuals. Several alternative structured decomposition techniques are permitted depending on the nature of the product and the skills and experience of the team members.

Section 5 provides an overall requirements definition process that is sketched with clearly defined responsibilities within integrated product development teams (PDT) and a program integration team (PIT). Verification planning is covered as a means to assure that product designs will satisfy approved requirements. Definition of verification requirements are developed concurrently with product requirements to encourage quantification of requirements and assure that they will yield to verification. Requirements traceability capability of company requirements tools is covered and how that capability will be employed on the program.

Section 5 describes methods for PDT synthesis of approved requirements and defines an effective trade study process for making difficult decisions between competing alternative design solutions where the selection criteria is complex and multi-functional in nature. Methods are presented to control the identification and development of interfaces with clear assignment of responsibilities of all interfaces tied to the PDT and the PIT. This document encourages the alignment of PDT with the product architecture and to associate needed functionality with these entities to support the evolution of the simplest possible cross-organizational interface development situation.

Section 5 provides an effective method for assembling the product architecture based on needed functionality overlaid by a sound work breakdown structure (WBS) for financial accountability purposes, configuration item or end item structure for management focus, specification tree telling what requirements documents must be prepared and in what format, and drawing and manufacturing breakdown structures for detailed development process organization. PDT responsibilities are correlated with this same structure.

Section 5 provides guidance in the capture of program decision traceability data such that throughout the program/system life cycle it is possible to recall the rationale behind all important decisions affecting the final product.

Section 5 defines how planned analyses will be managed and brought to bear on program problems in a timely way. Section 5 defines how testing requirements are developed and by whom. It covers the overall test and evaluation program and how the three major components (design evaluation, qualification, and acceptance) relate to other program activities. Finally, Section 5 provides the information required by MIL-STD-499A Part II.

1.6.6 Specialty engineering integration and concurrent development (MIL-STD-499A, Part III)

The key to excellent systems engineering work and program success is the effective and efficient communication of ideas and integration of the knowledge and skills of many specialized engineers and specialists with on-going program work accomplished in accordance with a plan and schedule. It is helpful to this end for all program personnel, but especially engineers responsible for systems engineering work, to understand the nature of the work accomplished by specialists; so, this section identifies a range of specialized disciplines, not all of which may be required on any one development, and shows how they all fit into a single pattern of work traceable to the overall process flow.

Section 6 describes how the various inputs into the systems engineering effort will be integrated and how interdisciplinary teaming will be implemented to integrate appropriate disciplines at the item and system level into a coordinated systems engineering effort that meets cost, schedule, and performance objectives. The section describes how the company organizational structure relates to program organizational structures and how this relationship supports effective product-oriented teaming on contracts while satisfying company needs for continuous process improvement through the functional organization.

This section defines how the company assigns responsibility on programs using integrated PDT and a PIT for product development and production. The program-specific planning document will identify the specific teams identified for the program, their range of responsibility with respect to the product architecture and process, and specific persons who are assigned principal leadership and specialty roles in those teams. Guidance is included for optimizing the application of the virtual team concept to programs and functional organizations. Section 6 satisfies the requirements of MIL-STD-499A Part III.

1.6.7 Section 7, notes

Section 7 includes a list of abbreviations, terms, and acronyms used throughout this document.

1.6.8 Appendix A

Appendix A provides detailed tailoring for any standards referenced in Section 2 to cause perfect alignment between the resulting document and the content of this document. This tailoring data must be included within any contract requiring compliance with listed documents in order to permit company personnel to apply the process defined herein in the interest of encouraging repetition and continuous improvement of company capability. Where MIL-STD-499 is a required compliance document, the contract should call for its replacement on contract by this plan or acceptance of the tailoring included within Appendix A. In either case, company personnel will be able to follow the standard practices contained herein.

Appendix A also provides high level traceability between MIL-STD-499A and its corresponding data item description for a systems engineering management plan as well as ISO-9001. Traceability was established with these standards to ensure that the content of this plan was complete with respect to potential customer requirements and to provide benchmarking results for the systems engineering metric defined in Appendix B.

1.6.9 Appendix B

Appendix B provides a means to audit a program for systems engineering effectiveness. This may be applied by a program periodically for self-assessment or by the Enterprise Integration Team for company management purposes. It includes a traceability map between 20 criteria that define the minimum acceptable system engineering process and the content of this document in the interest of demonstrating coverage completeness.

1.6.10 Appendix C

Appendix C consists of the company generic system life cycle process diagram including expanded coverage of the company system development process. This is in the form of a process diagram composed of blocks representing specific processes that must be accomplished in the development of products and processes for programs and interconnecting directed line segments that suggest process sequence. These diagrams are annotated with concurrent development bonds where processes entail intense cooperative efforts using concurrent development techniques described in the body of this document.

1.6.11 Appendix D

Appendix D includes a matrix that lists each block on the Appendix C diagram and provides additional information on these processes that could not be included on the diagrams without excessive clutter detracting from the diagram purpose.

section two

Applicable documents

This section lists the government documents and non-government documents applicable to use of this SEM/SEMP in the generic situation. It includes documents appropriate to major company product lines based on past experience. Some listed documents include tailoring that adjusts the listed document into perfect alignment with the content of this or other company generic plans. Only those documents listed in the contract statement of work along with any tailoring included there are compliance requirements for the particular contract. Proposal and program managers are encouraged to tailor contractually required applicable documents for agreement with internal company procedures wherever possible to permit employees to gain advantage from repetition of the same procedures incrementally improved based on program experience. Documents identified with a "T" in parenthesis include tailoring statements in Appendix A.

2.1 Government documents

2.1.1 Specifications

2.1.2 Standards

MIL-STD-499A	Systems Engineering	(T)

2.1.3 Handbooks

2.2 Non-government documents

Company Manual 01	Company Standard Practices Manual (SPM)
Company Manual 02	Engineering Practices Manual (EPM)
Company Manual 04	Quality Assurance Practices Manual (QAPM)
Company Manual 05	Production Practices Manual (MPM)
Company Manual 06	Integrated Logistics Support Manual (ILSM)
Company Manual 07	Material Management Practices Manual (MMPM)
Company Manual 08	Software Development Manual (SDM)

Company Manual 09	Configuration Management Manual (CMM)
Company Manual 10	Data Management Manual (DMM)
Company Manual 11	Test and Evaluation Manual (TEM)
ISO 9001	Quality systems — Model for quality assurance in design/development, production, installation, and servicing

section three

Development environment

3.1 *Program development environments*

This section explains the possible environments within which program work may be accomplished. The company recognizes 24 specific development environments organized in Figure 3-1 in the form of a three-dimensional Venn diagram. Any of these combinations of phasing correlation, product composition, and sequence model alternatives may be applied on a program but some are more appropriate to specific kinds of programs than others. Program personnel must evaluate the program situation and select the environments to be applied with care.

It is not required that a complete program apply to only one environment. At the system level on a grand system program (includes a complex combination of hardware, software, and people), rigorous phasing under the waterfall sequence may be applied at the system level. Within the product development team responsible for computer software, a rapid prototyping

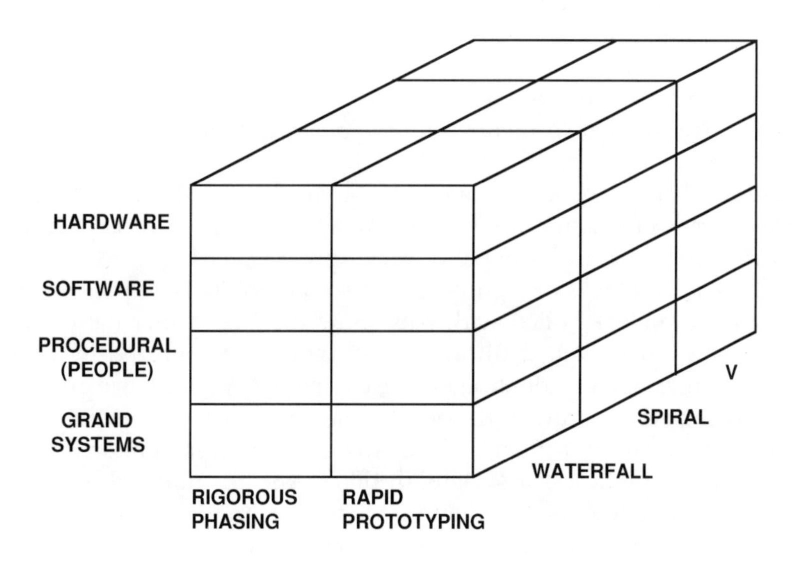

Figure 3-1 Program environment possibilities.

spiral sequence may be applied while a rigorous phasing V may be applied on the avionics subsystem. These terms will be made clear in paragraph 3.2 below.

The PIT is responsible for approving all environments applied on a program. Prior to approving any environment at lower system levels different from the environment applied at the top level, the means for correlating schedules, cost, interfaces, and work status shall be determined.

3.2 Phasing correlation

Two phasing correlations are defined, as noted in Figure 3-1. They are called rigorous phasing and prototyping. In the former, specific milestones are established and a narrow range of related work (for example, requirements definition, design, or testing) is accomplished, leading to a milestone decision on readiness to proceed to a subsequent phase based on the work in that phase. In prototyping, all of the phases may be in progress together with intense interactions leading to a version of the final product reflecting current knowledge.

3.2.1 Rigorous phasing

Appendix C provides a detailed generic process diagram depicting how the company accomplishes development and production programs. Appendix D defines each block on the process diagram illustrated in Appendix C in terms of work performed, principal functional charter responsibility, and references for accomplishment (commonly back to the content of this document by section or paragraph numbers). The development work depicted in Appendix C may have to be accomplished in customer-defined phases focused on specific milestones and goals punctuated by major decision-making reviews that determine the degree to which the company has satisfied the success criteria for the phase. Figure 3-2 illustrates the generic company phasing diagram that correlates with the phasing definitions of every member of our customer base. The tasks defined in Appendix C and listed in Appendix D are mapped in Appendix D to the phases shown in Figure 3-2. Table 3-1 provides a summary level map and correlates the steps in Figure 3-2 with DoD project phasing and NASA project phasing.

The shaded blocks of Figure 3-2 correspond to the DoD phasing definition while the blocks filled with hats show the NASA phasing definition. Blocks with cross-hatched fill are not specifically recognized by either of these customers but are identified in the company generic phasing definition.

Note the cyclic nature of the development/acquisition process. Dissatisfaction with a current capability leads to a new development activity and a phase-out of a predecessor system. In this sense, it is unusual when a new system is completely unprecedented. Most development programs entail a mix of precedented and unprecedented work. It is important to understand where this boundary condition lies in a new program.

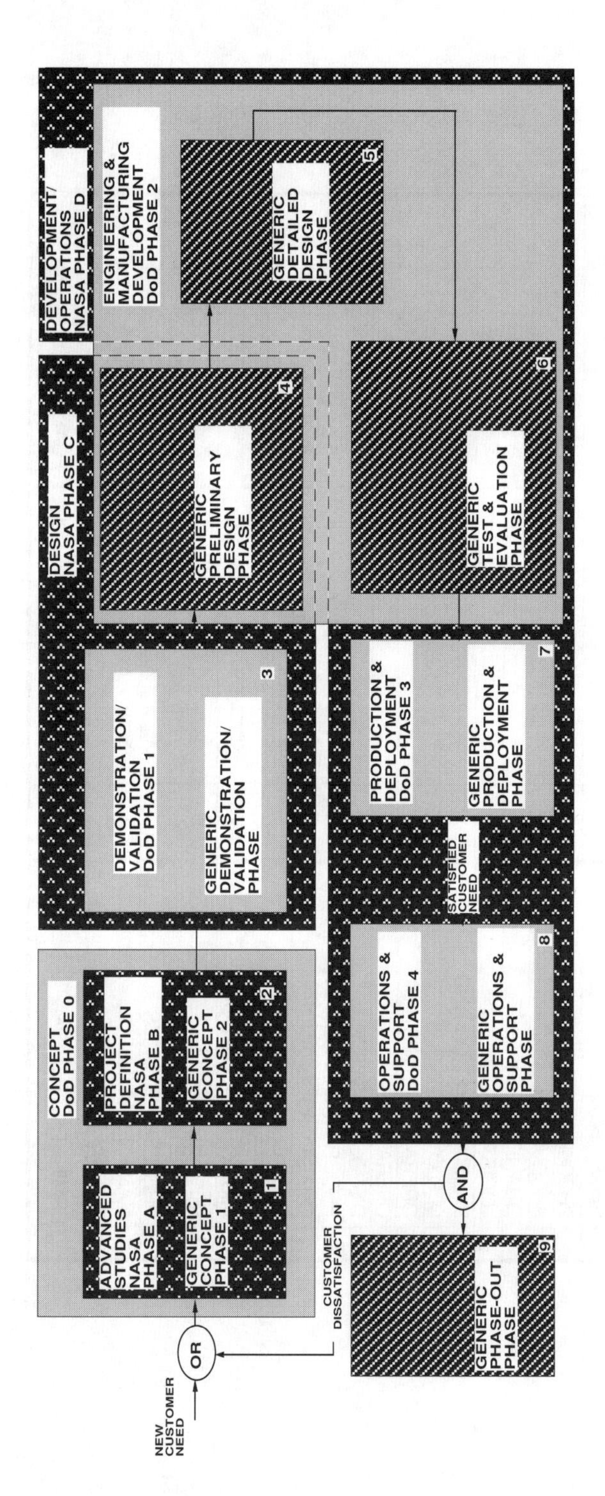

Figure 3-2 Generic company phasing diagram.

Table 3-1 Process-Phasing Map

GENERIC PROCESS (APPENDIX B)		PHASE									
		DoD	CONCEPT (0)		DEM/ VAL (1)	ENGINEERING & MANUFACTURING DEVELOPMENT (2)			PROD (3)	OPS & LOG (4)	
		NASA	ADV STUDY (A)	PROJ DEF (B)	DESIGN (C)		PRODUCTION & OPERATIONS (D)				
		COMPANY GENERIC	1	2	3	4	5	6	7	8	9
#	NAME		STUDIES	PROJECT DEF.	DEM/VAL	PRELIM DESIGN	DETAIL DESIGN	TEST & EVAL	PRODUC-TION	OPERA-TIONS	PHASE OUT
1	PRODUCT AND PROCESS DEVELOPMENT		X	X	X	X	X	X			
2	PROCUREMENT						X	X	X	X	
3	PRODUCTION								X		
4	QUALIFICATION TEST							X			
5	LOGISTICS SUPPORT							X		X	X
6	BASE ACTIVATION AND DEPLOYMENT									X	
7	OPERATIONAL TESTING							X			
8	OPERATIONS									X	
9	DISPOSE OF SYSTEM										X
P	PROGRAM MANAGEMENT		X	X	X	X	X	X	X	X	X
A	QUALITY ASSURANCE							X	X		
B	ENGINEERING CHANGE PROPOSAL						X	X	X	X	
C	MISSION ADAPTATION		X	X	X	X	X				
D	RFP AND PROPOSAL		X	X	X	X	X	X	X	X	
E	MARKETING AND PREDESIGN STUDIES		X	X	X						
F	MANAGE BUSINESS ACQUISITION		X	X	X						
TERMINATING REVIEW			SRR		SDR	PDR	CDR	FCA			

It is possible to become confused between the generic process diagram illustrated and defined in Appendices C and D and the program phase-oriented diagram in Figure 3-2. Appendix C shows the sequence of steps leading from a customer need through the whole evolution of the product development and manufacturing process. Commonly, this work is actually accomplished in program phases, each terminated at a major milestone. The difficulty and complexity of the development work demands that we periodically collect our aggregate knowledge about the product and process under development and expose it to critical review at these milestones. These major reviews offer an opportunity to demonstrate the depth and breadth of the work to date and gain acceptance of that work product on the part of customer and company management personnel. Table 3-1 shows the nature of the principal relationships between the program phases and the nine generic process steps. Figure 3-3 illustrates this same relationship in fundamental terms, but for the sake of illustration simplicity, only four program phases are included. In each program phase, we accomplish a subset of the generic processes. In subsequent phases, we accomplish another subset with some repeated work, where the overlap corresponds to needed iteration, and some new work.

The Terminating Review row of Table 3-1 identifies the major review that terminates the indicated phases. These acronyms are defined as follows:

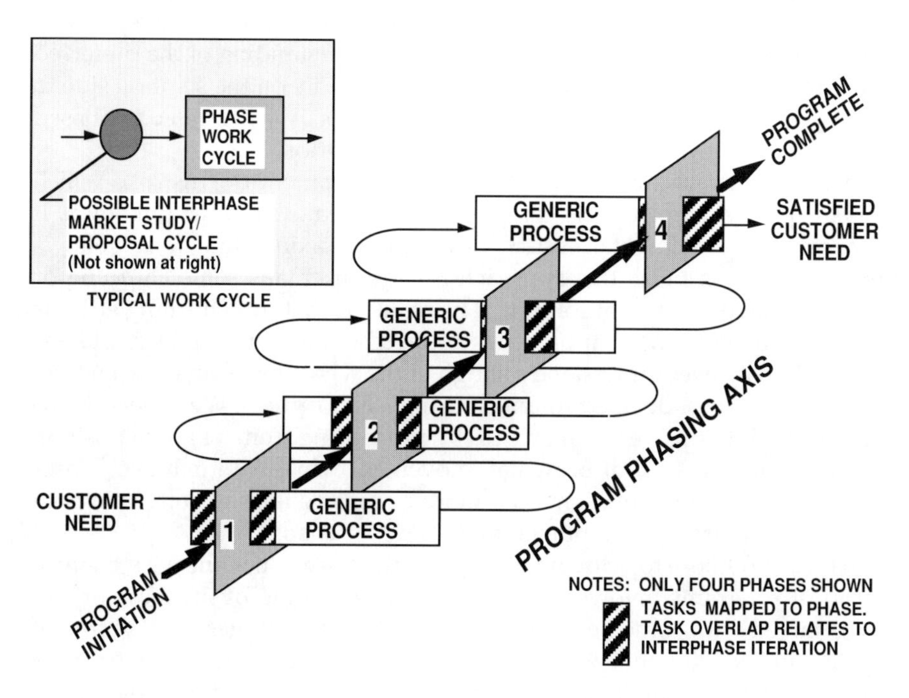

Figure 3-3 Correlation of program phasing and generic process steps.

SRR System Requirements Review where the customer and company reach a clear understanding and agreement on the customer need and system level requirements.
SDR System Design Review where the system design concept is disclosed and shown to be compliant with system requirements. Technologies are defined that will play a part in the development process and the maturity explained. Risks, including any immature technology, are identified.
PDR Preliminary Design Review where the major elements of the system are described in terms of their allocated requirements and complying design, analysis, and test approaches.
CDR Critical Design Review where the design solution is essentially complete (drawings 95% released) and described in the context of design features and the rationale supporting requirements compliance.
FCA Functional Configuration Audit where proof of compliance with development requirements is presented in the form of test and analysis reports. This same technique may also be applied to the physical product in the form of a physical configuration audit (PCA) to show that the first manufactured article satisfies product requirements.

The major phase-terminating reviews also mark the major iteration or recycling points focused on two different perspectives. First, we iterate the problem decomposition process as we penetrate deeper and deeper into the problem definition. This increasingly detailed understanding of the problem space is allocated to an increasingly detailed understanding of the preferred architecture (the solution space). Secondly, we iterate on the solution side to improve our definition of the designs for the items in the architecture based on improved knowledge about the optimum solution.

This manual encourages a top-down development downstroke which works from the simple to the complex, from the known to the unknown. It endorses the fundamental system development axiom that form follows function. Some criticize this approach because it entails some churning or iteration to arrive at a final solution. The reality is that this churning effect is necessary no matter what approach is applied because on complex problems we can almost never understand enough of the whole problem early enough to conceive the best detailed solution baring a lucky guess. We have a choice to put order into this iteration process by proceeding from a known position or to encourage a kind of Brownian movement while we lurch ever closer toward the optimum solution via a more expensive and less effective path called unstructured development, autonomy, or chaos.

Thus, we choose top-down development as the norm while we recognize that most programs involve a degree of re-engineering of predecessor elements mixed with some new development. As a result, there is a degree of middle-out development work that must be accommodated on most programs. In each program phase illustrated in Figure 3-1 we apply some subset of the process diagram shown in Appendix C. Some of these tasks appear for the first time in particular phases while others are repeated, as iteration leads to more refined understanding of problem and solution spaces.

3.2.2 Rapid prototyping

This plan emphasizes the top-down development of new products applying one or more structured decomposition models. For whatever reason, however, it may be determined that the rigorous phasing approach will not encourage a sufficiently rapid response. Rapid prototyping has the advantage of moving faster than rigid phasing by recursively applying a combination of analysis, design, and build where the design and build process stimulates rapid recognition of potential solutions. Rapid prototyping does not await final development of the requirements for an item before developing design solutions. This stream of solutions takes the form of actual physical entities or software structures that have increasingly capable functionality. Requirements are identified corresponding to desirable features of a given prototyping cycle and used as a basis for the next cycle.

While this approach can be accomplished rapidly, it can also lead to a leap to point design solutions that may not be optimum. There is a tendency for the team to fall into the group think psychology with an increasing strength of commitment to the current path. The team leader must be alert to the absence of critical comment and alternative ideas that are symptoms of this phenomena.

One or more teams on a large program may be given authority to pursue rapid prototyping while other teams are required to apply the rigorous phasing approach. In this case the PIT must make a determined effort to maintain respect for interfaces defined across the team boundaries.

Rapid prototyping is valid where there is relatively little information available on the customer's needs and minimal history of successful development. This technique is especially suitable where human operator actions are involved. It is very difficult to define requirements for human actions; yet, features of prototypes can be experienced by people with good and bad features determined as a result. Rapid prototyping is most effective within the spiral development sequence described in paragraph 3.4.3.

3.3 Product composition

The products produced by the company fall into four specific classes. The company prefers to be the prime contractor for development of grand systems poorly defined by our customers initially in terms only of a customer need expressing a difficult problem the customer wishes solved. These systems will commonly evolve into a solution entailing some combination of hardware, computer software, and human activity covered in procedures. It is possible that specific programs may entail only one or less than a complete set of the three component product types as well.

3.4 Sequence model alternatives

Three specific models related to development sequence are identified for use: the waterfall, V, and spiral. They are all reflective of the same process of providing teams of analysts and engineers an expanding knowledge base in an orderly way. But, each focuses on different characteristics of that process in ways that are useful in emphasizing particular process strengths. As noted above, a specific program need not apply only a single sequence model across the complete architecture or even at different times in the program development period. Different teams with different problems may be able to make a good case for different sequences.

Also, at different stages of the overall program, one or another sequence model may be more appropriate. For example, it may be very efficient to initially develop a computer software product using the spiral model to the point that the program is convinced that a viable solution is possible and that the requirements for that solution are fully understood. At that point it may also make sense to switch to the waterfall model to encourage schedule compliance.

3.4.1 Waterfall sequence model

Of the three models permitted by this plan, the waterfall model is the original and almost universally accepted model of system development. As such it is characterized by some as old fashioned or outmoded. On the contrary, the company accepts this sequence as the normal model barring sound reasons to apply the V or spiral model. The model derives its name from the common Gant chart portrayal of a series of linear task bars with the next task bar beginning at the time the prior one ends and illustrated below the earlier task as shown in Figure 3-4. One can picture the water cascading from the earlier task down to the next and so forth to the last task.

Some people prefer to illustrate the chart building in the opposite fashion where the later task is shown higher on the chart than an earlier one. This is still a waterfall model, but the connection to the sequence name may be harder for the uninitiated to fathom.

Contrary to the process depiction in the pure waterfall model, it is more commonly the case in actual situations that one can begin one development activity prior to completing an earlier activity. For example, we can begin the design activity on parts of a system where the requirements have been developed prior to the end of the complete requirements task. Beyond a certain point of overlap, the program environment moves away from rigorous phasing and closer to the rapid prototyping.

3.4.2 V sequence model

The V model extends the waterfall model to illustrate a finer granularity of a complex system composed of end items, subsystems, and components, and

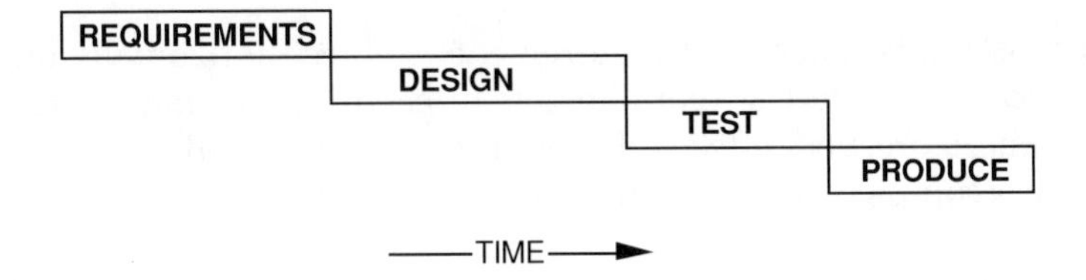

Figure 3-4 The waterfall development model.

it makes clear the very special relationship between the requirements work accomplished on the downstroke of the V and the testing or verification work done on the upstroke of the V. Figure 3-5 illustrates the V model.

The downstroke represents the front end top-down decomposition process working from the customer's need to a more expansive understanding of the problem represented by that need in terms of requirements for progressively lower architecture tiers for the system. At any level of architecture indenture, the V is essentially a waterfall sequence and it may be so illustrated on item team schedules.

The upstroke corresponds to integration work related to the items exposed through downstroke decomposition. This includes optimization as well as physical integration, assembly, and test. During the downstroke, we define test requirements as well as product performance requirements and

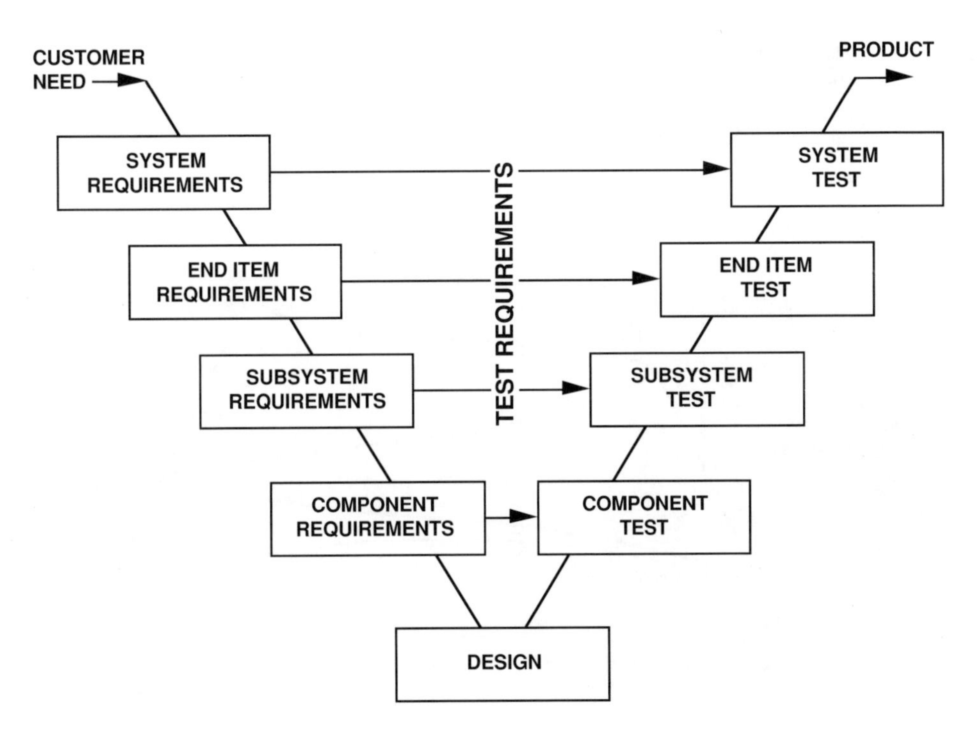

Figure 3-5 V sequence model.

design constraints. Those test requirements connect horizontally on our V diagram to define the test requirements at each level of indenture. The V diagram, therefore, very effectively communicates the need for test program hierarchical completeness as a function of the product requirements hierarchy.

3.4.3 Spiral sequence model

The spiral model illustrated in Figure 3-6 emphasizes the growing knowledge base from program inception with the customer's need (possibly not well defined at the beginning) through completion as an expanding spiral base as time progresses. It also clearly shows the iterative nature of the intended process. The development team applying the spiral sequence cycles through the same activities as in the waterfall or V model but does so at a higher repetition rate and with less perfectly well-developed ideas at each

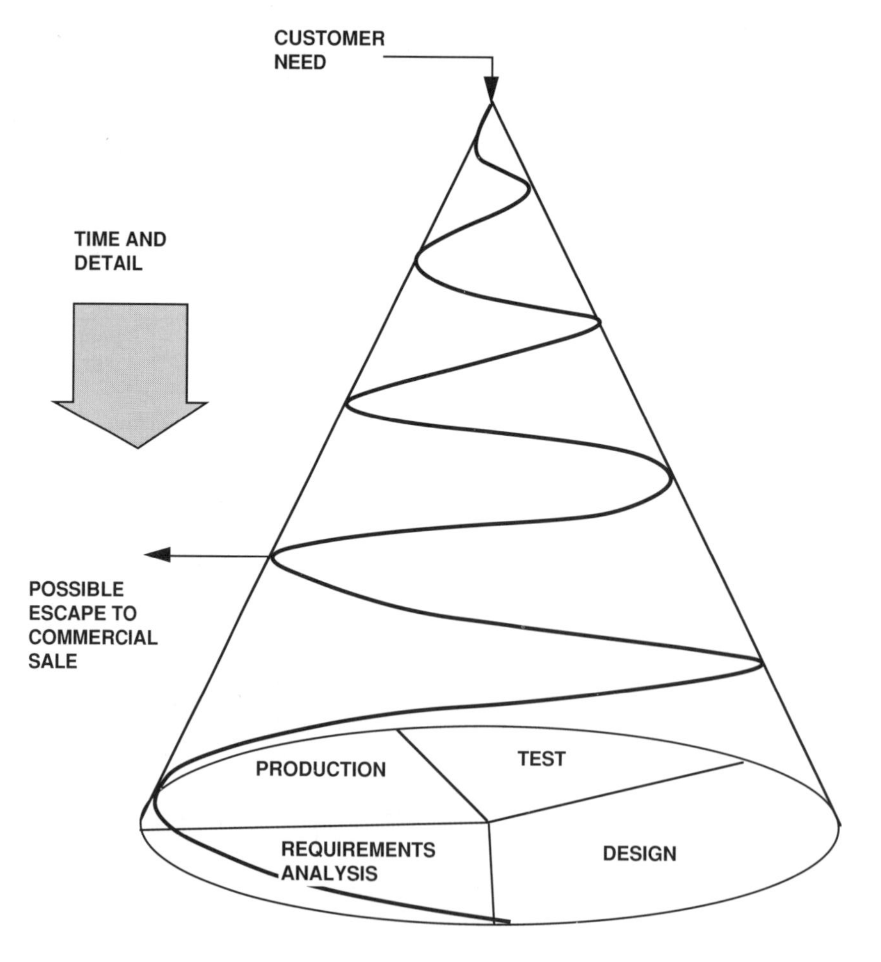

Figure 3-6 Spiral development model.

cycle. This model is appropriate when insufficient knowledge exists to characterize the item requirements as a prerequisite to beginning design work.

This notion appears to be in conflict with the declared intent in this plan of defining requirements prior to design and it is to some extent. But there are narrow situations where an acceptable solution can be more rapidly derived through rapid construction of an imperfectly functioning model and subsequent improvement of that model. Also, in the spiral model, we do not forego all requirements work. We make an attempt to understand the requirements as a prerequisite to initial design work, but we do not hold up attempts to create a working prototype until we have perfected the requirements. Once the working model has been constructed and tested, the team uses the accumulated knowledge to better phrase the requirements defining the problem space and cycles through the sequence again with a working model closer to satisfying the need. This process continues until the team is satisfied that an acceptable product has been created or certain predetermined customer guidelines have been achieved.

The spiral model is especially appropriate to computer software development entailing a heavy human interface requirement as in operation of a computer terminal. In these situations it is very difficult to forecast what the human will consider desirable characteristics. It is very easy for the humans to tell analysts on a working model what "works" and what does not work. The results can then be folded back into the product on the next cycle as the team spirals in closer to the goal. Note that the spiral is commonly illustrated as an opening spiral in time to emphasize the expanded knowledge in time. But, it is also helpful to think of the spiral development process as closing in on the desired solution in time. Each cycle of the spiral provides better information about our goal and we progressively reduce the error to closure.

The spiral also illustrates an idea of interest on commercial products. The development of a product can continue indefinitely with periodic product escapes from the spiral at points where the product is believed to be good enough to market in a competitive situation. At each product release point work continues to progressively improve the product based on new knowledge, including customer feedback from previous releases. This notion is not useful where each release would be characterized by an expensive physical replacement or modification of hardware previously delivered, but it is very useful where the product is software that can simply be replaced or upgraded with no physical change to the apparatus it is used with.

It is also obvious that the spiral model is compatible with the rapid prototyping approach described above in paragraph 3.2.2. Where the program applies the rapid prototyping approach, it should also apply the spiral generally.

3.5 Encouraged environments

For grand systems, the rigorous phasing waterfall environment is encouraged. For hardware systems, the rigorous phasing approach is encouraged using any of the three sequences. Computer software not requiring intense human interaction should be developed using the rigorous phasing waterfall environment. Computer software entailing human responses should be developed applying the rapid prototyping spiral environment.

3.6 Program information environment

Programs are responsible for establishment of effective means to retain information created to satisfy program needs. Commonly this will entail computer databases of one kind or another. There are some items that may only exist in a physical sense but these are no less important. Examples of these are physical models and sketches executed in a physical media. Some information and physical objects may require special handling and storage as a function of its security concerns either because it has a customer-defined classified nature or it has a very sensitive proprietary nature with respect to potential company competitors. All programs are required to comply with company security manuals with respect to proprietary material and customer security requirements for classified material.

3.7 Management style and the working environment

Those responsible for program management are responsible not only for meeting customer technical, cost, and schedule requirements but also for doing so while avoiding damage to and improving the company's most vital resource, its employees. These goals are not in conflict. Quite the contrary, employees who are well led and knowingly and vitally involved in program activity will produce quality work. Program Managers are responsible for company employees during the period they are assigned to their program. When employees are no longer needed on a particular program they return to their functional department, for which the manager should already have negotiated a new assignment.

As these transitions occur, the employees should have derived some benefit from their past program experience and therefore be of greater value to the company. Whether this occurs is more a matter of the management style of the program than the nature of the work on the program. Good managers will establish a good environment within which employees can efficiently apply their specialized skills toward achieving program objectives with minimum negative reinforcement.

3.7.1 Decisiveness versus participatory management

Programs should be organized using teams populated by many kinds of specialists appropriate to the item for which they are responsible. The leaders of these teams should encourage active participation by all members as a means of sharing specialized knowledge toward team goals. In team meetings, team leadership must seek out the opinions of those who do not actively participate and respectfully restrain the efforts of the more forceful team members to dominate meetings and team affairs.

While it is very important that teams gain full benefit from the knowledge and work of all team members, it is also vital that teams reach timely decisions or conclusions where alternative courses of action are presented. Where active team participation is encouraged, there is a tendency for rule by committee which can lead to paralysis and missed schedule events. The team must seek out and carefully study alternatives, but the leader must act decisively to encourage a conclusion and a movement forward toward team goals.

3.7.2 Order versus creativity

Systems engineering is sometimes described as the science of order and discipline. It is more accurately described as the science of providing the needed balance between order and creativity. In early program phases characterized, by relatively little knowledge about the product, it is important to encourage a balance point far over in the direction of creativity. As the system solution becomes clear and the commitment to the preferred solution builds, the work represented by that solution must be protected by a progressively increased appeal to order. Subsequent to CDR, the resistance to change should be very strong.

section four

Technical program planning and control (MIL-STD-499A Part I)

4.1 Precedence of controlling documentation

Section 2 of this document lists all standards that are respected in the normal implementation of the company systems engineering process. The authority on standards for a given program is commonly the Statement of Work and the system specification and the standards referenced therein. The specific precedence of controlling documents for a program is defined in the contract which is generally the ultimate controlling document on any program. Customer documentation is intended to be tailored, if necessary, such that this document may be used as the program authority for application of the systems engineering process.

4.2 Company organizational structure

The enterprise is managed through a matrix structure with a lean functional staff and one or more strong program structures which are required to perform program work in accordance with a common process defined by the functional organization through an enterprise integration team (EIT). Each unique program is managed by a program manager reporting to the enterprise executive. Programs employ personnel provided by the functional departments and are responsible for managing those personnel. Within the functional structure, a manufacturing facility organizational arrangement is employed to manage the potentially conflicting demands of the programs. Figure 4-1 illustrates this construct, and the elements of the organization are described in subordinate paragraphs. Temporary proposal organizations are also established from time to time to encourage transitions into contract wins and program starts.

4.2.1 Company functional organization

The functional organization is structured with departments arranged in a vertically thin hierarchical network as a function of the way the company has

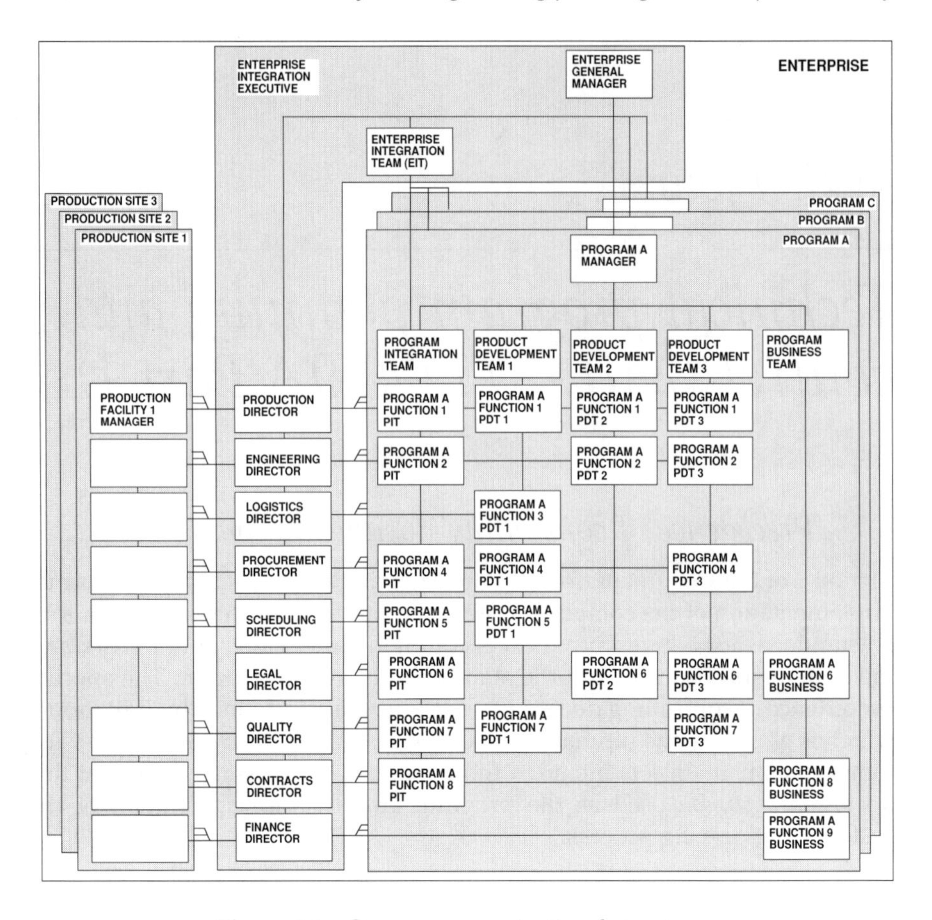

Figure 4-1 Company organizational structure.

decided to partition the knowledge base needed for the company product line and customer base. The functional managers reporting to the executive are members of the enterprise executive board responsible for enterprise management through the EIT and their own efforts within their own departments. Functional department managers are responsible for four specific activities defined in subordinate paragraphs.

4.2.1.1 Personnel

Functional management shall establish and continuously improve the personnel staff responsible for accomplishing the work defined in department charters. They shall make assignments of personnel to program teams and accept returning personnel from program teams for reassignment. Functional management acts as the personnel clearinghouse between programs. In the process of accomplishing this responsibility, functional managers are encouraged to make assignments that have the effect of improving the experience base of their department. Functional department managers are responsible for hiring and separation decisions.

4.2.1.2 Company training program

The company has a responsibility to provide for the continuing education of its work force. This is accomplished through the company training program administered and conducted by the EIT in accordance with content direction from functional management.

Recurrent systems engineering process training is provided for company personnel in their specialties in classes on the company facility and through an approved system engineering certificate program at a local university. In-house training uses the company procedures manuals as text books and these documents are available throughout the facility via a computer networked library.

The in-house training program also includes a company product component to maintain personnel knowledge about the technical aspects of common product subsystems. Where a program depends on new technology not in the existing knowledge base of its employees, special courses are offered to reinforce this area.

Where company or customer procedures require maintenance of a prescribed certification status for a specific task through initial and refresher training, records shall be maintained by the responsible functional department to ensure that all personnel performing those tasks are currently certified.

Customer and supplier personnel are invited to participate in company training activities when they are working within the facility or in close proximity to the facility.

On certain contracts, the company may also have a contractual commitment to train customer personnel on specific facets of the product. These product training programs are under the overall responsibility of the particular PIT with specific elements possibly assigned to PDTs.

The program Statement of Work identifies any specific training responsibilities for customer personnel for the purpose of operating and maintaining the product system. The specific training program will identify customer performance and behavior deficiencies or shortfalls and offer specific training opportunities to achieve required proficiencies.

4.2.1.3 Enterprise practices

Functional management is responsible for preparation and continuous improvement of company practices used by all programs. Common practices encourage task repetition and corresponding personnel skills improvement even though personnel move between programs. They ensure predictable performance of program teams.

Each functional department is responsible for preparing one or more department practices manuals covering all of their charter tasks. This manual is the functional manual for systems engineering. The content of these manuals must relate to the standard process diagram included in Appendix C of this manual and be coordinated with the department task estimating approach.

Functional management is responsible for an external interface with personnel from other companies, related literature, and professional societies to benchmark company practices for the purpose of ensuring that internal practices represent the best possible way to perform company work. In addition, functional management will seek out program review assignments for the purpose of maintaining contact with the ways that their practices are applied. Where evidence suggests a need for an improvement, the changes will be coordinated with EIT and prepared for EIT review and approval. Programs may be used to test practices changes and permanent changes made or not as a result.

4.2.1.4 Preferred tools

Functional management is responsible for providing programs with effective computer tools for use in implementing standard practices. Tools must be consistent with standard practices. Functional management must maintain contact with tool sources, related periodicals, and other professionals knowledgeable about tools to stay abreast of the latest developments.

4.2.2 Enterprise integration team

The EIT performs both functional and programmatic tasks. It is responsible for integrating all enterprise functional activities involving two or more functional departments. It is also responsible for integrating the demands of two or more programs where common enterprise resources are in conflict. The team leader reports to the executive. The executive, all program managers, and all functional managers form the executive board that provides the EIT with guidance and direction. The EIT is, therefore, the day-to-day working authority that operates the company within the guidelines and priorities set by the executive. In many companies this work would be accomplished in an ad hoc way by the several functional managers working together where necessary.

4.2.3 Program organizations

All customer work is accomplished by program organizations. Each program is the responsibility of a program manager installed by the executive. All major program managers report to the executive. The managers for some programs that are derivative from other programs may report to the related major program manager. The generic term manager is applied to these posts even though in a particular situation the post may be occupied by a person in a pay grade named Director or Vice President.

4.2.3.1 Program staffing

Programs are staffed by the functional departments based on the work that must be accomplished as defined in the Statement of Work. The numbers of specialists in each functional discipline are determined based on the planned schedule and work intensity during particular periods of time. Program planning should focus on satisfying customer needs, but, if it is possible to do that while also smoothing customer demands across all programs, that goal should also be considered. Personnel assigned to programs shall be qualified to perform the tasks identified for the department providing personnel that are required on the program. Functional management is responsible for ensuring personnel qualification through training classes, on-the-job training, assignment sequencing, hiring of new personnel, and separation in accordance with company practices.

4.2.3.2 Development program organizational structure

A program is organized into several teams all reporting to a Program Manager or Director. One of these teams is a Program Business Team (PBT) responsible for program administration, contracts, legal matters, and finance. A second team is called the Program Integration Team (PIT) responsible for the technical development and management of the product and related program processes. In addition, each program includes several integrated Product Development Teams (PDT). Each PDT is responsible for the development of some specific assigned portion of the product in accordance with a team budget, schedule, work statement, and specification. This structure is illustrated in Figure 4-1 in the context of the complete company, but Figure 4-2 extracts only the portions related to a single program for the sake of simplicity.

The personnel to staff these program teams are provided by functional departments through negotiation between the program and functional management. Team personnel are supervised within their program team structures once assigned. Functional management provides programs with qualified personnel, good tools, and sound standard practices (such as this generic SEM/SEMP). Functional management is encouraged to follow the progress of their program personnel through participation on program review boards

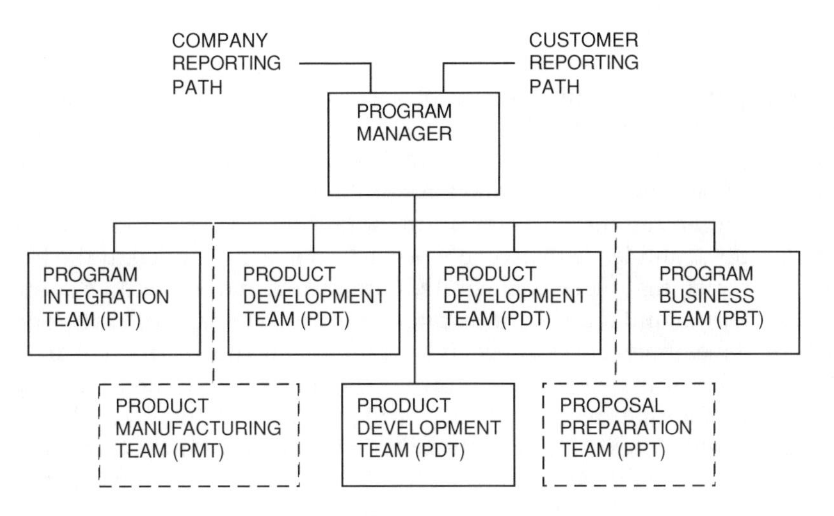

Figure 4-2 Generic program organizational diagram.

and to provide refresher training and professional support for difficult tasks but is prohibited from exercising supervision on a day-to-day basis.

The program may establish other special purpose teams to work specific projects but should avoid excessive use of special purpose teams since it is a symptom of bad program and product planning or management, suggesting that each new stress must be addressed by a special team. If a program is properly planned and organizationally structured, most new stresses should have been foreseen, expressed as program risks, and subjected to mitigation action precluding a resort to crisis management. The vast majority of program work should fit very neatly into the PDT, PIT, and PBT pattern if the program has been properly planned and executed.

Where it can be shown that the company will realize a significant cost savings from pooling program work at the functional level in isolated cases and the programs will not suffer in terms of loss of synergism, functionally managed work may be authorized. Where this is necessary, the program may apply virtual team concepts to encourage concurrent participation by team members collocated by function. The virtual team concept entails use of networked computer resources with real time audio/video capability to link personnel together in virtually the same experience they would have if physically collocated.

No matter how well implemented, however, the virtual concept is not as effective as the real physical collocation environment. It is important to align the physical collocation selection with the organizational axis that entails the most complex inter-personal relationships. The product development teams represent the most complex relationship; so, the emphasis must be placed on program managed co-located work by cross-functional teams. Functional departments are encouraged to apply the virtual team concept to hold brief periodic cross-program meetings to discuss problems, tool use, current practice problems, and other matters of interest to the discipline.

A program generally has its beginning as a proposal team as indicated in Figure 4-2 with the dashed box. Many programs are implemented in phases requiring proposals for each phase or groups of phases. In this case an on-going program may have to establish a temporary proposal team several times in the life of the program. These teams are staffed from program personnel possibly augmented by additional personnel supplied by the functional organizations; they transform the customer's request for proposal into a proposal and leave the scene when selection is concluded. If the team wins a contract, that work is integrated into the existing program stream.

The diagram in Figure 4-1 illustrates an organizational structure appropriate for development of product given that the company has more than one program in house at one time. If and when the company production capability includes multiple manufacturing facilities, these facilities can best be managed in the context of the facilities. During development work, manufacturing personnel will be members of the PDTs and, during production, engineers must be assigned to the facility-oriented teams. Generally, the right time to switch from PDTs to facility teams is after first article acceptance. The program organization must coordinate with one or more of these teams as indicated in Figure 4-2.

As a product matures and passes into a production status, the PDTs migrate to facility teams, and the program organization collapses to the PIT and business team. As engineering change proposals, major modification programs, or mission-peculiar work expands, the PIT may be able to accomplish this work directly or may have to assign specific teams to specific change work during the duration of time required to fold the results into the normal production process.

4.2.3.3 *Program management*

The Program Manager/Director is responsible for overall program activity and has the authority to act for the company in accordance with the contract and company policy. The Program Manager/Director will be supported by a staff composed of leaders of the several teams established for programs as illustrated in Figure 4-2.

4.2.3.4 *Program business team*

The PBT is responsible for program administration, contracts, finance, estimating, program office management, and customer formal interface. The team leader reports to the program manager.

4.2.3.5 *Program integration team*

The PIT is responsible for system requirements down to the level that includes the top set of requirements for each PDT and any interface control documentation with associates or the customer; product and process integration and optimization at the system level; and all other product and process technical issues. The PIT is the system level PDT.

4.2.3.6 *Product development team*

The PDT is responsible for the development of an assigned element of the system in accordance with a set of product requirements, a clear statement of required work, and a schedule. A program may have several top level PDT and the leader of each shall be on the Program Staff. A top level PDT may include one or more subordinate PDT, but those leaders shall report to the parent PDT. The PDT is responsible for lower tier requirements analysis and staged development (concept, preliminary design, and detailed design) of the preferred design solution coordinated with a system level process for manufacturing, testing, quality assurance inspection, logistics support, and operation. On programs using manufacturing facilities serving multiple programs, the PDT will commonly disband after first article inspection and responsibility for continued product development assigned to facility-oriented teams called Product Manufacturing Teams. Otherwise, PDT may continue to serve the program through the manufacturing process.

4.2.3.7 *Principal engineers*

PDT may assign lower tier responsibilities for elements to persons called Principal Engineers, responsible for development of the assigned item. The

Principal Engineer draws on the personnel base of the responsible team for specialists. Normally, a separate organization is not built up around the Principal Engineer.

4.2.4 Transition from development to production

Figure 4-2 illustrates the organizational arrangement for the development period and does not address all of the transition to production possibilities. Figure 4-3 shows two of several possibilities. If a program product is manufactured in several company manufacturing facilities, each with multiple program responsibilities, the company will have one Product Manufacturing Team (PMT) for each facility it employs. This arrangement prevents any one program from dominating production capacity and optimizes resources for the aggregate demand of all using programs.

In the one case of Figure 4-3, all company manufacturing will require facilities with multiple program responsibilities. Upon completion of first article inspection of the item for which each PDT is responsible, the PDTs will be disbanded, leaving the PIT and PBT to handle any continuing program responsibilities for these products. Manufacturing responsibility is picked up by PMT based on company facilitization. In this example there were three PDT and these products will be manufactured in two facilities shared by other programs. Some combination of the three product elements will be manufactured in each of the two manufacturing facilities.

In the second example of Figure 4-3, there remain three teams after the transition, but two of them entail program-dedicated manufacturing facilities,

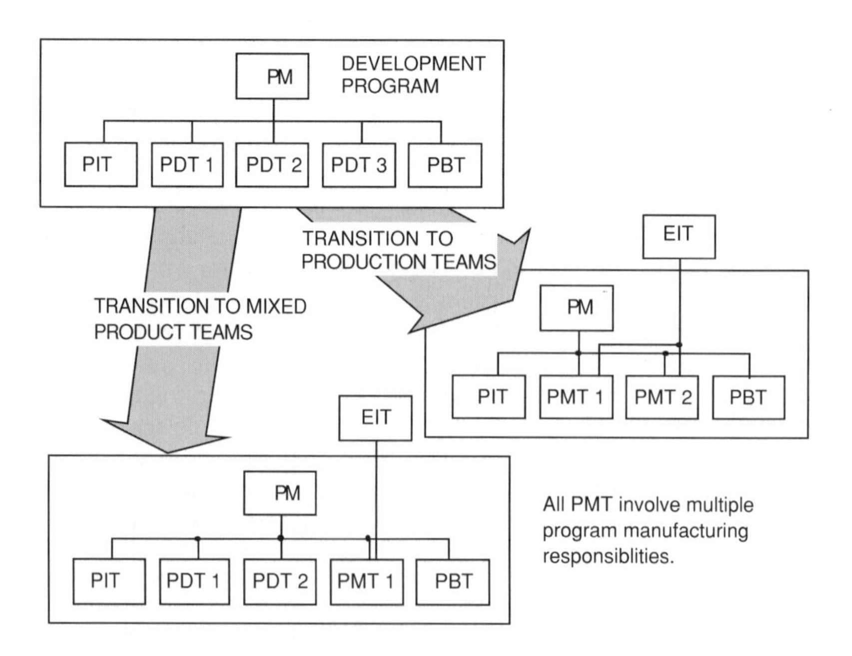

Figure 4-3 Team transitions.

while the third is manufactured in a facility with multiple program responsibilities. One or both of the PDT could correspond to a major procurement item or an associate item as well as to items that are manufactured in-house in program-dedicated facilities.

Each program is responsible for structuring their organizational relationships to accomplish development and manufacturing within the context of company manufacturing resources. There are many possible arrangements besides those illustrated in Figure 4-3. The following rules summarize company-mandated program organizational relationships:

a. During development, programs must organize in cross-functional teams aligned with the way the product is organized.
b. Any company manufacturing facility responsible for manufacturing products for multiple programs will be managed in their joint interest with representation from all participating programs.
c. A manufacturing facility dedicated to a single program may be managed as a program-peculiar PMT unless and until it undertakes manufacturing tasks for another program. The EIT shall periodically review cases of program peculiar manufacturing to ensure that all enterprise facilities are being optimally employed.

In cases where there is a team realignment at first article as explained in the previous paragraph, it is essential that there be manufacturing representation on the PDT during the development process to ensure that the manufacturing process is concurrently developed with the product design. Similarly, when the product responsibility transfers to a PMT, there must be engineering representation for liaison purposes to cover any needed engineering changes. Some of the manufacturing people and design engineers involved in the development process should be retained in the related PMT to ensure that lessons-learned knowledge transitions into the production process.

The rationale for the transition between a product-oriented organizational structure during development and facility-oriented structure during manufacture, where multiple program responsibilities exist, is that manufacturing facilities can best be managed, and the personnel motivated to perform, as a facility with specific inputs, outputs, and processes defined by the several programs. This arrangement moves the management of all material acquisition into a common framework. The program source for an item is managed much the same whether the source is a supplier or an internal team.

4.2.5 Proposal organizational structures

Proposal teams are temporary organizations that act as transitioning instruments between a desire to win a particular program and achievement of a contract status. They are applied to that part of the company business where customers notify prospective contractors of their needs and the contractors offer proposals telling how they will satisfy those needs.

Given a decision of the EIT and company executive to bid on a particular program, the company executive appoints a proposal manager who is then responsible for all work focused on responding to the potential customer's needs within the constraints of company policy.

The program manager is responsible for acquiring the support of functional management for staffing the proposal effort. These personnel will be organized to respond to the customer's request for proposal (RFP). Commonly, the proposal team must be organized in accordance with the information that must be produced for inclusion in the proposal as defined in the RFP.

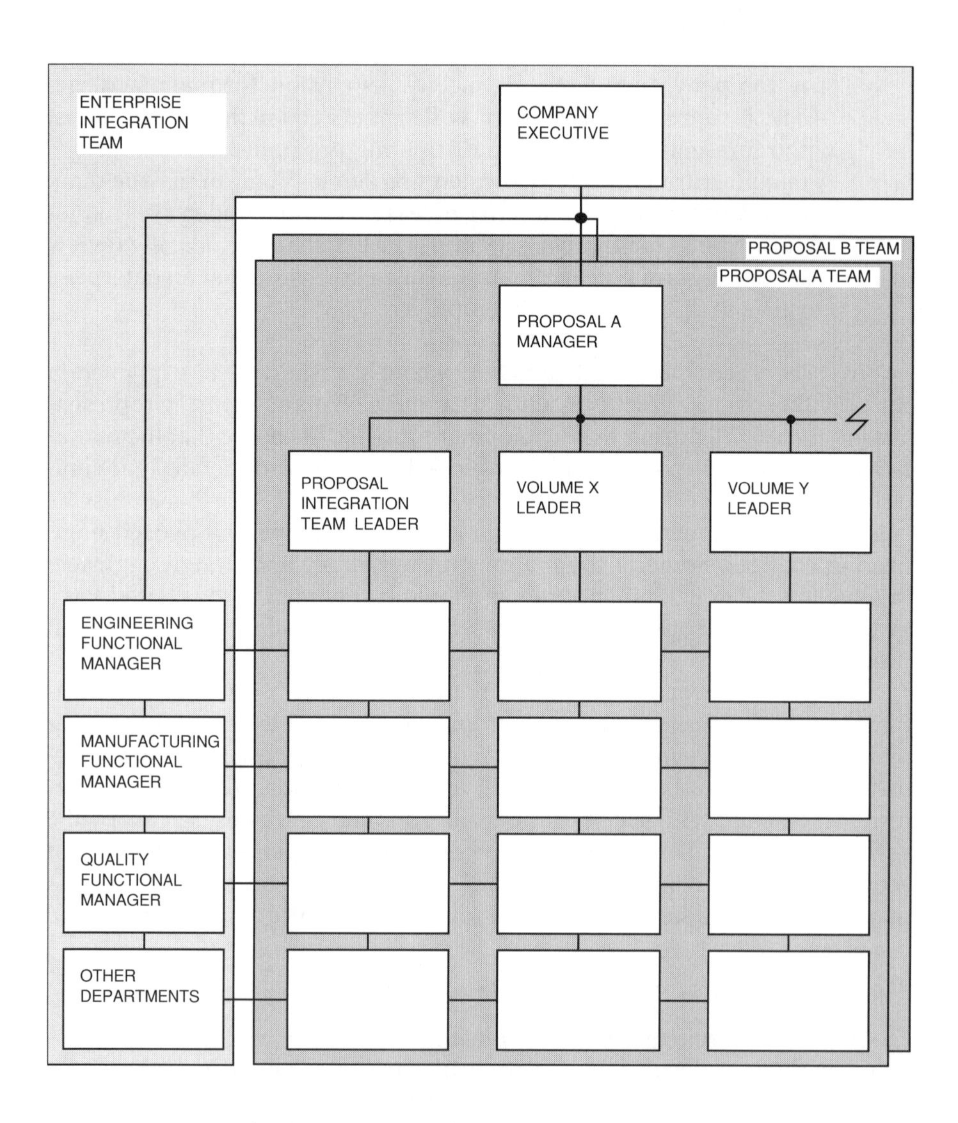

Figure 4-4 Proposal organizational structure.

Teams shall be defined, selected, and staffed for major elements of the proposal and their work integrated by a proposal integration team. Figure 4-4 shows only a representative structure. An actual program must determine the detailed structure based on the customer's RFP structure and needed functional disciplines.

4.3 Generic company process documentation

This SEM/SEMP provides the foundation upon which all enterprise systems engineering planning is based. It identifies and describes the systems engineering practices that shall be applied on programs consistently with program needs. The companion enterprise process diagram (Appendix C) identifies the tasks that must be accomplished to develop products. In response to this generic definition of the process applied by the company, all functional departments with systems engineering responsibilities are required to develop a set of supporting systems engineering task statements that clearly identify the specific tasks that in the aggregate are required to assure enterprise success in the process. These generic task statements have been reviewed by the EIT, as are recommended changes to them, and found to be consistent with a least-cost effective method for implementing the systems engineering process.

4.4 Input-process-output

4.4.1 Major outputs, deliverables, and results

While performing the systems engineering process as described in this manual, several major technical outputs or deliverables must be produced and made available to program personnel and to the customer as a function of their identification as Contract Data Requirements List (CDRL) or data accession list (DAL) items. These items are described in subordinate paragraphs. Commonly these items will be developed on programs whether they must be delivered to the customer or not, but some may not be required on specific programs or contracts due to the nature of the program and the kinds of risks that influence the development effort.

4.4.1.1 Planning data

Program-specific plans are prepared prior to and during the program to clearly define needed work, associate budget and schedule with those tasks, and coordinate related tasks. These plans must be maintained throughout the program.

4.4.1.2 Requirements analysis data and specifications

The results of the requirements analysis process in the form of specifications and all of the data created to gain insight into appropriate requirements for system items are used as a basis for the design effort, qualification and verification work, and many other program activities.

4.4.1.3 Design and analysis data

Engineering drawings, analysis reports, and other design data are created through the conceptual, preliminary, and detailed design processes described elsewhere in this IMP/SEMP. They shall be the result of an integrated team effort coordinated with process designs created for procurement and material planning, manufacturing, quality assurance, logistics support plans, and test planning documents. All of these data are internally reviewed by PDT supervision during informal in-process reviews and after completion. They shall be selectively reviewed by the PIT for system optimization and consistency with system and program requirements and compliance with item requirements.

4.4.1.4 Verification planning and reporting data

All requirements in customer specifications must be verified through an appropriate verification method (test, analysis, demonstration, inspection), resulting in collection of compliance evidence that must then be subjected to review for suitability. Reference to surviving evidence shall be recorded in a verification management matrix that correlates requirements with the evidence and status. Evidence shall be accumulated and retained for reference by company and customer personnel upon demand. Customer access may be conditioned by DAL constraints but should be encouraged between company and customer specialist counterparts to ensure the earliest possible identification of disagreement between company and customer positions relative to the adequacy and accuracy of verification evidence.

4.4.2 Process inputs

Effective implementation of the systems engineering process requires specific minimum resources and information that encourage cooperation and communications between the customer and company and the many people staffing a program. Some of these process inputs must be supplied by the customer (or associates identified by the customer) while others must be supplied from within the company or from suppliers under contract to the company.

4.4.2.1 Customer inputs

The systems engineering process requires the following external inputs and supporting resources and conditions from the customer and any associates identified by the customer:

a. A clear customer definition of their need and guidance on their meaning associated with that need;
b. Inputs on their system requirements (in the form of a previously prepared system specification or energetic participation with company personnel in development of one) and their prioritization of needs and requirements;

c. On-going timely critical technical feedback from the customer on requirements, design concepts, verification plans and evidence, and planning data;
d. Associate and customer data making clear the decisions reached across interfaces for which other parties bear the responsibility to develop; and
e. Easy access to customer standards and other customer documentation referenced in contractual documents.

4.4.2.2 In-house inputs

A program team requires the following inputs from company resources to create and sustain an effective system engineering capability:

a. Qualified personnel skilled in their specialty areas, knowledgeable about company standard practices, skilled in the use of available tools, and experienced on the company's product line;
b. Effective standard practices for tasks that commonly must be carried out in the performance of company business;
c. Computer resources, applications (tools) and services for word processing, database applications, spreadsheet work, graphics, special analyses, simulation, and modeling work, computer-aided hardware and software design tools, and other applications;
d. Effective voice communications capability (coordinated with computer workstation placement) including voice mail and conference calling within and without the program work space;
e. Office materials (paper, writing implements, etc.);
f. Furnishings and facilities including one large unbroken wall expanse (8 feet by 10 feet long preferred) or meeting room in close proximity to the area allocated to each PDT plus one each for the PIT and PBT;
g. Office equipment including viewgraph machines, some in meeting rooms equipped with computer data adapters for projection of that data from network resources, FAX (as appropriate to the program), and reproduction machines;
h. Cross-functional information resources including computer networking and communication resources such as inter- and intra-team teleconferencing capability, electronic mail, and a physical collocation environment supporting direct contact by voice and body language within a team space; and
i. A clear definition of the desired program organization structure from the program manager.

4.4.2.3 Supplier inputs

The company must have access to the following supplier inputs (commonly by requiring them as SDRL items) in a timely way as defined in the subcontracts with the suppliers:

a. Supplier data called for in the subcontract;
b. Access to supplier management and design and specialty engineers involved in the development of supplier components for the purpose of routine, informal status, and technical conversations; and
c. Formal management and technical reports as prescribed in the contract.

4.5 Program planning transform

Upon receipt of a customer request for proposal or enterprise identification of an approved marketing opportunity, the person assigned responsibility for exploiting the situation must begin a systems engineering planning exercise specific to that project. The key elements of that process are briefly described below:

a. Understand the customer's need. Either through reading customer prepared material describing their need, talking to customer representatives about their need, or conducting studies of customer needs, determine exactly what the customer's need is in terms of products related to the company's historical product line or future directions. Determine the schedule and available customer resources in time in order to frame the planning response within achievable boundaries.
b. Through an organized analysis of the customer's need, determine an appropriate mission scenario and matching system architecture, environmental definition, and logistics support concept. Prepare a system specification or precursor document or review and edit a customer supplied specification.
c. Define a WBS based on the system architecture defined in the system specification respecting customer guidelines to the maximum extent possible consistent with item boundaries responsive to manufacturing, procurement, logistic support, and test concerns.
d. For each product WBS element, prepare one or more statements of work required to satisfy the customer need. Do the same with all process-oriented WBS elements. Collect all of these statements into a SOW. If the customer prepares the SOW, review for consistency with the planned approach.
e. Assign personnel derived from the functional organizations to planning teams related to the architecture (WBS). For each SOW element, the responsible team must identify the generic, functional task statements that implement the statement. The team leader should be the person who will lead the integrated product development team during program implementation and that person is responsible for integration of the tasks to ensure elimination of excess and redundant entries and filling of any detected omissions.
f. The detailed work definition data is related to specific program events and tied to significant accomplishments — essentially the individual

task objectives with clear definitions of task completion criteria. The results are included in an integrated master plan and scheduled in a coordinated integrated master schedule.

g. This SEM/SEMP provides the narrative guidance on accomplishment of the systems engineering tasks so listed. Other company standards provide similar narratives for other functional work areas. Augment the program planning data with identification of program specific risks.

4.6 Program planning documentation

Figure 4-5 illustrates the relationship between the several program and work planning documents. Subordinate paragraphs describe these documents and define their purpose. Note that these documents provide the top end of several major program document series commonly arranged in trees.

4.6.1 System need statement

The need statement may or may not be supplied by the customer. Even when it is supplied by the customer, the company must be very careful to validate with the customer that the statement expresses their real need. In cases where the customer need was never written or has been lost in the process of

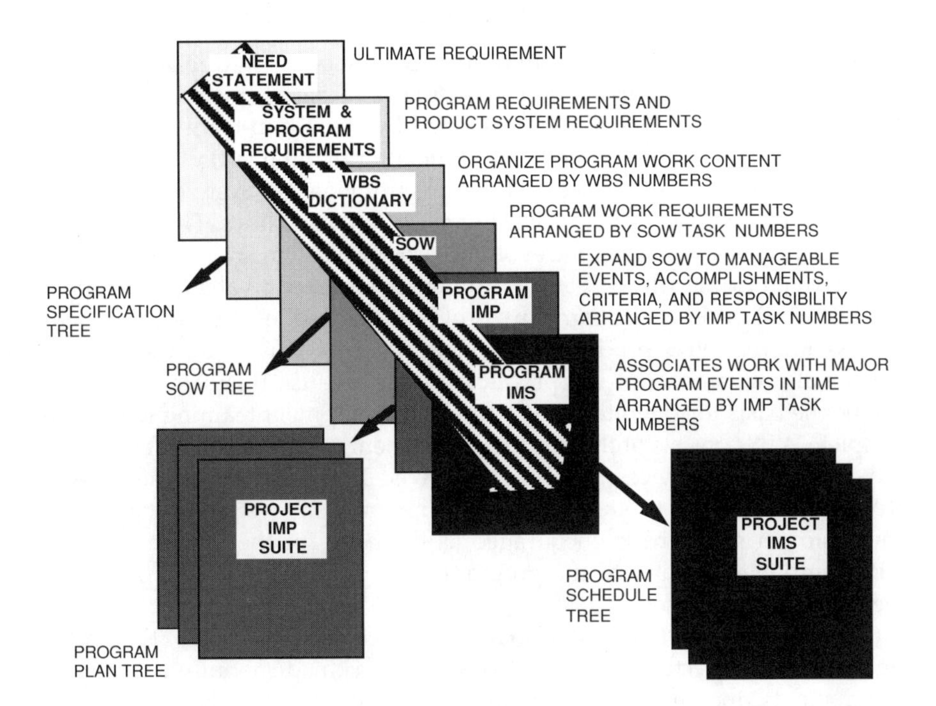

Figure 4-5 Systems engineering planning documents.

customer development of a system specification, the company, in coordination with the customer, must re-create it in order to be certain that the correct problem is being worked. This will commonly be a very difficult undertaking encouraging one to stop work on the statement, but the more difficult it is to characterize the need to the mutual agreement of customer and company, the more need there is to complete the task. The need statement must be completed very quickly as well because impatience to get on with the program and the urgencies of schedule compliance will eventually force the program to move onward regardless of the program team's state of readiness to do so.

4.6.2 *System specification*

The system specification (very early in a program this may be a precursor document such as a system requirements document or operational requirements document) defines the product requirements based on an initial decomposition of the customer's ultimate requirement called the need statement, a mission analysis, and other analyses possibly involving some simulation. The system specification also defines the top level system architecture selected to satisfy top level system functionality decided upon in these same analyses. This architecture is the basis for the work breakdown structure (WBS).

4.6.3 *Work breakdown structure (WBS) dictionary*

The WBS collects product elements and needed services into related groups for the purpose of organizing cost, schedule, and management concepts for the program. The WBS shall be developed by a PIT sub-team early in the program (initially on the proposal and thereafter maintained) with membership including Program Office, PIT leadership, Finance, Systems Engineering, Software, and the principal hardware design disciplines. The WBS shall be developed by overlaying the evolving product system architecture with a WBS derived from MIL-STD-881, the other customer equivalent, or an effective historical or on-going company program cost model. The development process for the WBS product component shall be driven by the needed functional architecture of the product system because this same structure will be the basis for the integrated product development teaming logic. The complete WBS consists of the product component created as indicated above and a process or services component derived from a study of the needed process to create the product defined in the program system specification. The customer standard is encouraged as the basis for the service-oriented WBS components. Refer to the program WBS dictionary for the program-specific WBS structure.

The specification tree shall also be developed as an overlay of the evolving product architecture with conscious decisions made about what kind of requirements documentation is required for each item on the architecture diagram. The WBS and customer configuration/end item specification selections shall be coordinated to ensure that there is a level of WBS that corresponds to each integrated PDT such that the team may be provided with an overall product requirements document on the element for which they are responsible.

The PDT shall also be provided with task definitions in the form of one or more WBS elements and corresponding SOW and IMP content focused on the work necessary to satisfy the item requirements. Where major items are to be procured and development work is required, those items shall have a procurement specification and SOW prepared as part of the contract between the company and the supplier. Each team, whether a company PDT or major supplier company, shall have a top level requirements document and planning data to define their expected product results and associated work. Lower level specifications may be necessary as well.

The PIT shall regulate the passage of each PDT, sub-team, or principal engineer in the sequential development of appropriate requirements followed by development of a design concept, preliminary design, and detailed design or procurement strategy and procurement action. Each of these steps shall be accomplished as an integrated product development team activity under the management of the product development team leader or a person designated by the PDT leader. The PIT shall maintain a work control matrix coordinated with the master schedule that defines the team authority to engage in each of the major steps of the development sequence listed above.

The PDT leader is responsible for cost, schedule, and technical control of his/her team in their performance toward goals defined in the team item specification and work planning data. Since the product WBS is aligned with the product structure and teams are assigned based on the product structure, budgets allocated to WBS elements are aligned with the PDT structure.

The methods of time phasing of work breakdown structures, development and control of work packages, development of planning packages and their conversion to work packages, size of work packages, resource use traceability to work breakdown structure and organizational breakdown structure, and integration to scheduling and critical path identification shall be included.

4.6.4 Statement of work (SOW)

The SOW identifies high level work elements required to accomplish each WBS element. This is the level of task identification that the customer or internal sponsor wishes to use in managing the program. Each SOW element maps to one and only one WBS element at some level of indenture.

4.6.5 Program plan or integrated master plan

The program may use a Program Plan and subordinate hierarchy of subplans or an integrated master plan (IMP) to define required work. In either case, the plan will refer to this SEM/SEMP as the authoritative source of information about the systems engineering process to be applied on the program. The plan identifies a series of specific program events that will be used to accomplish formal reviews of progress and establish baselines for future work. It then correlates SOW tasks to these events and expands each to identify detailed work that must be accomplished by the several PDT and the PIT to realize specific accomplishments that are paired with specific measurable completion criteria by which customer and company management may clearly determine if the tasks have been completed successfully. Each task listed in the plan must be carried on the master schedule which gives timing requirements for these tasks.

4.6.6 Integrated master schedule (IMS) or SEMS and SEDS

Customers will normally call for a systems engineering integrated schedule and a series of systems engineering detailed schedules to define when each planned task shall be accomplished. The U.S Air Force integrated management system calls for an IMS. On commercial programs, the master schedule will simply be called the master schedule. In any case, the master schedule aligns task accomplishment with the calendar. Each planned task defined in the SOW must be shown on the master schedule. If the schedule entails significant schedule risk, or if the customer requires it, a critical path shall be defined and managed as a means to mitigate schedule risk.

All planned tasks are correlated with the WBS through the responsible team and architecture relationships since the team and WBS are coordinated with the system architecture. All tasks are coordinated with a set of predefined program events commonly aligned with the major reviews and program terminal phasing points. Each planned task, at some level of indenture, shall have associated with it a clear definition of what it means to complete the task in terms of accomplishments and a clear definition of how to determine successful accomplishment in terms of a success criterion. These will commonly be contained in the program plan keyed to the SOW paragraph and assigned master schedule events.

The PIT is responsible for laying out the master schedule at the top level and laying in the major events through which all work will be timed. This schedule will show a level of detail sufficient to define the overall schedule requirements for each PDT for requirements definition and design development. Each PDT will first determine if the allocated time is sufficient for what they perceive the job to be and interact with PIT if they conclude it allows too much time or insufficient time to accomplish the team responsibilities. Given that there are no problems in this regard or that they are resolved, the team shall refine this top level portrayal on a coordinated team schedule, showing a greater degree of detail and intermediate milestones of value in managing the team activity.

The PIT is responsible for integrating the detailed team schedules into the master schedule and for determining the critical path through the master schedule supplemented with detailed team schedule data. The PIT will maintain, in clear view for those working on the program, a portrayal of the critical path and current schedule status throughout the program period. Should re-planning be necessary, it shall be coordinated by PIT.

4.7 Program work definition and authorization

All program work shall be pre-defined with clear goals and criteria for completion in an approved plan with an associated budget and schedule span. Each work item shall be clearly assigned to a program team and within that team to a specific person responsible for accomplishing the work. All work shall be specifically authorized to begin in accordance with the approved plan by a person in the chain responsible for that work.

4.7.1 Product definition

The product shall be organized using a hierarchical architecture structure. The elements of this structure should be determined as a result of allocating needed functionality to objects. These objects should be assembled in the architecture using a series of overlays to ensure an optimum structure is defined. The first overlay is financial in the form of the WBS explained below. The manufacturing breakdown and drawing breakdown structures will also be considered. Interfaces between the objects will be studied to ensure that the simplest cross-organizational interface situation evolves as the responsibility overlay is established.

4.7.2 The work breakdown structure

Ideally, the program product-oriented WBS shall be defined by overlaying a finance view onto the functionally derived product architecture. The reason for this is that programs must apply the PDT organization oriented about the product architecture and the management and budget structure must be set up to encourage team leader responsibility for cost, schedule, and product performance.

The customer may demand that the WBS be defined otherwise and fail to understand the sound basis for the company policy that would make their job of managing the program easier as well. In such cases, every effort should be made to communicate the rationale for the company policy. If it cannot be done, the program must map the WBS to the product architecture and make every effort to allocate budget in accordance with the functionally derived architecture without violating customer reporting rules.

The process- or service-oriented WBS element structure should respect the company process definition defined in Appendix C.

4.7.3 The program statement of work and work responsibility

The SOW includes one or more work statements for each WBS item. These statements expand upon the corresponding WBS item and tell at the program level what work must be done. The SOW work items should be allocable to specific teams in a one-to-one relationship if the WBS and responsibility architecture overlays have been properly established.

4.7.4 Program budget definition

Program budgets shall be established and aggregated within the definition of the WBS and SOW task statements. They shall be time phased as a function of task scheduling in the program schedule. Each budget item at some level of indenture should allocate to a single program team.

Budgets are initially determined during proposal efforts or equivalent commercial activities by mapping company generic budget estimates for functional department tasks supportive of SOW tasks to the WBS and SOW definition. These are reviewed by team representatives and the PBT finance function and integrated for identification of any conflicts, omissions, and mismatches. The surviving budget estimates must be reviewed and approved by the prospective program team leaders and subsequently by the program or proposal manager and company executive.

4.7.5 Program schedule definition

During the initial program planning period, commonly occurring during the proposal, PIT and PBT will cooperate to integrate schedule needs identified by teams to achieve customer-defined milestones or events. Each task at the SOW level shall appear on the program master schedule linked to major program events. Each master schedule task should be allocable to one specific team at that level.

4.7.6 Work authorization

All required program work is defined in the program SOW, plan, and schedule as described above. This work is coordinated with the WBS which is further coordinated to the preferred system architecture to which teams have been assigned responsibility. All budgets are coordinated with the WBS; so, budgets are coordinated with the PDT structure defined for the program by default. In keeping with DoD and NASA C/SCSC requirements and sound business practices, these budget elements are also coordinated with the company functional departments, but the functional departments are not involved in working supervision of the team tasks. This is all accomplished on the program organizational side to preserve the sound practice of holding a single person (a team leader) accountable for process cost and schedule and product performance. Functional department allocations are used internally to determine staffing needs.

All tasks appear on the master schedule. When the time nears for tasks to be initiated, the PIT coordinates with the responsible team leader and ensures a state of readiness to accomplish the task in terms of availability of qualified personnel, prerequisite task readiness, availability of needed information, and supporting resource availability (workstations for example). Plans are set in motion to resolve any task support actions. At the time the task must begin according to the schedule, the required budget is released to the appropriate team or teams and the PIT begins tracking task accomplishment and status.

Should it become necessary to change the plan or schedule, the impacts are evaluated and the changes are either approved and implemented or ways resolved to continue with the current plan. If additional budget is required, it is made available from reserves, or parts of the existing plan are deleted to allow completion within the existing budget. If additional time is required, it is either added to the schedule if not on the critical path, task components deleted, or overtime authorized to preclude schedule changes.

4.8 Performance tracking and reporting

Programs will apply the company cost/schedule control system to track and report all program tasks and aggregate task reporting for teams and functional disciplines as well as program management. The PBT is responsible for setting up budgets for program teams and monitoring performance to that plan. Budgets are assigned at team levels and team leaders are responsible for performing in accordance with planned profiles. Where a team cost and schedule performance indicates a significant deviation from the planned track (defined on each program by the PBT and customer), that team is required to provide details on the problem and information about how the problem will be solved and cost or schedule performance placed back on track.

4.9 Development controls

The PIT is responsible for exercising overall control over the program development process. Each PDT is, however, responsible for controlling all work for which they are responsible. All work that must be performed is defined in the program SOW. This work will be accomplished in accordance with the program schedule. This work shall produce the products called for in the program plan. System engineering work defined in the SOW shall follow the guidance provided by this document. Specific control techniques are described in subordinate paragraphs.

On programs applying the integrated management system, all SOW tasks will have been broken down to a finer level of detail and these tasks coordinated with specific achievements and completion criteria providing management with a very clear definition of task goals and the conditions that must be satisfied to claim completion.

4.9.1 Sequence expectations

The company applies the concurrent development or integrated product development approach to the development of systems. The first step in this process is to determine what teams are required and to populate those teams with appropriate personnel. As the functional decomposition process fills in a preferred architecture, the PIT will review the options for team assignment, make initial assignments for team leadership, and empower these team leaders to staff and begin work on their assigned system elements in accordance with the program schedule and detailed work definition planning. This process entails concurrent development within a prescribed sequence as follows: item requirements analysis and concept development, preliminary design, detailed design, and product manufacturing and test support.

Concurrent development does not encourage concurrent development of requirements and designs, rather the sequential concurrent development of requirements followed by concurrent development of compliant designs. Requirements must be approved by PIT, for top level PDT requirements, or PDT leadership, for all lower tier PDT requirements, before preliminary design may be authorized or undertaken. Some concept development work may be authorized concurrent with requirements development as a way of validating that certain requirements values and combinations can be achieved. PDT and PIT management must be alert to avoid overuse of this technique, resulting in a leap to point designs before the problem space is fully characterized.

4.9.2 Baseline management

The development of products shall be regulated through imposition of technical baselines in accordance with the program plan. Major baselines shall be defined at major program events and intermediate ones may also be established by the PIT to more finely regulate the development process. A baseline is a freeze of requirements and/or designs to capture the definition of a particular configuration of merit at a particular point in the program. Once a baseline has been established, it is used as a briefing basis and a point of departure for subsequent work toward the next baseline.

Baselines shall be defined and enforced by the PIT and respected by all PDT. They will be defined in terms of a specific list of information products in particular versions. All items on the baseline list will be provided in electronic media to the PIT and placed under a controlled condition. These items may not be changed by anyone except the PIT based on a formal decision reached in a Development Review Board with capture of the rationale.

Early in programs prior to development of formal engineering drawings, baselines may be managed informally by the PIT through computer databases or development information grids (DIG). At some point in a program the PIT shall place baselines under formal control through a configuration management function within the PIT and the baseline shall be managed in accordance with the configuration management plan.

4.9.3 Decision database

All decisions that have a bearing on the product configuration or related processes shall be traceable to the source of those decisions in terms of the person who made the decision, the circumstances and rationale surrounding the decision, and the forum (meeting identification, for example) within which the decision was reached. The database may, but need not, be an integrated computer database composed of tables or fields and records. The information may be retained by any means so long as it is retrievable in a reasonable period of time to support other decisions.

Ideally, depending on program scope and available resources, the database will be retained in an integrated computer database with imbedded traceability to requirements, design concepts and features, risks, and process steps. As a minimum, the following decision database components are encouraged:

a. **Requirements, Traceability, and Verification**. This system will retain all decision data pertaining to requirements definition, vertical traceability information between requirements, and traceability of requirements to test and analyze responsibilities for verification of requirements.
b. **Meeting Management**. This system will retain the identification, purpose, and results of all formal meetings, reviews, and audits in the form of minutes with reference to decisions reached during the meeting and direction given for current and future work.
c. **Action Item Tracking**. This system will retain the identification of all formal actions required from formal meetings, audits, and reviews, the responses to those required actions, and initiator acceptance of actions taken.
d. **Trade Study Management**. This system will retain the definition of the candidates considered, selection criteria and requirements defined, candidate values or ranking in the selection criteria, and selections for all formal trade studies.
e. **Design Rationale**. This system will retain the reasoning and logic behind design decisions reached in the PDT and PIT in the form of a running log.
f. **Validation, Verification, and Assessment Record**. This system will retain a clear definition of the correlation between product configuration features and product representation configurations including test articles, models, simulations, mock-ups, and related data acquired from these representations.
g. **Program Task Definition**. SOW and Program Plan or IMP tasks shall be identified and available to map to decision-making events.

h. **Program Schedule**. Program events, dates, status, and other time-critical information shall be available for mapping to decision events.
i. **Program Document Listing**. All program documents will be listed and correlated with document type information (specification, plan, etc.), author, purpose and brief description, planned and actual release dates, and status. These documents will fall into one of the following four categories.
   (1) CPI, Company Private Information — Documents the content of which must be retained within the company to protect a competitive position in the market place. Commonly these documents contain information of a proprietary nature. This document type will not be used to withhold information from a customer that they rightfully must have in accordance with the contract.
   (2) DAL, Data Accession List — All such documents may be requested by the customer in accordance with the contract requirements.
   (3) CDRL, Contract Data Requirements List item — Documents that the company is obligated under the contract to deliver to the customer in accordance with a prescribed schedule and content definition.
   (4) SDRL, Supplier Data Requirements List item — Items prepared by a supplier that must be delivered by the supplier in accordance with a contract with the company that stipulates content and schedule requirements.

4.9.4 Design reviews and audits

Design reviews and audits will be held at key points in the development of products and processes, permitting exposure of recent work and approval of the results of that work and near term future planned work. The purpose of these reviews is to maintain contact between the process and customer and company management expectations. Major reviews are scheduled at critical decision points in the program where significant changes will occur in terms of the magnitude of budget commitment or rate of consumption of budget. Both product and related processes shall be reviewed with the expectation that they are together in a condition of optimum relationship.

4.9.4.1 Major customer reviews and audits

Major reviews are held at the termination of program phases to review status of planned work completion and to review readiness to proceed to the next phase. These reviews must include customer participation. In commercial situations where no clearly definable customer exists in the form of a single representative or organization, a customer advocate should be selected (ideally from outside the company) to represent the customer's views. The result of the review should be a clear decision on whether or not the goals of the program phase have been satisfied and, if not, a specific list of deficiencies should be developed that must be cleared to the satisfaction of the reviewing official before further work is accomplished. This should be in the form of action items with specific deliverables and dates agreed upon by the actionee.

Minutes of these reviews must be prepared as part of the design rationale information stream. These meeting minutes must identify attendance and roles of attendees, a copy of any presentation materials used, decisions reached and the rationale for those decisions, a clear statement on continued work authorization, and action items assigned. Minutes may be captured in paper or electronic media, but the latter is preferred if the program has the resources to provide on-line access to these documents. PIT is responsible for facilitating and conducting system level major reviews including minutes and follow-up. A PDT may be assigned these responsibilities for a major item incremental review.

Table 3-1 identifies the timing of these reviews with respect to program phasing. Generally, these reviews are a phase-terminating action that reviews phase progress against phase goals and specific accomplishment criteria in terms of the products produced. If the program truncates on the front end eliminating phases normally concluded by particular reviews, the objectives of these early reviews should be considered in the earliest review that is held. The logic for early phase deletion is based on the balance in system development needs between precedented and unprecedented portions of the new system and the resultant risks. The early phases of multi-phased developments are implemented to pace the expenditure of development resources (including money) commensurate with improving the knowledge of development risks.

The theory is that by spending a relatively small amount of money to fully study and characterize the problem, one can avoid obligating future fund expenditure on a solution to a problem not yet defined. The clear message from our history is that this route will require many design shifts and changes as improving knowledge of the problem exposes the inadequacy of design solution features based on incomplete knowledge. It requires a great patience and management skill to restrain a team of goal-oriented engineers bent on producing solutions to problems they believe they understand until such time that they truly do understand them.

Where the problem is well characterized based on precedented system implementation and operation, the development process may proceed without some early phases and reviews. Many of the products commonly defined in these early reviews should still be produced in a front-truncated program. For example, we should have a system specification, possibly one modified from a previous system implementation, even though the program entails a major modification rather than a new development. We should also understand any need and mission differences. The first major review held on a front-truncated program should quickly review the kinds of products commonly reviewed at those reviews omitted with their corresponding phase.

4.9.4.1.1 *Alternative system review (ASR).* This review occurs at the close of company phase 1, but on a DoD program could be called at the close of company phase 2. It offers an opportunity to down-select from two or more system concepts to one or more for further study or subsequent demonstration and validation (company phase 3) of the one selected concept. Accomplishments reviewed should include the following:

a. Concept study complete with one recommended solution. If the problem is very complex more than one solution may still be viable and another cycle needed to resolve the final recommended solution.
b. System architecture for the preferred concept complete. If multiple system concept survive at this point, multiple architecture may also survive.
c. A top level analytical assessment should be complete that demonstrates the preferred solution meets established needs, requirements, and/or objectives.
d. Cost, schedule, and performance objectives (figures of merit) and threshold estimates for the preferred system concept complete.
e. Product and process technologies and a means to verify their readiness defined for all preferred concept elements.
f. Product and process risks identified and a risk management approach complete.
g. Draft specification tree overlay on a program WBS complete.

Upon completion of the review, there should be a single surviving concept that passes into the next program phase or two or more surviving concepts that are recycled into a repeat of the initial phase to resolve the relative advantages. Multiple contractors may be involved in this and later early program phases. The customer may use early phases to focus their future budget on not only the best concept but the best contractor measured in ways that they may or may not share.

4.9.4.1.2 *System requirements review (SRR).* The SRR should be held at completion of company phase 2 or early in company phase 3. The purpose of the SRR is to reach mutual agreement between all parties to the development of the requirements for the system. This should include a review of the customer need, the functional decomposition process used to identify lower tier functionality, the allocation of the functionality thus exposed to lower tier items, performance requirements derived from system level functions, and system level constraints. The applicable documents should be reviewed for completeness, necessity, and tailoring appropriateness. All of the requirements should be captured in a system specification or precursor to such a document. The ASR and SRR may be merged into one review on programs that can move rapidly due to a high proportion of precedented system composition. The following accomplishments should be reviewed:

a. Draft system specification complete with all TBD items clearly identified with closure responsibility, planned completion date, and nature of any associated risk.
b. Draft system architecture and external interfaces defined.
c. Identification of any non-viable technologies needed and associated risk abatement plans complete.

Completion of the review must signal agreement on system requirements between company and customer and a definition of a draft functional baseline in terms of a specific list of controlled documents. Future changes of this baseline will require capture of traceability to the rationale for the change.

4.9.4.1.3 System functional review (SFR). The purpose of the SFR, alternatively called the system design review (SDR), is to establish an appropriate set of functional and performance requirements for the system. This review would, in a fully phased program (a development program requiring technology validation prior to establishing a functional baseline), occur at the conclusion of company phase 3. Where the program uses mature technologies not requiring a demonstration validation activity, the review may be scheduled during preliminary design. Where the program entails modifications, upgrades, or product and process improvements, the review should be conducted early in the initial work to establish a clear understanding of any requirements changes from the precedent system. Accomplishments that should be reviewed include:

a. All prior scheduled subsystem level reviews complete and unresolved issues assigned with clear responsibility and due date.
b. All TB's resolved in the system specification and system requirements complete for product and process. System specification ready for approval as part of the functional baseline.
c. All system level trade studies and those needed to identify/select end/configuration items complete and results documented.
d. System architecture complete down through end or configuration items with specification tree overlay complete to that level as well. Draft WBS overlay of the system architecture complete.
e. Draft end or configuration item development specifications complete to a level needed to inspire confidence in the completeness and achievability for parameters established in the system specification and the associated costs. These documents form the draft allocated baseline.
f. Specialty assessments complete at the system level in support of the system specification.
g. Configuration management process defined and in place on the program in readiness for starting preliminary design.

h. External interfaces defined in draft ICD where appropriate.
i. Mature risk management process in use on the program with a list of specific risks defined and evaluated.
j. PDT leaders selected, team composition defined, and team responsibilities defined for all architecture.

Successful completion of the review should signal agreement on a functional baseline defined by a specific list of documents in particular versions/revisions. Subsequent changes will be based on this definition and be traceable to it.

4.9.4.1.4 Preliminary design review (PDR). The purpose of the PDR is to determine if the predetermined conditions have been met to authorize detailed design work involving a considerable increase in program manpower and rate of cost accumulation. All product development requirements must be complete as well as credible design concepts responsive to those requirements. A single PDR may be held on a relatively simple development program or a series of partial reviews each focused on a particular end or configuration item may be held on a more complex development program culminating in a system PDR. It terminates company phase 4. The accomplishments that should be reviewed at the system PDR include:

a. All lower tier PDRs complete and any unresolved issues clearly defined in terms of closure responsibility, scheduled closure, and the nature of any associated risk.
b. All end or configuration item development specifications complete.
c. Functional decomposition and system architecture complete, terminating the downward stroke of the system development process. Make-buy decisions finalized.
d. Preliminary design, including any lower tier trade studies, complete in the form of layout drawings, design evaluation test results, and design support analyses reports.
e. All internal and external interfaces characterized. All external interface development documentation (ICD) finalized. Interface compatibility between all end or configuration items established analytically.
f. Verification rationale complete that is supportive of the notion that all design approaches are compliant with their requirements. Formal requirements verification planning complete.
g. Preliminary allocated baseline complete for all end or configuration items in the form of a specific list of documents in particular versions.
h. A review should be completed of any long lead procurement items and authorization obtained to proceed with that procurement with possible list revision stimulated by the review.
i. A review of remaining risks and their mitigation.

4.9.4.1.5 Critical design review (CDR). The purpose of the CDR is to review the completed design signified by release of at least 95% of all planned engineering drawings. The decision must be made at this review whether or not to proceed to the manufacture of sufficient resources to accomplish needed testing activities. The review may be accomplished as a single event or through a series of end or configuration item reviews capped by a system review. The following will be reviewed at the system level CDR:

a. All planned lower tier reviews complete and action items under control in terms of a responsible person identified, a required delivery date established for the answer, and an assessment of risk potential.
b. All development specifications complete and released including in-house and procurement types down to and including parts, materials, and processes.
c. Draft product specifications, if required, complete. The product baseline shall be defined in terms of a list of specific documents in defined versions including product specifications and drawings.
d. Evidence of compliance in the designs with the corresponding requirements. Design presentations should include a discussion of the requirements preceding a discussion of the design features where it is shown how the requirements are satisfied. Analytical evidence of compliance is also referenced or presented.
e. Updates of functional and allocated baselines complete for work since PDR.
f. The designs and manufacturing methods complete for any required qualification test articles. CDR should provide authority to fabricate test articles needed to verify requirements compliance in preparation for FCA where the evidence of compliance will be reviewed.

4.9.4.1.6 Functional configuration audit (FCA). The purpose of the FCA is to prove that the design solution for a given end or configuration item will satisfy previously approved requirements by making available evidence of that proof in the form of test and analysis reports linked to the verification requirements defined in the item specification Section 4, in turn traceable to the requirements defined in Section 3 of the item specification. To support this review, all planned qualification tests and analyses must be complete. This evidence must be studied by the customer and the PIT prior to the review and any differences of opinion resolved. The review should conclude whether a product that satisfies the engineering definition contained in drawings and reports will satisfy the requirements defined in specifications.

4.9.4.1.7 System verification review (SVR). Where a product requires several item FCA, a SVR is held to support a decision that the system is ready to enter production phase. Progress toward accomplishment of the following goals should be reviewed:

a. All test and analysis activities prescribed by all of the development specification Sections 4 complete and resultant evidence reviewed for compliance verification. All lower tier FCA complete with no outstanding action items.
b. Verification that each qualification test article conforms to its design. If the requirements or design have changed since or during qualification tests, it must be ascertained how the test results relate to the final requirements and test article configuration. Special analyses may be required to show how this data should be used.
c. Verification that the analyses were conducted on the baseline product definition.
d. Updates of functional and allocated baseline complete.
e. Proofing of the support, training, and deployment processes complete.
f. Availability, capability, capacity, and readiness of the procurement and manufacturing process verified. Production readiness for a prescribed rate assured.

4.9.4.1.8 Physical configuration audit (PCA). The purpose of the PCA is to demonstrate that the product that results from the manufacturing process then in place will satisfy the engineering, quality, and manufacturing design embodied in engineering drawings and reports, quality inspection planning, and production planning documentation. The PCA may be accomplished as a single event for a single end item or incrementally for several capped by a system PCA. It must also be shown that the results will be repeatable from one article to another. As a result of the PCA, the customer should approve the product baseline. The following accomplishments must be verified at the PCA:

a. The as-built or as-coded version of the item compares with the requirements in product specifications and/or design documentation.
b. All acceptance test requirements complete and approved and effective test practices in place.
c. All lower tier PCA, if any, complete and action items closed.

4.9.4.2 Special reviews

Additional reviews may be held on programs as a function of the nature of the program, customer preferences, and the phase of the development efforts.

4.9.4.2.1 Software reviews. Software requirements reviews shall be held prior to committing to design of the code to ensure that all requirements have been identified and the interfaces of the software entity are fully characterized. Refer to the Software Development Plan for details.

4.9.4.2.2 Training reviews. Special training reviews may be held to permit customer review of the requirements and design for customer training materials. Refer to the Integrated Support Manual for details.

4.9.4.2.3 Supportability reviews. Special integrated logistics reviews may be held to track the development of supportability requirements, designs, and issues. Refer to the Integrated Support Manual for details.

4.9.4.2.4 Test readiness reviews. Prior to beginning major test events, the test plans and procedures shall be reviewed to assure full readiness to commit test resources. In many cases, test events will result in the destruction of or damage to costly test resources that would be difficult to replace should a test not achieve its goals. Every reasonable effort must therefore be made to assure readiness. Refer to the Test and Evaluation Manual for generic details and the program integrated test plan for programmatic details.

4.9.4.2.5 Manufacturing and quality assurance reviews. Special manufacturing and quality assurance reviews may be held to review requirements and designs for manufacturing and inspection activities. Facilities, tooling, and material handling plans may also be included. Refer to the Manufacturing Practices and Quality Assurance Practices Manuals for details.

4.9.4.3 In-house reviews

Throughout the development period, the development teams will reach many decisions within their team meetings and informal discussions that have a great bearing on the final product configuration. It is important to capture these decisions in what is called the product design rationale record that will later permit traceability of particular product features to the source rationale. This will be useful in understanding why features were selected and may act as a basis for retaining features subjected to later challenge or permitting a change based on conditions different than the original rationale. It is very important that the program have a way of capturing the results of formal decisions reached in system or program level meetings and in particular at meetings involving the customer.

PDT are responsible for conducting reviews within the scope of their authority. The same information provided for major reviews applies for lower tier reviews.

4.9.4.3.1 Development review board. The principal formal decision-making activity during the development period (outside of major and internal reviews and audits) through first article inspection is the Development Review Board (DRB). The DRB is a formal meeting characterized by an agenda; attendance by a chairman responsible for any final decisions, advisors, and presenters; and minutes that capture the decisions derived and action items for continued work not defined or differently scheduled in program planning documents.

The DRB may be called for any of the following reasons:

a. To review a set of item, and related process, requirements in database or specification format for the purpose of in-process review or to determine team readiness to proceed to design activity.
b. To review the requirements or a design for the purposes of in-process review; as a test of readiness for transition between concept development and preliminary design or preliminary design and detailed design; or in preparation for a major customer review.
c. To review process plans, documents, and results. By way of example, the following are included here: integrated test plans and results of tests, planned major analyses and results, requirements verification planning and results, the manufacturing flow, the quality assurance overlay of the manufacturing process, and the tooling list.
d. To resolve a specific problem in the development of the system, an item, or a process.
e. Review the results of trade studies and select the preferred alternative.

The DRB is chaired by the team leader with responsibility for the item or problem being reviewed. At the system level this is the PIT Leader. At the lower level this is the responsible PDT Leader. If two or more teams are involved in the decision-making process a joint meeting shall be held with attendance by all integrated Product Development Team Leaders and chaired by the PIT Leader.

The PIT is responsible for facilitating all DRB held by the PIT and in maintaining records of all DRB whether held by the PIT or PDT. The PDT that is responsible for a DRB shall facilitate that meeting. Whichever party is responsible shall publish a report of the meeting for inclusion in the decision rationale information component of the program information library. Facilitation includes:

a. Gain access to physical space and resources within which to hold the meeting.
b. Prepare and distribute a meeting agenda, brief the agenda at the meeting, and provide the meeting coordinator responsible to the Chairman for keeping the meeting on the agenda while the Chairman focuses on the agenda topics.
c. Capture all action items identified in the meeting, assigning responsibility, and getting agreement on a due date by the assigned party.
d. Prepare meeting minutes including meeting purpose, attendance, Chairman/decision-maker's name, action item summary or list, alternatives discussed, and decisions reached.

4.9.4.3.2 Internal reviews and audits. Any of the reviews and audits described in paragraph 4.8.4.1 may be held internally either as preparation for a major review on a program where the customer requires it or as the final review on a commercial program. Refer to the criteria noted under the appropriate review above.

4.9.5 Technical information retention and communications

All documented technical communication shall be retained for later access. Ideally, this information should be retained in electronic media and placed in a data structure on a computer server for program access from available microcomputers and workstations. Paragraph 4.9.3 defines the minimum required information that must be retained. Programs are encouraged to retain technical information in structures called interim common databases early in the product development through PDR. Between PDR and CDR, this information will migrate to final forms under formal configuration control.

Figure 4-6 illustrates the interim common database concept. It consists of six elements. The DIG is simply a folder or directory structure on a computer server to which are connected all of the workstations of the program. The folders/directories are organized into development data packages (DDP) corresponding to each team item and each DDP contains the same folder/directory set (the discipline row), providing a place for all team personnel to place their information product (information cell). This information is created and edited within tools installed on the workstations that can access the DIG on what is called a tool ring. Initially, the empty cells are operated in an open door mode where specialists may place their product in the cells without impediment through the DIG/Tool interface. As a baseline condition approaches, the PIT should switch to closed door mode where cell owners must drop their information products in the appropriate dropbox cell

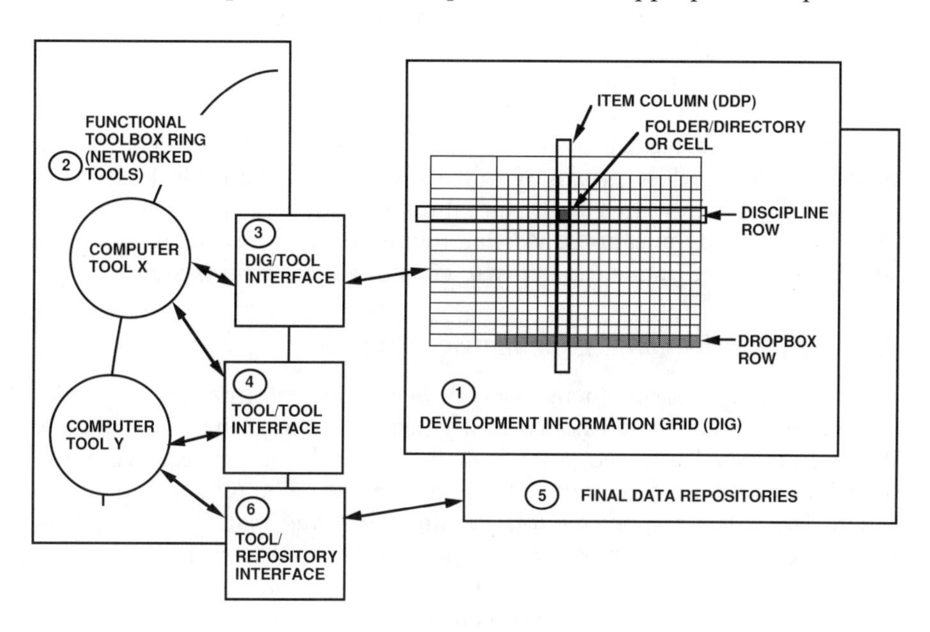

Figure 4-6 Development information grid.

(for their team). A person on each team is assigned the responsibility to review dropbox material and move it to the appropriate cell updating the baseline.

Some tools may have the capability to interact to produce aggregate products through a Tool/Tool interface. Over time this DIG concept may evolve into a true common database where selected information products may be produced directly through these tool-to-tool interfaces. As a program matures, much of the information retained in the DIG gradually migrates to final repositories such as specification libraries, engineering drawing libraries, and other electronic media.

Any information in the DIG may be called up in read-only mode by any person on the program, providing for easy access to everyone's in-process work and avoiding a fundamental problem in communication between engineers reluctant to expose their unfinished work. All program personnel are required to place their current work product in the DIG and to maintain that information up to date. Any information in the DIG may be projected in meetings using computer projection, and within the context of a team meeting, this information may be changed with all team members present and contributing to the discussion enabling true concurrent engineering.

Each team must also be served by a section of wall space, within a meeting room or otherwise, upon which it may place large drawings through which team members may understand the correlation of several views. This wall space should also be served by a computer projection space such that the team may use the space to provide the large view and computer projection from the DIG for details. This facility feature is called a war wall.

The information in the DIG may be used directly for meeting materials or assembled into a presentation using a presentation tool. A complete major review can be assembled in this way, making it unnecessary to spend considerable time preparing for major reviews. The DIG should contain at any point in time exactly the information any reviewer would be interested in understanding.

The teams must also be served by good telephone communications with phones located in close proximity to work stations. It should be possible to hold meetings with virtual attendance through shared computer visuals and phone voice communications. Meeting rooms and war wall areas should be served by speaker phone capability to encourage this arrangement.

4.9.6 Action item and issue management

An action item is a statement of need for specific information or activity leading to a specific result not currently defined in program planning data. It is characterized by a requester, a definition of the action required and a date when it should be completed, and an actionee who must complete the action. The PIT is responsible for tracking status and managing the action item system. Periodic reports should be provided to team leaders showing their performance. The goal is zero overdue action items. If this figure rises above ten of all open action items for any PDT, special action should be taken to quickly drive that number back to zero and keep it there.

Action item data shall be retained in a computer database accessible from common networked workstations available in the program work space. The action item tracking database is part of the design rationale component of the program information library and will be maintained by PIT.

Issues are concerns expressed by management for specific possibilities or realities that have or may have a detrimental effect on the program. An issue is a precursor to a risk if it cannot be resolved quickly and simply. Issues raised at formal program meetings are tracked and reported by PIT until resolution so long as they are not transferred to the risk category, in which case they are tracked as a risk.

4.9.7 Resource allocation

The company employs an integrated product development process to each program and the PIT is responsible for identifying the resources needed to accomplish program goals within prescribed cost and schedule limits. Each PIT is represented on a cross-program EIT at the company management level which seeks to balance company resources to serve the aggregate of program needs.

Where this team identifies near or far term conflicts, efforts are first made to overcome any negative effects by re-balancing existing resources within the time and money constraints of the programs. Failing this alternative, an effort is made to understand the most critical needs of each program in conflict and to identify a compromise that has minimum adverse impact to all programs. All program managers, as well as functional department managers, are members of the EIT and so have a voice in deciding these issues. Decisions on company capital expenditures are largely driven by the results of the analyses performed by the EIT.

4.9.8 Program constraints

Program constraints include those things a program cannot or will not do. Early in a program the company and customer should develop a list of constraints based on program requirements and available resources. The company and the customer must then work to reduce this list to a null condition by including the necessary wording in the SOW and system specification to fully define contractual requirements and limitations. There should be no long-surviving separate list of things that shall or shall not be accomplished. Where, for example, there are funding limitations imposed by the customer or the agency from which those funds are acquired, the SOW and system specification must be made consistent with those limitations for the phase under contract.

In addition to changes to the SOW and system specification, deviations from these documents may be appropriate where the difficulty of achieving a specific provision will be avoided or delayed in a very specific way for a specific time period after which the original requirement will be fully satisfied. A third approach is to define preplanned product improvements that will be accomplished at predetermined times in the future.

4.10 Risk management

Risks materialize on programs as a result of a failure to foresee detrimental eventualities. Not all eventualities can always be foreseen but to the extent that they can, program risk is reduced as a function of the effectiveness by which the program works to avoid their occurrence through good program planning and sound technical development of the product.

4.10.1 Program discontinuity

Programs are planned and requirements defined for products to ensure a smooth development process devoid of discontinuities that act to introduce and increase cost and schedule problems. It is especially important that a risk program detect and avoid causes of program discontinuity.

4.10.1.1 Discontinuity detection

Perfection is seldom attained. If a program was perfectly planned and executed with good fidelity to a good plan and everything unfolded as conceived by the planners, everything would go according to plan. Programs are planned, however, with imperfect knowledge of the future and this reality practically guarantees that everything will not go precisely according to plan. The presence of uncertainty in our future forces us to consciously evaluate the possibilities for adversity and its effects. Adversity will commonly take the form of a discontinuity in our planned path which was based on predetermined conditions and assumptions about the future.

A program discontinuity is an interruption in planned activities. It is a condition that precludes continued efficient execution of work in accordance with the predetermined plan. There are three principal factors involved in determining whether conditions of discontinuity will develop during program execution:

a. Planning quality
b. The degree of correlation between the assumed conditions that will be in place during execution while planning and the reality during execution
c. The fidelity of plan execution

Figure 4-7 is a three-dimensional Venn diagram illustrating the possibilities from these three factors. In the simple case of binary possibilities for each variable, there are 2^3, or 8, outcomes. The preferred outcome during execution is, of course, that we had planned well, executed the good plan faithfully, and conditions in effect during execution were those for which we had planned. All of the other combinations will yield less desirable results.

We generally think of discontinuities as being bad, but it is possible that conditions can change for the better and we may wish to change our plan to gain full benefit from these changed conditions. For example, our plan could involve using a very expensive technology because another is thought to be

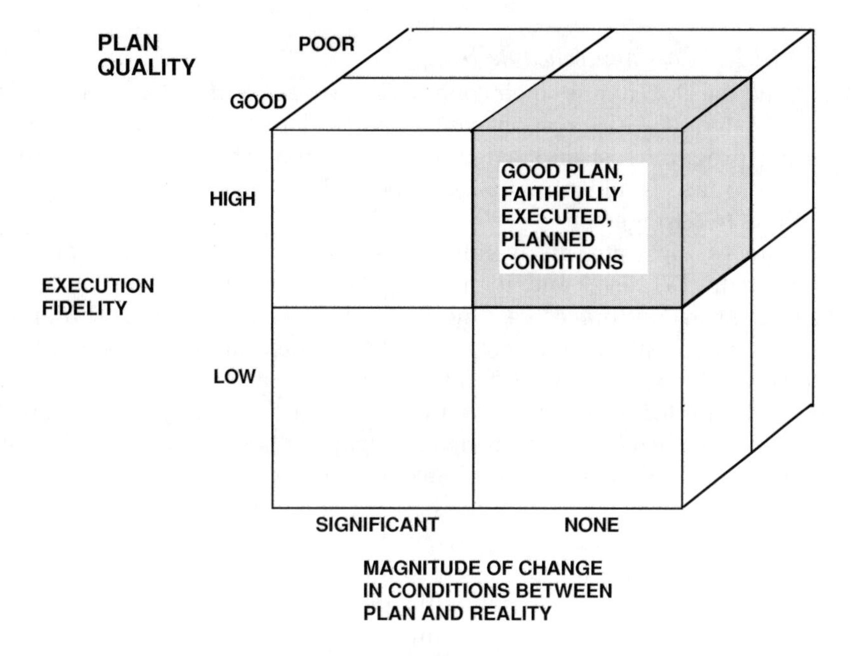

Figure 4-7 Discontinuity cause cube.

too far out in development. During our program execution, that other technology might become available in practical terms making it possible for our customer to save hundreds of thousands of dollars. Certainly, we would want to embrace this new technology no matter what our current plan called for.

More commonly, discontinuities are associated with bad news. A program discontinuity can arise from several sources suggested from Figure 4-7 and we need a detector tuned to each of these sources. The three principal program planning focuses are technical or product performance, cost, and schedule. Most everything that can go wrong in terms of plan quality, execution faithfulness, and changed conditions can be expressed in these terms. We add technology to this list even though technology unavailability can be expressed in terms of cost, schedule, and performance. A new development program that is pushing the technology will benefit from this added trigger. A program that can rely on existing technology need not include it. This gives us four triggers for which we must arrange detectors.

The detectors we need are really tuned to what commonly are called risks. Risks are potential program discontinuity triggers that should be avoided, mitigated, or corrected for. Wherever possible, we should avoid them by detecting their potential appearance in the future and taking action in the present to ensure that the conditions needed to bring them into existence never occur. When a risk is upon us, but not yet disrupting our plans, we need to be actively mitigating it to prevent it from becoming an open discontinuity. A risk that is fulfilled is a program discontinuity that must be removed through replanning.

4.10.1.2 Cost and schedule triggers

NASA and the DoD require their contractors to use a certified Cost/Schedule Control System (C/SCS) responsive to a particular criteria for tracking and reporting program cost and schedule information. The information these systems produce is adequate to detect possible or realized discontinuities from these two triggers.

Figure 4-8 illustrates an overall process for accomplishing system integration at the PIT level and it includes discontinuity detection and risk management as key processes. One function of the PIT is to be Program Cost/Schedule Control agent for the PDT and this is accomplished within task 1083 (DTC, LCC, and C/SCS Integration). This activity must constantly monitor the performance of all of the PDT and the PIT against planned expenditure and achievements using the C/SCS database as a data collection and reporting medium. Deviations beyond predetermined boundary conditions are used to trigger recognition of a potential risk. This risk must then be studied by PIT personnel responsible for program risk analysis.

During the earliest program phases the company must come to a clear understanding with the customer on the planned program spending profile and any risks to customer ability to maintain that profile. The funding profile may require that the company hire new personnel and this should not be done without assurances of the customer's ability to implement its plans. Where a planned profile is not compatible with current company personnel resources in time, the program may propose a different spending profile that smoothes company personnel demands and may even reduce customer financial burdens. If, however, the customer's need is very time sensitive, a particular spending profile may be required and the company will have to adjust to it, identifying a corresponding risk.

4.10.1.3 Product performance trigger

A technique called technical performance measurement (TPM) shall be used to detect problems in satisfying the key technical product requirements. TPM involves customer and contractor agreement on a small list of key quantified parameters, parameters the condition of which signal the general health or illness of the complete program and product system. Weight is a common parameter chosen where this is an important characteristic and historically difficult to satisfy.

The assigned principal engineer will track the value of these parameters in time against the required value and make predictions, based on planned work, how the value will change in the future relative to the required value. The history of past values and future predictions of these parameters is reported periodically to the company and customer management in graphical form. If a TPM parameter is outside predetermined boundary conditions, the customer will expect the contractor to explain the reasons and what they are going to do about getting the parameter back on track with its planned value. As noted in Figure 4-8, the PITTPM function feeds its concerns to the person responsible for risk analysis.

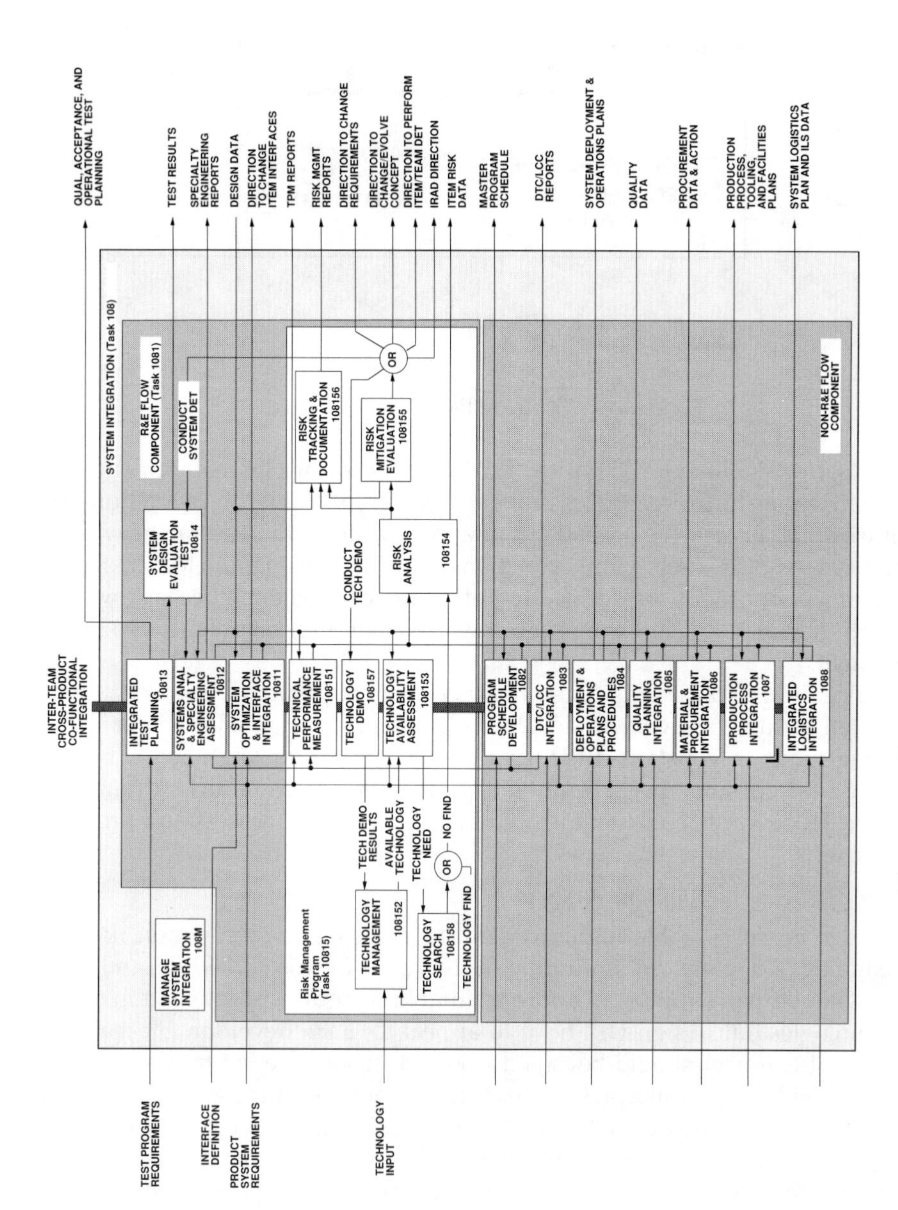

Figure 4-8 Discontinuity detection network.

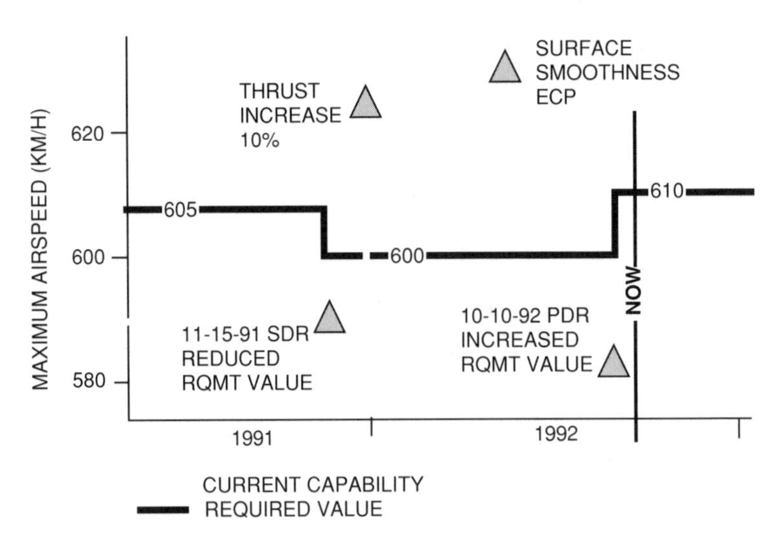

Figure 4-9 Typical TPM parameter chart.

Figure 4-9 illustrates a typical TPM chart showing the requirement and current value historical traces. When the current value or predictions of future value exceed the agreed boundary conditions, the contractor will be required to brief their planned actions to bring the parameter on track. Boundary conditions are defined to permit the contractor some slack within which they may manage the program without undue customer micro management.

The TPM parameters may shift as the program matures, exposing some selected parameters well under control and others not selected for TPM management in constant difficulty. The list of parameters should be reviewed for currency periodically and changes made to reflect current and future needs for management insight into the evolving process and product features.

4.10.1.4 Technology trigger

There is no comparable highly organized mechanism in common use for technology-triggered discontinuities as there are for cost, schedule, and control. The company applies an approach that first requires identification of the technologies our design will have to appeal to. Each development team or principal engineer should be called upon to survey needed technology for the planned design concept. Concurrently, the manufacturing, quality, logistics, and operations people should be doing the same thing for their own fields based on the current product and process design. We then must compare the content of these lists against a list of technologies available to us. These technologies may be available to us because: they are freely available for anyone's use, we know how to gain access to them on the open market, or we hold them as proprietary properties.

PDT Leaders are required to identify the technologies they intend to appeal to for each design concept during preliminary design. The PIT should review this mix against available technologies, as indicated in the Technology Availability Assessment block of Figure C-5. The program must first have a way of knowing what technologies are available that might have a program relationship, and this may require a specific technology search or simply the application of an edited list from a comparable prior program. We must realistically assess the availability of all needed technologies. Any immature technologies we intend to depend upon should be recognized in our risk management data and accompanied by risk mitigation plans corresponding to the case where they do not come on line at the needed time.

In all cases where there is a match between required and available technologies, there is no risk-triggering condition present. Where it can be demonstrated that a technology needed by an PDT does not now exist on the program list or there is a likely probability that it will not be available in a practical sense at the time it will be needed on the program, there exists a technology discontinuity trigger event and corresponding risk. The appropriate response is to conduct a technology search followed either by adding a found technology to the available list or signaling a technology risk.

The outcome of the risk analysis process may be direction to the designer to change the design to avoid the unavailable technology or a technology demonstration to develop and report the new technology needed for the existing design concept. The latter path should carry with it a periodic monitoring of the status of this new technology as the time approaches when it must be mature. If there is a risk that it may not be available in a timely way, it may be necessary to run parallel development paths to mitigate the risk of the technology not coming in on time. In this case, it may be possible to convince the customer that the alternate technology will be adequate for initial capability, even though it does not meet the system requirements, and that it can be replaced at a particular cut-in point in production. The customer may agree to a deviation from specification requirements for a particular number of articles that will be corrected when the needed technology becomes available.

4.10.2 Risk identification and management

Risk identification and avoidance is the responsibility of management at all levels in company programs. Risk identification is the responsibility of all program personnel in their specialized disciplines. This is accomplished by comparing planned activity with anticipated or possible eventualities. There is a creative nature to this work involving thinking through things that can go wrong and understanding the consequences of the resultant problems. There is a lot of commonality between risk identification work, failure modes and effects analysis, and safety hazard analysis.

Risk management is an inseparable part of good management. Each program will manage risk through its PIT. The PIT is responsible for identification and prioritization of risks through cooperative work with the PDT and PBT. Assigned personnel are responsible for mitigation of risks and closure.

4.10.2.1 Risk assessment and abatement

When any of these triggers are activated for a specific risk, they should stimulate the PIT to a particular course of action. First the risk should be studied to determine the nature and scope of exposure. What specifically will happen if the risk becomes a reality? Then we must try to ascertain the probability of the risk materializing. A risk with a combination of very serious consequences and a high probability of occurrence should be dealt with as a very serious matter. A risk with minor consequences and low probability of occurrence may be set aside. Under different conditions at a later time in the program, this risk may take on a more urgent concern; so, it should be logged and retained for continuing review. In between these extremes we need a policy for selecting risks for mitigation. The problem is that we have limited resources and cannot afford to squander them on every possible problem.

It is helpful to have a special form or worksheet for evaluating risks that encourages the person most knowledgeable about the risk to provide the information needed by those who will decide whether to spend scarce resources on mitigating it. Figure 4-10 offers a form for this purpose useful throughout the risk life cycle in summarizing the status of the risk.

Each risk identified on one of the forms shown above and accepted by the PIT shall be added to a formal list of risks under active management. This list will be used to communicate identity, responsibilities, and status of all active risks.

The program should also have a means to communicate in a graphical, summary way what risks are being managed and what their status is. Figure 4-11 illustrates a fragment of one graphical way of doing this. Each major program area is identified in a hierarchical structure mapped to the WBS, with subordinate areas in each case listed. For each block on the diagram, three blocks correspond to the three major risk types (cost, schedule, and technical) and offer the analyst a place to enter the corresponding risk probability. The latter are given in colors as: GREEN = Low Risk (L), YELLOW = Medium Risk (M), and RED = High Risk (H). Letters are used here to avoid the need for colored pages and can be used in databases.

The PIT and program management use the status summary to determine on a program scale where the most serious risks are and how to allocate available budget in the most intelligent way for the best overall effect. Each active risk must have a principal engineer assigned to work issues associated with that risk and take appropriate mitigation actions. Periodically, the PIT should meet to assess the current status on all active risks and provide direction for future work on those risks. Where a risk has been fully mitigated, energy should be focused in other directions. As new risks become identified and determined to be serious, they should be formally accepted into the set being actively managed.

4.10.2.2 Formal discontinuity identification

At any one time, a program may be managing 25 risks using the assessment summary illustrated above. If the risks are well mitigated and conditions

<table>
<tr><th colspan="4">COMPANY XYZ
RISK IDENTIFICATION</th></tr>
<tr><td>RISK ID NUMBER</td><td></td><td colspan="2">ID DATE</td></tr>
<tr><td>RISK NAME</td><td colspan="3"></td></tr>
<tr><td rowspan="2">RISK TYPE</td><td colspan="3">COST SCHEDULE PERFORMANCE TECHNOLOGY</td></tr>
<tr><td colspan="3">ITEM PROCESS</td></tr>
<tr><td>WBS</td><td colspan="3"></td></tr>
<tr><td>RISK STATEMENT</td><td colspan="3"></td></tr>
<tr><td>RESPONSIBLE TEAM</td><td></td><td colspan="2">RESOLUTION NEED DATE</td></tr>
<tr><td>PRINCIPAL ENGINEER</td><td></td><td colspan="2">PLANNED COMPLETE</td></tr>
<tr><td colspan="4">PLANNED ACTION</td></tr>
<tr><td>CURRENT STATUS</td><td></td><td colspan="2">DATE</td></tr>
<tr><td>TEAM LEADER</td><td colspan="3"></td></tr>
<tr><td>PIT APPROVAL</td><td colspan="3"></td></tr>
</table>

Figure 4-10 Risk identification form.

permit, none of them may ever become program discontinuities requiring program re-planning. At the same time, if we have done a good job of monitoring the discontinuity triggers and converting them into risks to be mitigated before reaching a discontinuity status, no discontinuities should befall us that do not flow from our on-going risk management program. An

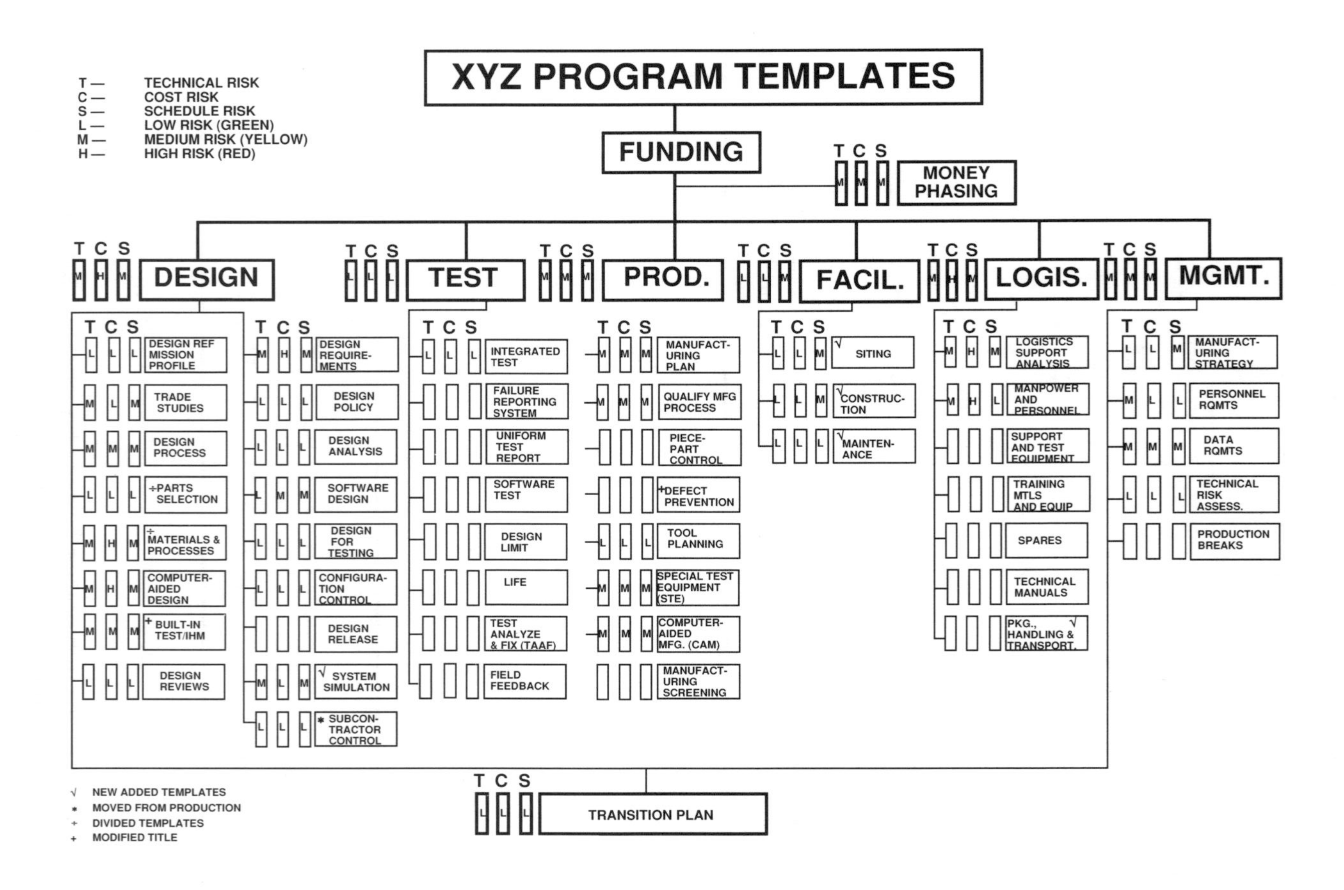

Figure 4-11 Program risk assessment status summary.

exception would be an act of God like an earthquake tearing our factory in half or a flood carrying away downstream the factory of a prime supplier.

These are cases where a discontinuity materializes instantly without passing through our best efforts to prevent them from happening. If we include every possibility, however remote, in our risk mitigation work, there may be insufficient resources to do the actual program work; so, some unforeseen risks will remain despite our best efforts.

Our periodic review of the risks that are being managed should ask if any of these risks have reached a condition where we can no longer manage related progress in accordance with our current plan. This could be another indicator that is tracked on our chart. When we must answer this question with a yes, we must accept the need to re-plan part or all of the program depending on the scope and seriousness of the discontinuity.

4.10.2.3 Program re-planning

We can apply the same techniques in re-planning that we applied in the original planning activity. Clearly, we must identify the scope of the tasks impacted by the discontinuity. These task changes must be defined in the plan as they relate to major program events. The interfaces between these tasks and other program tasks must be studied to determine if the critical path is influenced, possibly triggering other schedule risks not previously identified. If possible, the original major events schedule should be respected unless the customer indicates otherwise. This may require parallel rather than serial work performance or overtime to avoid overall program schedule impact. Unless the schedule was very optimistic in the first place or prior re-planning has absorbed all of the slack, there may be some margin in the schedule that precludes major program rescheduling.

4.10.2.4 Risk closure

The goal of risk management is the identification of all potential risks and closure of all of them through purposeful action. Risks that have been thoroughly mitigated to the extent that they cannot occur or the worst effects are not detrimental to program goals may be closed. Risk records should carry the history of the risk but status reports need not include them.

4.11 Configuration management

The PIT is responsible for configuration management on a program. All PDT and the PIT are responsible for implementing the requirements of the company Configuration Management Plan with respect to product elements for which they are responsible. On a DoD program, MIL-STD-973 (tailored to match the company internal Configuration Management Plan) will be used as the authoritative reference source for configuration management. NASA Centers apply their own configuration management requirements, which will be tailored to equivalency with the internal plan to the maximum extent possible. Commercial programs will simply apply the content of this plan augmented by internal configuration management practices. Configuration management focuses on four specific areas described in subordinate paragraphs. Refer to the company Configuration Management Plan for details not supplied here.

4.11.1 Configuration identification

The product shall be organized for development using a WBS overlay of the system architecture for cost and management purposes. Key items in this structure will also be identified as end items or configuration items as a function of the customer terminology. These items provide a focus for management of the developing program. They are selected by PIT and approved by program management and, if appropriate, by the customer. They are selected to form a development control band across the architecture at some level of indenture such that all items below this band requiring development are subordinate to one of these items. They are selected based on the degree of risk and management intensity that will be applied to the program. A low risk, highly precedented program should have fewer high level items selected while a high risk, highly unprecedented program should have more lower level items selected. In the latter case, it is possible that two layers of items (called prime and critical by DoD) could be selected, but avoid this if possible.

An end or configuration item is identified by assignment of an item control ID and listing on the program configuration or end item list. All items on this list must have appropriate specifications published for them and major reviews will focus on their development condition and status. All of the representations (specifications, engineering drawings, mockups, test articles, test and analysis reports, simulations, and models) of these items as well as the items themselves must be controlled to the degree that it is always clearly understood what the content or composition of the representation or item is at any point in time. The degree of identification for which the company is responsible subsequent to delivery is a function of the contract.

4.11.1.1 Product architecture

Configuration management shall be accomplished upon items defined in the product architecture. This structure shall define the existence of objects in the system and their hierarchy and family structure relative to other items.

4.11.1.2 Work breakdown structure

The architecture is overlaid by a WBS structure by the PBT and PIT to indicate the way cost planning, implementation, accumulation, and tracking shall be accomplished.

4.11.1.3 Specification tree

The configuration management representative on the PIT shall determine the items on the architecture tree that shall require specifications and the kinds of specifications that shall be prepared. A specification may only cover one item identified on the architecture diagram. That specification may include subordinate item requirements on the next tier so long as it is comprehensive for that tier.

4.11.1.4 Configuration item identification

The configuration management representative on the PIT shall define, in consultation with the customer and program management, the items that will be used for formal program management and selection for major reviews. Each of these items must be represented on the architecture diagram by a single entity.

4.11.1.5 Drawing breakdown structure

The configuration management representative on the PIT shall define end items for the purpose of organizing the drawing breakdown structure overlay of the architecture diagram or list. Drawing numbers shall be assigned such that no one drawing will contain content on two or more architecture items unless they are subordinate to only one item and all of the subordinate items for that item are illustrated or defined thereon. Refer to the Configuration Management Manual and Drafting Manual for details. Customer requirements may be more restrictive than this requirement, imposing mono detailing where only one part may be shown on one drawing.

Each team is responsible for creating a drawing list corresponding to the end items for which they are responsible.

4.11.1.6 Software breakdown structure

The team responsible for an item of software shall consult with the PIT for agreement on the breakdown of software entities into programs and modules. The software development manual provides detailed guidance on this.

4.11.2 Configuration control

It is required on all programs that the company be able to tell the configuration of product items and their representations unequivocally at any time. This is done through assignment of part numbers to product items (hardware or software) and correlating the permissible combinations of the corresponding part numbers for particular end or configuration items. These permissible combinations provide baselines for these items for use in engineering changes (and the resulting updated configurations) consisting of specification and/or engineering drawing changes and preparation of deviations and waivers.

Process or product requirements and design changes that involve changes in contractual commitments in the SOW and/or system specification will not be accomplished until the full effects on the product and contract are understood. Customer requirements for contractual changes will be followed to explore contract impacts prior to contract changes. This will commonly involve a preliminary assessment to a determination of the responsibility for the need of the change, either on the part of the company or customer. Subsequently, the full impact and solution will be defined and prepared in the context of a contract or engineering change proposal. Upon approval, of the change, the work required to implement must be integrated into the stream of normal program work so as not to conflict with predetermined goals and milestones.

In addition to maintaining product configuration control, the configuration of the many representations of the product must be controlled to ensure that the degree of correlation between those representations and the product baseline is always known. Product representations may include test articles, test and analysis results, models, and mock-ups. An inventory of all such representations should be prepared and maintained. This matrix should also include a definition of responsibility and a correlation between representation configurations and product configurations.

4.11.3 Configuration status accounting

As a minimum, the program shall maintain a list of end or configuration items correlated with specific articles in each case, the current configuration of those articles, and the approved configuration of those items. Additional accounting requirements may be defined in a contract.

4.11.4 Configuration audits

The three principal configuration audits are FCA, SVR, and PCA described in paragraphs 4.9.4.1.6, 4.9.4.1.7, and 4.9.4.1.8, respectively. These audits are conducted on end or configuration items and summarized at the system level. Together these audits verify that the product configuration described in the engineering drawings will satisfy its requirements and that a product is built to the engineering drawings and process controls. The PIT is responsible for system level audits and PDT are responsible for item audits.

4.11.5 Baseline management

Baselines provide clearly defined plateaus of progress at which program accomplishments may be assessed and compared with intermediate goals and adjustments made for future work. Baselines are composed of sets of documents in specific versions or revisions.

4.11.5.1 Product requirements baseline

The requirements baseline for an item is defined by the content of a specification for that item that has been approved by the appropriate authority. The internal authority for a system or configuration item specification or interface document is the PIT Manager. The internal authority for all other specifications is the responsible PDT Leader. A document that satisfies this criteria is the current requirements baseline for the item.

4.11.5.1.1 Specification existence. The PIT configuration management person defines items for which specifications shall be prepared and the type to be prepared. PIT defines a responsible team and determines a schedule in consultation with the responsible party for its completion. The specification exists and is placed in the requirements baseline upon initial release.

4.11.5.1.2 Specification type and form. All customer specifications shall conform to the customer definition of type, format, and content standards. All company specifications shall be prepared using the standard 6 sections defined in MIL-STD-490A, with the possible exception of some software specifications (see the software development manual), so all company specifications will conform to form 1a as defined in MIL-S-84490 with respect to format control.

4.11.5.1.3 Specification responsibilities. The responsibility for a given specification shall be clearly assigned by PIT to a team and within the team to a specific person. That principal person is responsible for the timely development of the specification, coordination with all needed specialty engineering persons for content, and husbanding it through the approval process.

4.11.5.1.4 Specification schedules. PIT shall define all specification schedules in consultation with the responsible persons named to prepare them. The analysis work needed to define specification content shall be scheduled and accomplished prior to design work on the item. The process of formatting and releasing the specification may extend into the design period so long as the requirements for the item are readily available to the responsible team.

4.11.5.1.5 Specification approval and authentication. All system and configuration item specifications shall be approved by the PIT Manager. All customer specifications must then be approved or authenticated by the customer before they are fully approved. The PIT Manager and responsible PDT Leader may, with the Program Manager's approval, proceed with design work based on internal approval only where the risk of customer disapproval is determined to be small. Internal and procurement specifications shall be approved by the responsible PDT with possible PIT Manager interference by exception. Prior to PDT Leader approval, the specification should be subjected to a traceability review by PIT.

4.11.5.1.6 Specification changes and revisions. Once approved and released, any changes to a specification must be approved by the same office approving the original release.

4.11.5.2 Product design baseline

A product design baseline is defined in a specific set of engineering drawings, drawing lists, and reports. The content of this set of data and its specific version must be defined as part of that baseline. The current released engineering package forms the design baseline for an item.

Design work on items shall not be initiated until the requirements for the item are understood and approved. Changes to a released design baseline may only be accomplished with the approval of the office that originally approved the release.

4.11.5.3 Product representation baseline

Product representations reflect selected characteristics of product items for the purpose of evaluating product features or possibilities. The team responsible for the development and use of a particular representations is responsible for maintaining the configuration of that item with respect to a specific design baseline. Examples of product representations are

a. Models that mathematically or physically represent the design solution.
b. Prototypes that include selected features of the intended design.
c. Special test articles used to test specific features of a design.
d. Test benches that allow controlled operation and test of functional systems.
e. Simulations that produce output results based on specific input conditions.

4.11.5.4 Production process baseline

The production representative on the PIT is responsible for maintaining a product production process baseline coordinated with the product design and quality process baselines. This baseline includes production facilities, procurement plans, tooling, process flow, material control and flow, and work station definitions.

4.11.5.5 Quality process baseline

The quality representative of the PIT is responsible for maintaining a product quality process baseline coordinated with the product design and manufacturing process baselines.

4.11.5.6 Logistics support baseline

The logistics representative on the PIT is responsible for maintaining a logistics support baseline coordinated with the product design baseline.

4.11.6 Change control

4.11.6.1 Engineering changes

Engineering changes during the period leading up to initial release of a design package simply come under the heading of original design work no matter how many times the development team may change direction. Subsequent to initial release, any changes must be specifically approved prior to accomplishing detailed work on those changes. Each such change shall be categorized in terms of its motivation, impacts, and optimum schedule for accomplishment. Each change shall also be associated with a change classification, class I or II, described below.

4.11.6.1.1 Change motivation and impact analysis. The motivation for a change shall be defined in one or more of the following categories:

a. Current design will not satisfy requirements.
b. Cost reduction.
c. Schedule improvement.
d. Performance improvement (current performance not in violation of requirements).
e. Process optimization.
f. Customer direction (with compensating cost coverage, if appropriate).

The full impact of a change shall be determined in terms of the product and the process. These impacts shall be evaluated in terms of cost and benefit relative to the status quo as part of the process of decision making on the change. Changes shall not be made without a clear benefit to the customer.

4.11.6.1.2 Change timing and grouping. The effects of a change shall be evaluated in terms of schedule relationships and the relative difficulty and cost of accomplishing the change at one time vs. another. In the absence of a reason demanding urgency, a specific change may be delayed until it can be accomplished in combination with one or more other changes in the interest of lower cost and less aggregate customer activity interference.

4.11.6.1.3 Class I changes. Class I changes affect the customer requirements baseline and require customer approval before they are implemented. These changes may require submission and approval of an engineering change proposal defining in some detail the precise work that shall be accomplished and the effects of that work.

4.11.6.1.4 Class II changes. Class II changes have no impact on a customer approved baseline and depending on the specific contract may not require customer approval to implement. In all cases, the impact of such changes will be clearly defined before an implementing decision is made.

4.11.6.2 Deviations and waivers

Deviations and waivers are agreements between the company and a customer that permits continuation of a current design that is not in compliance with approved requirements on a temporary basis. The agreement will specify what articles are covered and in what ways it fails to satisfy item requirements. A deviation covers a problem first identified through acceptance testing of an article and permits that one item or a specific series of articles with the same problem with the understanding that changes will be made that correct this problem on subsequent items. A waiver is identified ahead of time, generally during the requirements or design period. An agreement is made that the design may temporarily fail to satisfy one or more requirements for a specific reason while work to resolve the problem continues.

4.11.7 Product interface management

Program, PIT, and PDT leadership share in responsibility for management of product interfaces. No management responsibility carries with it more significance for cost and schedule control than interface management because it has such an expansive influence on the whole technical development effort. Interfaces well defined between the product elements under the responsibility of two teams will yield compatible designs on both sides of the interface that survive development challenges. This same condition reached across all system interface pairs leads to program success when combined with sound management of the internal affairs of all teams.

4.11.7.1 Interface identification

Interfaces are defined by the needed relationships between items in the system architecture. They are pre-determined by the way that functionality is allocated to things in the system. Interfaces unlike the items in a system, the responsibility for which can be assigned to individual people or teams, generally have a responsibility shared by two people or teams. This increases the complexity of management of the development of interfaces over items.

It is commonly and correctly held that the development process most often breaks down at interfaces where different parties are responsible for the interface terminals. These are referred to as cross-organizational interfaces. The development team should seek to minimize these interfaces but cannot eliminate them since they also provide the system with richness that distinguishes it as a system. Cross-organizational interface can be minimized by the way functionality is allocated to the system architecture, the way that the architecture is grouped into subsystems, and the way responsibility is assigned for the architecture to the PDT.

The use of functional N-Square diagrams to show correlation between functions during the function allocation process helps to understand and avoid complex relationships that may eventually materialize as difficult cross-organizational interfaces on the solution side of the allocations. The functionality should be allocated to architecture by the PIT with participation by manufacturing, logistics, and hardware and software design personnel with careful attention to where the functional subsystem and physical boundary conditions are drawn in order to avoid different physical and functional boundaries where team responsibility boundaries will be defined.

Figure 4-12 illustrates the preferred relationship between the product interfaces between the architecture items and the communications paths that must exist between the teams responsible for development of the product. Note the perfect alignment between cross-organizational interfaces and the team communication paths that commonly break down in the development of these interfaces. This interface alignment results in the simplest possible definition of interface responsibilities and helps to encourage communications where it is most needed.

Two kinds of interface exist: internal and external. External interfaces exist between our system and outside entities. Internal interfaces exist between

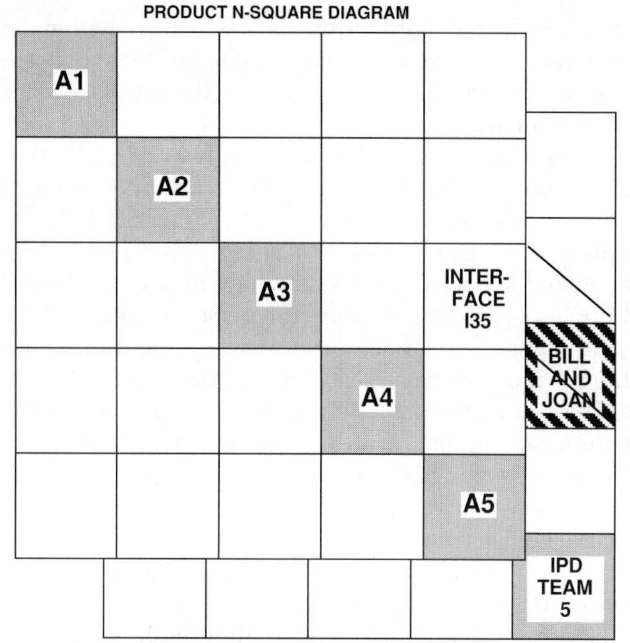

Figure 4-12 Interface alignment in product entities and team communications.

items within our system. One of the PDT is responsible for one terminal of each external interface, but the PIT is responsible for ensuring compatibility for all system level external interfaces. This same notion is carried down into team items where the internal interfaces are an internal team responsibility and the external interfaces between team items are a joint responsibility between two teams with PIT responsible for ensuring compatibility.

4.11.7.2 Internal interfaces

Internal system interfaces shall be developed under the guidance of the PIT through cooperation between the PDT. The PIT shall maintain a system schematic block diagram (in graphical or database format) that identifies all needed interfaces and their terminal items defined in accordance with the system architecture notation. Since each architecture item has a clearly defined development responsibility, the pairs of persons or teams responsible for development of each interface are therefore also clearly identified. In each case, the person or team on the receiving (destination) end of each interface shall be the principal responsible party and shall take the leadership in development of the interface. In cases where the sending/receiving sense is unclear, PIT will arbitrarily define the sense for the purpose of principal responsibility determination.

The teams and personnel should view these internal interfaces from one of three perspectives based on their degree of responsibility for the interface, and this is the basis for PIT definition of interface development responsibility.

a. **Interface**. An interface element both terminals of which are within the development responsibility of the team in question. The team in question is totally responsible for the development of the interface since they control both terminals.
b. **Outerface**. An interface element neither terminal of which is within the development responsibility of the team. The team in question is not responsible in any way for development of this interface but may have some indirect interest in it. Refer to paragraph 4.11.7.4 for a discussion of extended software interfaces which may color the degree of interest in these interfaces for software personnel. All of these interfaces are in someone's innerface or crossface category because every item in the architecture has a responsible team assigned.
c. **Crossface**. An interface element for which the team in question is responsible for the development of the architecture item at only one terminal. The team in question has a shared development responsibility with the team responsible for the other terminal, but the receiving team is the principal held accountable for the proper, compatible, and timely development of the interface. The interface requirements shall be developed in each case through cooperative work between the responsible teams during the requirements definition period for the items themselves under the leadership of the identified principal team or engineer. These interface requirements will be checked across the interface in the requirements for each item by PIT where crossfaces are involved before design authorization is given to either team. These two teams will thereafter cooperate to develop compatible design solutions for these interfaces with audit by PIT.

4.11.7.3 External interfaces

All external interfaces shall be identified and their development managed and accomplished by PIT in cooperation with the PDT responsible for our terminal. These external interfaces will be characterized by an interface control document (ICD) prepared by PIT of one of two types:

a. A living ICD retains the external interface requirements for the life of the system. External interface requirements in this case should not be carried in the system specification to avoid redundant requirements capture.
b. An interim ICD is used as a tool to initially identify and bring about a condition of mutual agreement between the parties to the interface. It is a management mechanism. The agreed upon requirements are included in the terminal specifications. When the interface is fully characterized, the ICD is dropped and the interface is defined by compatible requirements in the two terminal specifications.

Where the external interface connects to an associate item under contract independently with a common customer, the company will reach agreement with that common customer and associate on the way that the interface development will be managed in a letter or memo of agreement or understanding that tells who is responsible for interface documentation and outlines the broad programmatic principals of the relationship. The company will also reach agreement on a common interface management plan with the customer and associate contractor that will guide the detailed work between the associates. Commonly, this agreement will call for one of the two parties to prepare the ICD with the approval of the other and the common customer. Any of the three parties can thereafter offer changes to the ICD that have to be studied by each party and approved or rejected for cause. Where agreement cannot be reached, the items are discussed at an interface control working group (ICWG) meeting or preparatory technical interchange meeting (TIM).

4.11.7.4 Other interfaces and clarifications

The external interfaces between the product and the natural environment will be treated as environmental requirements and solutions. Any hostile and non-cooperative system interfaces will be treated the same way since there is no one with whom the company can coordinate development work. Cooperative system interfaces will be handled as external interfaces. All humans responsible in some way for operation or maintenance of the system and parts thereof will be treated as elements of the system (sometimes called the human subsystem) and the interfaces between the hardware/software elements and the human subsystem are internal at the system level. There may be both external and internal human interfaces at the PDT level.

Software personnel involved in interfaces entailing hardware and software planes may choose to recognize extended interfaces where the immediate physical interface may be the input/output port in a processor, but they have interests in the dynamics of a whole control loop beyond that plane.

4.12 Data management

The PIT is responsible for product-oriented data management and the PBT is responsible for programmatic data management including contracts, finance, and program administration. These teams will cooperate to provide a common and simple approach to information access balanced with a degree of protection as appropriate in response to customer security demands, company proprietary rules, and potential for willful or unintentional destruction, loss, or alteration. Computer network solutions are encouraged that are compatible with available networked workstations and software applications.

All product data taken together forms the program product data library. It falls into one of five categories with different rules associated with each category:

a. **CDRL Items**. These are documents that the company is obligated to supply to the customer in accordance with the provisions of the contract. They may include system and configuration item specifications, engineering drawings, analysis and test plans and reports, schedules, major review data, plans, and status reports. As information is passed through the PIT data management person for delivery to the customer, one or more copies are placed in the CDRL section of the program product library for internal use.

b. **DAL Items**. All data developed by the program that is not on the CDRL is generally accessible by the customer for a fee defined in the contract. These items may include engineering drawings, procurement and in-house specifications, test procedures and reports, memos, and internal meeting minutes. Each team manager is responsible to provide a copy of any internal data to the PIT data management person for inclusion in the DAL section of the program product library for internal use and submission to the customer in response to a specific request.
c. **SDRL Items**. SDRL items are supplied by a supplier under the terms of the contract with the supplier. They are formal contractual inputs called for in the supplier contract. Commonly these items must be reviewed within the company and formally approved. Therefore, PIT data management must provide a means to route these documents (preferably electronically) for internal review and approval as well as maintain track of the status of the review and approval process. All supplier data is a subset of the DAL but treated separately because of its special routing needs, volume, and uniqueness. A copy of all internally approved SDRL items is placed in the SDRL section of the program product library.
d. **Associate Items**. The PIT persons dealing with associate contractors on interface development are responsible for providing the PIT data management person with copies of all related correspondence for inclusion in the associate section of the program product library for internal use.
e. **Program Company Private.** The program may produce some limited set of data that is properly defined as company private. This data must be safeguarded from exposure to any outside parties including the customer. The PIT must be very careful to avoid use of this category for information that program persons wish to improperly withhold from the customer. The proper content of this section of the product library include bidding strategy and proprietary information related to future business opportunities not covered in the current contract. The content of this section should be reviewed periodically by personnel from the company's legal and contracts departments.

4.13 *Parts, materials, and processes standardization*

Design teams shall maximize the use of standard parts, materials, and processes to control costs, reduce logistics complexity, and simplify and standardize production. The PIT is responsible for implementing a parts, materials, and processes (PMP) standards process on a program. Generally, this is done by making approved lists from which designers should select needed items. During the concurrent design process, the PMP representative should check for compliance. A computerized or manual drawing check process shall formally verify compliance.

4.14 Supplier technical control

Wherever possible, suppliers will be accepted as members of PDTs and these suppliers will be controlled by the responsible PDT. The PIT will include procurement and material specialists to manage the overall program subcontract technical effort and the Program Business Team will provide contracts support for all supplier activity.

4.14.1 Responsibility

PIT is responsible, with the cooperation of PBT and PDT, for formation of a make-buy team which is then responsible for definition of make or buy selection for each item in the system architecture. For all items that will be purchased, the PBT is responsible for management of all supplier and subcontractor relations. The PBT shall rely upon the PIT and PDT for technical support in these relations. The responsible PDT shall prepare the technical definition for the item in one of the following forms:

a. **Development situation.** Product requirements defined in a procurement specification and process requirements defined in a procurement SOW.
b. **Build-to-print situation.** Product design definition in an engineering drawing package and process definition in a procurement SOW.
c. **Off-the-shelf situation.** Product definition in a source control drawing that clearly defines the specific product required and any alterations that must be made prior to delivery.

The team responsible for a given procurement action must ensure that item requirements are clearly conveyed to the supplier in a form that will be understood, that all supplier questions are answered in a timely way consistent with any contractual relationship with the supplier, audit the product information provided by the supplier for compliance with requirements, and assure that the delivered product complies with requirements.

4.14.2 Supplier selection

The number of suppliers shall be minimized and favor suppliers with which the company has had positive experiences of delivery on time within cost targets with products that satisfy predefined requirements.

4.14.3 Long lead items

All items required for the development will be screened against their availability with respect to their need dates to support development and manufacture. Any items which are at risk of not being available will be listed on a long lead list and special provisions made for procurement in a timely way to mitigate the risk.

4.14.4 Specification items

Items that will require some development and testing by the supplier shall be contracted for, using a procurement specification and SOW.

4.14.5 Build-to-print items

Items designed by the company which require no supplier development and testing should be procured in accordance with a drawing package supplied to the supplier.

4.14.6 Off-the-shelf items

Wherever possible, items that can be procured off the shelf from existing production runs should be used. These items can be procured by a simple purchase order defining the item by its supplier identification. If such parts must be modified for company use, they should be procured with the necessary modification instruction provided to the supplier. If the modifications become too complex, the item will require special testing and possibly even qualification to the customer requirements.

4.15 Associate contractor relationships

Associate contractors are other companies responsible to a common customer for some other element of a system under contract to the company where the company has no contractual relationship with the other company except through the common customer. In these situations, the company should enter into an associate agreement with the other company and the common customer telling how matters of common interest shall be handled.

4.16 Customer furnished property

Property supplied by the customer to be integrated into the product system shall be acquired in accordance with contractual provisions by the PBT and its control passed to the responsible team during the development period.

section five

System engineering process (MIL-STD-499A Part II)

5.1 Process overview

This section of the SEM/SEMP describes the company systems engineering process activities as they are to be applied to a contract phase. Some of these activities will properly relate to a given program phase while others may not. The planner must select the activities that are appropriate to the goals of the program phase in question. This section, and appended data, provides guidance on those selections.

Appendix C provides a detailed generic process diagram depicting how the company accomplishes development and production programs. Appendix D defines each block on the process diagram illustrated in Appendix C in terms of work performed, principal functional charter responsibility, and references for accomplishment (commonly back to the content of this document by section or paragraph numbers). The development work depicted in Appendix C may have to be accomplished in customer-defined phases focused on specific milestones and goals punctuated by major decision-making reviews that determine the degree to which the company has satisfied the success criteria for the phase. Refer to Section 3 for phase and process relationships and Section 4 for controls applied to the systems engineering process.

5.2 System definition

Two possibilities exist when undertaking the development of new capability for a customer or market. The problem expressed by a customer need may be new to the company experience requiring an unprecedented solution. Alternatively, the problem may express a variation on past solutions with which the company is familiar. In the former, a clean sheet of paper approach is required applying the full range of development techniques. The precedented approach will require an initial analysis of some existing solution and what parts of it are appropriate to the new problem.

Whether the system is precedented or not, the earliest program work will define a common starting point for decomposition of the large problem represented by the customer need and expanded by this information into a series of smaller related problems that become the focus for concurrent development work. In precedented systems the subsequent work should be primarily focused on the areas changed or replaced, but it is necessary to keep re-evaluating whether you have done something that affects elements thought to remain unchanged.

5.2.1 Customer requirements credibility and stability

Customer requirements may be stated as simply as a terse customer need statement, an expression of a needed mission scenario, or more completely in the form of a system requirements document, operational requirements document, or draft system specification. In any case, the program must challenge all of these requirements and validate the need for them and associated quantitative values on the basis of the need that must be satisfied. This must not be approached from a position of arrogance, rather from a perspective of helpfulness assuring that the customer will receive exactly what they need as a result of company work. Requirements offered with the contract will be compared with those upon which a preceding bid was based and any differences resolved prior to final contract agreement.

Formal agreement must be reached at the earliest possible time between the company and customer on the system level requirements so as to form a binding definition of the technical work to be accomplished. This is normally accomplished at a system requirements review by whatever name.

Customers are commonly reluctant to authenticate a system specification early in a program due to the expense of formally changing it subsequently. That concern must be balanced against the valid need for a contractually binding definition of the problem the contract is attacking.

Subsequent to mutual approval of a system specification, company personnel must be very candid and disciplined in their conversations with customer personnel about additional capabilities desired. The approved system specification provides a basis for cost and schedule definition as well as technical capabilities. Any agreement to change system specification content or broaden the accepted meaning of existing content must be approved by the company program manager.

5.2.2 Unprecedented systems

On a new program where only the customer need has been established, the company will cooperate with customer personnel in the identification of an appropriate mission suite while also defining a corresponding basing concept, operations concept, logistics support concept, environmental envelope (including natural, cooperative, non-cooperative, and hostile as appropriate), and manufacturing scenario. System simulation and modeling will be applied as appropriate to assist in selecting preferred system concepts from among reasonable alternatives defined through team work and creative thought about the customer's need. The intent will be to determine a system concept characterized by a balance between least cost and greatest effectiveness for the customer to satisfy their expressed need. The result of this activity will be:

a. A preferred mission scenario with matching master functional flow diagram (or alternative diagrammatic treatment) and expansion of the operations and maintenance function one level,
b. A basing concept telling what major facilities will be employed in accomplishing the operational mission and logistics support functions,
c. A mission space definition telling within what space the system shall operate,
d. A system environmental definition,
e. A preliminary environmental impact statement, if appropriate,
f. A logistics support concept telling the maintenance levels and how they will be applied throughout the planned major facilities and mission scenario and supply support concept,
g. A manufacturing plan with major, high level material flow paths to include major vendors where known or history has shown a useful connection to similar product lines,
h. The outlines of the top level system architecture and interfaces (internal and external) to include identification of one or more alternative architectures that could be applied to solving the identified customer's need within the context of the surviving mission scenarios, and
i. Values selected for all system measures of effectiveness (MOE) and key parameters used in any modeling and simulation work.

Generally, the work required to produce these products will entail some kind of modeling and/or simulation work to permit a detailed examination of all of the alternatives within available time. More than one functional flow alternative may be composed to reflect different views of the optimum need solution and each characterized by one or more alternative architecture suites. In some cases the choices between elements of this two-dimensional trade may be easy to filter into promising and discouraging alternatives. Some remaining alternatives will be more difficult to determine their relative cost-benefit combination. It is here that models and/or simulations may prove effective in assessing the relative figures of merit for these different alternatives and accelerate the movement to a preferred mission definition and corresponding top level architecture.

5.2.2.1 The customer need

The customer need is the ultimate system function and requirement. It is the basis for derivation of all system functionality and performance requirements as well as the complete architecture for the system. Mission scenarios are developed as a systematic means to understand how the customer's need may be satisfied with one or more alternative approaches using architecture derived from alternative high level interpretations of needed functionality. These competing scenarios will impose different combinations of requirements on the system with consequent relative costs and effectiveness figures useful as a basis for selection of the most desired scenarios and resultant system requirements.

5.2.2.2 *Concept exploration tools*

In the early program phases, the development team shall take advantage of some combination of the tools described in subordinate paragraphs to obtain information from external sources and focus program understanding of the customer need.

5.2.2.2.1 Marketing analysis. Whether the customer be government or individual persons in a commercial sense, the program must understand the customer's need. The customer may or may not be able to articulate precisely what is needed. The customer may not even be aware of a capability that would be very useful and desirable if conceived. Company personnel must interact with the customer to gain an understanding of their views of their needs and to offer encouragement in directions that the company is prepared to support.

This interaction may take the form of direct discussions with persons in customer program offices or user commands. It may take the form of customer surveys or focus groups. The company must acquire an understanding of customer knowledge of their needs and their attitudes toward solving those needs. Where possible, the company should accomplish work to support potential customers in their efforts to understand their needs. This may be accomplished in the context of brief analytical studies of feasibility, technology evaluations, or information searches.

5.2.2.2.2 Feasibility analysis. A feasibility analysis is done to determine if a particular course of action has a chance of success and what impediments may stand in the way of success. It should consider enabling technologies and their availability or access or development difficulty. It should study alternative approaches and identify the relative difficulty and risk of each. Such studies should also determine the impact on company business from full implementation of a program along the lines studied.

5.2.2.2.3 Cost effectiveness analysis. A cost effectiveness analysis should determine the relative merits and difficulties of a particular course of action in satisfying a need. One of the principal aspects of this information is cost measured in current development dollars, life cycle cost, or some other measure. The other element is effectiveness or benefits. This may be measured in mission kill probability or some more peaceful measure, but it should be a measurable parameter that can be combined with cost in some way to relatively score competing solutions using cost/effectiveness ratios.

5.2.2.2.4 Performance measures and driving requirement identification. During the early steps of system definition, it is very important to determine ways to measure system performance quantitatively and to estimate or calculate these parameters accurately. Commonly some kind of mathematical model or simulation will be required for this purpose. In the process, the investigating team must work to focus on a small set of parameters whose values have a great deal of leverage in effecting system performance. These are called driving requirements and efforts to model system behavior must account for them.

5.2.2.2.5 Parametric analysis. Often it is difficult to determine the most desirable system based on a single parameter or even a collection of such parameters. A comparison of pairs of parameters is sometimes very helpful. This can be accomplished by forming parametric combinations where you state the amount of one parameter for each unit of another. For example, it may be useful to understand the cost per pound of payload while trying to determine the best payload capability. Such studies are called parametric analyses. Graphs help to illustrate useful combinations of these parameters and to make selection decisions.

5.2.2.2.6 Trade studies. Trade studies shall be used as a decision-making preparatory step where there are two or more alternative choices, the rationale for selection is unclear, complex, or difficult to understand, and the decision carries with it a lot of leverage for future action and cost. The trade study process applies to mission definition, functional decomposition, requirements definition, product design decision, and process alternatives.

A trade study need not require a great deal of time to complete, though complex ones often do. It is possible to complete a simple trade study in a day. Management must monitor trades study performance carefully to encourage timely completion of the work. These studies require careful study of the alternatives mixed with decisive leadership. The former is easier to accomplish commonly than the latter.

The PIT is responsible for maintenance of a trade study records database that retains and makes available for access information about past, current, and planned trade studies. This database is part of the design rationale component of the program information library.

The trade study process entails eight steps as follows:

a. Define requirements that must not be violated by any alternative.
b. Define the criteria by which a selection should be made and a way to assign quantitative or qualitative values to candidates in each criteria (utility curves, for example). Define how candidate values will be combined in the several criteria in terms of different weighting or priorities.
c. Define two or more alternative candidates and assure that they all meet the prescribed requirements.

d. Evaluate each candidate against the criteria and assign values within each.
e. Compare candidate scores and identify a preferred alternative.
f. Where there is uncertainty in one or more criteria or values assigned, conduct a sensitivity analysis to determine what changes would occur in the preferred selection if small changes were made in criteria or valuations. Consider alternative candidates based on new knowledge. Are there better candidates than originally conceived, perhaps that combine certain features of different candidates? If so, run the evaluation on those.
g. Collect all study data and include it in a final trade study report appropriate to the cost and difficulty of the study. Capture the decision rationale.
h. Make a decision on a final selection and subject the decision to PIT review (if PIT was not the decision-maker) and peer review.

5.2.2.3 *Mission analysis*

Given a well-defined customer need, that need must be expanded into a critical mass of knowledge as the basis for system development. Mission analysis offers an effective approach to this goal. Two or more alternative ways of satisfying the need are creatively determined and subjected to comparative analysis to determine a preferred mission definition. Mission scenarios are developed and evaluated. Alternative system resource mixes are considered, subjected to trade studies, and a preferred concept selected based on the results of the trades.

5.2.2.4 *Logistics support concept*

As the product concept evolves, it will be accompanied by a parallel development of the corresponding logistics support concept defining the basing scheme, maintenance levels, spares provisioning plans, and technical data and training plans.

5.2.2.5 *System boundary and environment definition*

During the initial development period, the system boundary shall be clearly defined along with the necessary interfaces with the system environment. The system environment is everything not included in the system. It is not necessary to identify everything in the system environment, only those elements that have some influence on the system. It is necessary to identify everything in the system architecture as part of the development process, but this is accomplished incrementally from the top down, layer by layer in each branch of the architecture, accounting for specific items the customer wants used.

The system environment should be defined in terms of: the natural environment, cooperative systems (those systems that have predetermined and controlled interfaces with the new system), non-cooperative systems (systems that may interact as a function of sharing mission space but have no predetermined or controlled interfaces with the new system), and hostile systems (systems with specific intent to do damage or otherwise defeat the new system). Special care shall be taken to determine what induced environmental effects apply. These are cases where energy sources within the system interact with the natural environment to produce environmental impacts, which would not be present without the presence of the system, that are applied to the system.

The environmental relationship is bi-directional. Natural environmental impacts by the system shall be defined and ways to mitigate any detrimental effects identified. Company policy encourages zero adverse environmental impact by company products. Where the systems are military in nature, environmental impacts are unavoidable but can be minimized within the constraints imposed by the contract.

5.2.2.6 Trade studies and preferred solution selection

The program shall arrive at a preferred selection for system definition through evaluation of alternative concepts in terms of the product configuration and mission, logistics support, technology availability issues, environmental, and cost characteristics. Where decisions cannot be easily and logically derived, trade studies shall be accomplished to clarify the most desirable cost benefit relationship maximizing customer value.

5.2.2.7 Reporting responsibilities

Programs must report the results of trade studies and mission analysis not only to support program design rationale database needs, but to provide the company with retrievable knowledge about the results which may be useful on other programs. Even if a customer does not require publication and delivery of a mission analysis report, one shall be prepared, telling the final preferred system definition in terms of a master flow diagram, top level architecture, top level schematic block diagram or N-square diagram, and one or more mission scenarios; alternatives not selected; and the important characteristics of the system environment.

5.2.3 Precedented systems

On systems based on prior system designs requiring some modification of product or processes or both, an analysis will be accomplished to determine if the previous mission definition covering the application prior to modification action is adequate. If those materials cannot be found or never existed, a rapid development activity should be undertaken to create a clear system problem statement based on discussions with users of the existing system. In either case, the resultant mission definition must be reviewed for potential changes and the above materials defined for the precedent system changed as a basis for subsequent system definition work.

The architecture items that are expected to remain unchanged should be identified as well as those that must undergo some change or replacement or new development. An effective way to do this that rapidly and visually conveys the problem scope is to place the architecture diagram (down to some level of indenture) on a wall and color the boxes to indicate change impact.

The program may have to work in a middle-out fashion, rather than the top-down fashion encouraged in this document, on such programs. The middle-out process first identifies all items that will have to be changed (product and process) or newly developed and then seeks to understand the requirements for these items that must be changed or newly created. These requirements are then focused downward (lower half of the middle-out process) from that point in the system architecture to develop lower tier items. Requirements traceability is encouraged as the means of establishing that the requirements for the changed items remain consistent with higher tier item requirements (upper half of the middle-out process) previously established.

Where higher tier requirements are shown to be inconsistent with changed item requirements, one or both must be changed. Changes to the higher tier item may force engineering change proposals with customer coordination to increase the scope of the contract work beyond that foreseen in initial program planning or a change in requirements to permit continued use of previously designed resources.

5.2.3.1 *Post-delivery changes*

Subsequent to company delivery of a product, the customer may become aware of changes that must be made to better satisfy their needs. These changes may be driven by new capabilities not included in the original contract or they may cover shortcomings in the delivered product. The cause of the problem may trigger a contractual analysis to determine fault and cost responsibility. Design rationale information retained on the development program may be useful in such an activity to clarify the situation.

Changes made subsequent to delivery carry with them the problem of not only making changes in an engineering sense, but of introducing them into the delivered system. The way the change is designed may have to be influenced by implementation as much as by the optimum technical change rationale. The change should minimize the burden on continuing customer operations while the changes are made.

5.2.3.2 *Single use mission adaptation*

Products characterized by a single use resulting in expenditure of the product offer a unique development environment. Each product operation may be tuned to a specific mission or all missions fairly standard. In the former case, each product article will have to be produced for use on a specific mission. Each mission is essentially a mini program involving a set of changes from some core set of standards. In the latter case, it may be possible to provide sufficient flexibility in the product design that the customer/user may make adjustments with the standard product to allow use on the full range of mission situations without special development work for each mission. The latter is in the customer's best interest generally where it can be done without significant relative cost or performance impacts.

5.2.3.3 Re-engineering existing systems

The company may become involved in work that changes systems originally produced by other companies. The fundamental difference in these kinds of changes is that the company will not have ownership and access to the detailed engineering data upon which the product is based. That data may not even exist or be accessible to the customer. Situations such as this are defined as re-engineering activities because it will be necessary to accomplish some work focused on understanding the current reality, work that would not be necessary for a company product delivered at a time in the past that is within the memory span of current employees.

5.2.3.3.1 Existing system evaluation. The first task in re-engineering is to clearly understand the existing system and application using such information that the customer is able to supply, information the original producer will part with at a reasonable cost to either the customer or company, actual product articles available from the customer, and the results of interviews with system operations, maintenance, and management persons. Special care must be taken to avoid this activity getting out of hand. It is interesting work and those doing it may become self trapped into pursuing more information than needed. Management must ensure that the team remains fixed on their objectives and avoids unnecessary work no matter how interesting the results may be for historical purposes.

5.2.3.3.2 Customer need evaluation. The second step in this process, which need not await the results of step one above, is to clearly understand the customer's or user's displeasure with the current systems and their precise needs for the system they wish it to become. This can only be done through careful discussions with the customer/user. Where the customer and user are different parties, both should be brought into this conversation. A special team should be formed of company, customer, and user to identify the need and top level requirements for the changed system and how they differ from the previous system.

5.2.3.3.3 Change impact analysis. Armed with the results of the study work suggested above, the company team must develop a clear definition of the design impacts which become the basis for an otherwise and subsequently normal development program. Each item in the system that will require some change must be identified and the nature of the change that must be made defined.

5.3 Functional decomposition

One of the fundamental principals of system engineering is that form follows function. This means that we should understand what a system must do before we decide what shall satisfy that functionality. The customer need is the ultimate function for a system. It states very concisely what the whole system must achieve. We require a means to expand the amount of information available to us about needed functionality and the process for doing that is called functional analysis.

By first working to understand needed functionality, we encourage a thorough study of the problem space before moving to the solution space defined by system architecture and item designs. We should recognize that there exist powerful forces attracting us to simply race to a solution to what we individually think the problem is. Since we are all specialists, such an unorganized approach will almost guarantee a less than optimum solution because we do not all have the same view of the problem. Engineers are action-oriented people who want to move quickly to a solution. They are creative people with little interest in order, discipline, and process structure. These characteristics do not make engineers bad people. These are traits that are badly needed to conceive the solutions that will have to be derived from a clear statement of the problem.

A structured functional decomposition process does impose order and discipline on the process of defining the problem space. We must give this process time to work but we must provide the teams involved in this work with people who are not only good analysts but effective decision-makers. This process can lure its practitioners into an almost endless study of possibilities. A responsible decision-maker must be identified who can encourage the process to completion and know when the right time has arrived to force a final determination of a preferred functional portrayal and selections in the allocation of exposed functionality to things that will populate the system architecture.

This process must move rapidly or anxiety will build in the design and management communities that the program will not have sufficient time to create the needed engineering drawings. Mr. Mac Alford, while Chief Scientist for Ascent Logic, popularized the poignant phrase "We must start the coding now because we have so much debugging to do." This is the powerful force that a thoughtful functional analysis is competing with in the early stages of an unprecedented program.

The number of drawing changes made subsequent to CDR is an effective metric for earlier requirements definition effectiveness. If the requirements are well defined in adequate depth early on a program prior to the start of detailed design and the development teams are staffed with qualified personnel and well led, it is probable that they will succeed in defining compatible product and process designs that will lead to very few drawing changes. Poor requirements work will almost always lead to discovery of serious problems during the integration process that require design changes. Unfortunately, this metric provides evidence of a past failure and can only be useful as evidence of a need for changes on future programs.

The functional decomposition process shall continue until the functionality of every item on the lower tier of every branch of the architecture has been characterized and the items along the lower fringe of the architecture diagram all satisfy one of the following two criterion:

a. The item will be purchased at that level from a supplier.
b. The item will yield to preliminary and detailed design by a company cross-functional team of specialists.

At this time, the needed product functionality should have been identified down through this terminal level of the product hierarchy represented in the system architecture and that functionality transformed into appropriate performance requirements. Therefore, the functional analysis process may be terminated and work continued on refinement of the architecture through continued interface analysis, performance requirements analysis, and specialty engineering constraints analysis in parallel with the preliminary design process.

5.3.1 *Preferred and alternative decomposition techniques*

The preferred functional model used by the company at the system level is functional flow diagramming unless the complete system is intended to be composed of only some combination of computer software and hardware. The ultimate function on any system is the customer need identified as system function F and diagrammatically displayed as a block enclosing the need statement (possibly paraphrased for brevity) centered in the block and identified with the function ID F in the lower right corner. The function F is automatically allocated to the architecture element identified with the word system centered in the block and identified by the architecture ID A in the lower right corner.

The ultimate system function F must be expanded into a master flow diagram defining the next layer of functionality. This level should embrace the complete system life cycle including development, manufacture, test, logistic support, operation and maintenance, and decommissioning or disposition. Both the ultimate and master flow diagrams are generally generic for all company products. The functions on the master flow diagram shall be further decomposed as a function of program-specific studies and techniques as shown in Figure 5-1 and explained below. The standard product system functions shown on Figure 5-1 are aligned with the standard top level company core system development processes defined in Figure C-2 of Appendix C as noted in Table 5-1.

As functions are allocated to hardware and computer software entities, the responsible team or principal engineer shall further decompose item functionality using an approved decomposition methodology and tool set. A PDT may, beginning at the top level of their area of responsibility, switch from functional flow diagramming to any other approved decomposition technique. The following techniques, in addition to functional flow diagramming, are pre-approved:

a. **Hierarchical functional block diagramming**. The needed functionality is assembled into a hierarchy and this structure used as the basis for the architectural arrangement. Commonly this results in a very

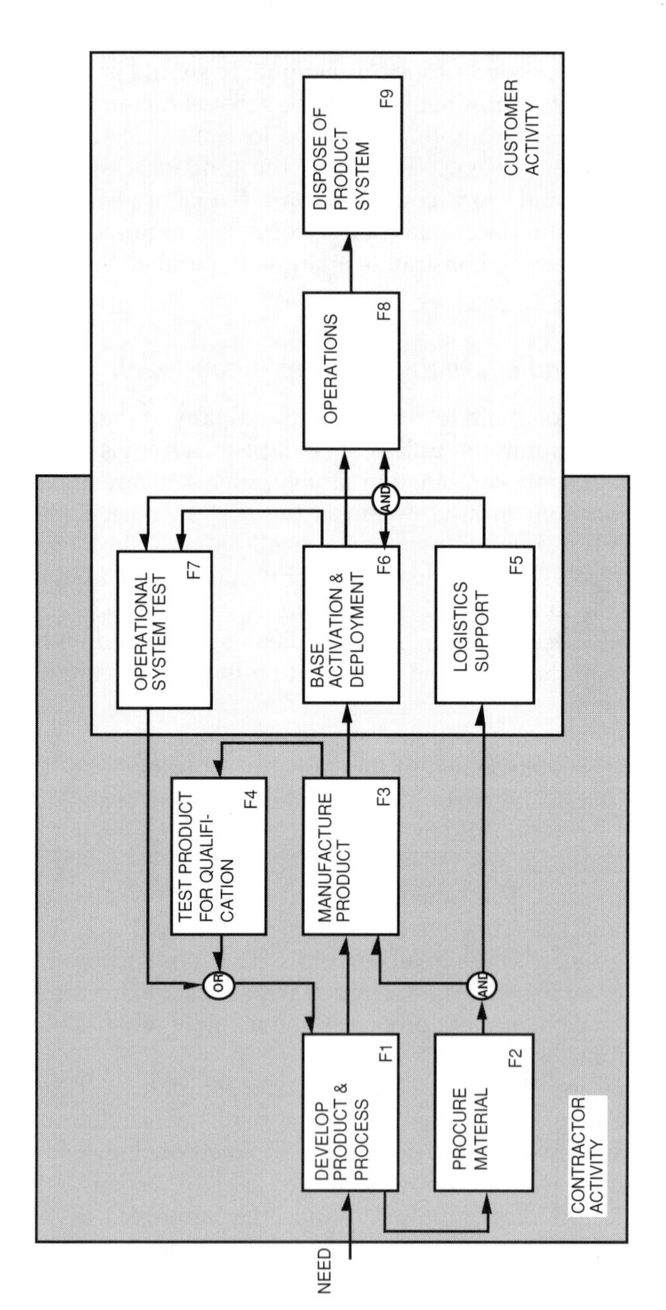

Figure 5-1 Generic master functional flow diagram.

Table 5-1 Process Correlation

Figure 5-1		Figure C-2 Major task	
F1	Develop product & process	1	Product & process development
F2	Procure material	2	Procurement
F3	Manufacture product	3	Production
F4	Test product for qualification	4	Qualification testing
F5	Logistics support	5	Logistics support
F6	Base activation & deployment	6	Base activation & deployment
F7	Operational system test	7	Operational system testing
F8	Operations	8	Operations
F9	Dispose of product system	9	System disposal

F1 Develop product & process. This activity is expanded with the results of a program planning study and resulting network diagram and corresponding program schedule showing the SOW and IMP task relationships including a critical development path.

F2 Procure material. Parts, materials, and assemblies are acquired for assembly and integration into the final product form in accordance with detailed procurement plans and contracts driven by content of engineering drawings and reports.

F3 Manufacture product. Manufacturing planning analysis yielding a detailed manufacturing process diagram to which manufacturing resources may be associated and from which manufacturing requirements may be derived. Quality Engineering shall work concurrently with manufacturing engineering personnel in selecting appropriate inspection points and planning their implementation. This includes test planning activities that will produce inputs to the manufacturing process to show how each produced item (or a sample of those items) will be tested to assure that it satisfies design and manufacturing requirements and will be produced with repeatability.

F4 Test product for qualification. A qualification test plan and process diagram expand this activity showing how the program will verify that the product satisfies development requirements.

F5 Logistic support. Process diagram expansion defining processes needed to maintain, store, transport, repair, and service product elements at all levels. Process blocks relate to human tasks that must be analyzed to define procedural requirements, test and servicing equipment needs and corresponding features, and human skill level requirements.

F6 Base activation & deployment. The product is introduced into its planned environment in accordance with a deployment process diagram.

F7 Operational system test. Where necessary, the principal product items are tested as a system (a flight test, for example). This activity is defined on a process diagram and clear plans associated with each activity.

F8 Operations. A functional flow diagram is developed in a mission analysis study or functional analysis activity as appropriate. Decomposition analysis continues to identify lower tier product system functionality and exposed functions are allocated to the system architecture with forced identification of associated performance requirements for the item to which the allocations are made.

F9 Dispose of product. A process diagram shows what actions must be accomplished to complete disposal of the product. This can be established as part of the original design activity but will generally require update when the system life cycle nears an end. New circumstances at that time may dictate application of resources to new uses not foreseen initially.

simple allocation process but sequence information is not provided. May be appropriate to a static entity with little dynamic activity. Commonly results in a one-to-one mapping from functions to architecture and susceptible to point design solutions.

b. **IDEF diagramming**. Takes the form of a standard flow diagram composed of rectangular blocks and arrow-headed lines connecting them on block ends. In addition, arrow-headed lines contact top and bottom surfaces signifying controlling influences and supporting mechanisms. This technique adds controlling influences not accounted for in functional flow diagramming but adds complexity to the portrayal of the functionality. An output from one step may provide a controlling influence on another, for example.

c. **Behavioral diagramming**. Technique used by Ascent Logic within the computer tool RDD-100 that presents a two-dimensional portrayal. A vertical functional flow diagram is connected to horizontal controlling influences. The resulting diagrams are very complex but can remain hidden within the computer model with outputs reflecting any one of several techniques including functional flow diagramming and IDEF. The benefit of this technique is that system operation may be simulated within the behavioral model using the requirements that are also included in specifications and for which traceability may be defined.

d. **Process diagramming**. A process diagram is similar to a functional flow diagram except that it is based on the physical model of the system (the architecture) and the blocks related to actual activities that must be accomplished with the system architecture items. In functional flow diagramming, the architecture is not necessarily known at the time the diagram is created. Also, a functional flow diagram reflects the problem space while a process model reflects the solution space. The process diagram is more appropriate to an existing system being modified or as an evolution of the functional flow diagram for use in logistic support analysis.

e. **Hardware state transition diagramming**. The system is perceived as existing in different states and those states are defined along with the transitions that cause a switch in state. This technique can be effective where the element can be characterized by time-dependent states with little movement in physical space.

f. **Yourdon-DeMarco computer software model**. Software processes are defined using circles representing processes and lines joining the circles to show sequences of processes. Other tools expand the definition, but the technique is most appropriate for process-dependent systems and has less utility for control-dependent systems.

g. **Hatley-Pirbhaa computer software model**. An extension of the Yourdon-DeMarco model to cover real-time applications as well as processing applications. Uses control flows as well as processing flows and two kinds of specifications.

h. **Object-oriented computer software model**. A software decomposition technique that associates all functionality with set theoretic entities and establishes the relationships between those entities.

5.3.2 Product-specific alternatives

There are many effective decomposition models, some most effective with hardware and others more effective with computer software. The responsible team must make a conscious decision what model to use in accomplishing its decomposition work.

5.3.2.1 Grand systems approach

Grand systems involve complex problems requiring an appeal to many different technologies for a suitable solution. They commonly entail hardware, software, and human activity. The company preference is to use functional flow diagramming as the decomposition aid for systems of this nature down to the point where further decomposition work is assigned to PDT.

5.3.2.2 Hardware continuation

Functional flow diagramming is effective for hardware involving a physical and functional dynamic. Static situations will sometimes quickly yield to hierarchical functional analysis, though care must be taken to avoid a leap to point design solutions. Functionality allocated to hardware with no physical dynamic but intense mode switching can often be further decomposed using state machines (state transition diagrams).

5.3.2.3 Computer software models

There are several effective computer software decomposition models. Those encouraged by the company include Yourdon-DeMarco, Hatley-Pirbhaa, and object oriented.

5.3.2.4 Personnel actions

At the lower tiers of the functional flow diagram we will often find that the functionality must be allocated to humans involved in operating or maintaining the system. It is helpful to apply process analysis in these cases. The fundamental difference between functional flow and process analysis is that in the former the blocks represent what the system must do while in the latter case they represent real-world physical analogs of work. In the latter case we should be able to associate with a block a certain facilitization, physical environment, and supporting infrastructure. Task analysis associated with these blocks involve use of the analysts' imagination to determine what actions the humans will have to take to accomplish the goals of the block (process). These human requirements flow into technical data defining how to accomplish tasks. This is essentially a logistics support analysis (LSA) activity.

5.3.3 Functional allocation

Given that the analyst clearly understands needed functionality through any of the decomposition techniques described above, this functionality must be allocated to architecture items that will form the physical model of the system. In most cases this will be a very simple assignment based on historical precedents from previous similar systems. Where it is a difficult decision, a trade study should be undertaken to clearly develop the relative benefits of alternative allocations.

When an unprecedented system is being developed it is advantageous to avoid allocation directly of functions while the top level character of the system is in a state of flux and trade. The functions may be further developed as functional requirements with quantification prior to allocation. With a precedented system with good historical precedents, it may be advantageous to allocate functionality with subsequent transformation into quantified performance requirements.

This plan makes a distinction between functions, functional requirements, and performance requirements as follows:

Function	A necessary system capability generally not quantified.
Functional requirement	A quantified function that has not been allocated to something in the system responsible for accomplishing the functionality.
Performance requirement	Quantified functionality that is associated with the whole system or a specific system entity.

5.3.4 Architecture synthesis

As items are identified in the system architecture through allocation of system functionality, the PIT shall integrate the organization of those items into families of things based on competing views of an appropriate product structure. The principal competing architecture views are listed below in decreasing order of precedence.

a. Manufacturing breakdown structure view based on bringing about compatibility with the planned production facilities, tooling capabilities, personnel skills, and historical precedents.
b. A desire for interface simplicity should drive the PIT to organize functionality into items and assign responsibility for those items so as to minimize the cross-organizational interface defined as interfaces with different organizations responsible for the development at the two terminals.
c. Engineering drawing view based on how engineering would prefer to capture the design in drawings.

d. Finance view identified in terms of the WBS overlay.
e. The responsibility model must respect the family structure of the architecture and WBS such that teams may be made fully responsible for cost, schedule, and product performance.

It is extremely important in all of these system composition views that all parties to the system development activity have precisely the same view of the system composition and structure and this view is embodied in the system architecture block diagram. The PIT is responsible for ensuring that all overlays are representative of the common system architecture.

5.3.5 *Interface identification*

As a prerequisite to allocation of functionality to system architecture items, the PIT and PDT may elect to identify needed interfaces between functions for the purpose of encouraging the simplest possible interface configuration. The interfaces between items in the physical model are largely predetermined by the way functionality is allocated to the architecture items; so, functional interface analysis can influence the evolution of simple interfaces in the physical model. Alternatively or subsequently, the teams may identify needed interfaces in the physical model between architecture items and feed back to the architecture identification such changes that appear prudent based on minimizing the cross-organizational interfaces. Cross-organizational interfaces may be influenced both by changing development responsibility assignments for the architecture items and by moving items between architecture structures.

The PIT shall define a means to assign clear accountability for interface requirements analysis and design responsibility. Interface responsibilities differ from item responsibilities in that they often entail a pair of persons or teams rather than a single person or team. Responsibility may be assigned in accordance with any one of the following three rules where the interface is cross-organizational in nature (different organizations responsible for the two interface terminals).

a. Shared responsibility — The two parties responsible for each of the two items at the interface terminals share interface development responsibility.
b. Source responsibility — Both parties share responsibility but the person responsible for the item on the source end is accountable for ensuring that the interfaces on both ends are compatible.
c. Destination responsibility — Both parties share responsibility but the person responsible for the item on the destination end is accountable for ensuring that the interfaces on both ends are compatible.

Where both terminals of an interface are the responsibility of the same organization, that organization is fully responsible for development of the interface requirements and designs.

The requirements for interfaces within the boundary defined by company responsibility should be captured in specification pairs with careful attention to ensure that these pairs remain consistent. Interfaces between items under company design responsibility and items under the control of associate contractors should have their requirements captured in an interface control specification or document (ICD) prepared by one of the associates in accordance with a letter of understanding signed by both parties and the common customer. In the latter case, the document should be used on one of the following two scenarios:

a. **Living ICD**. Include all interface requirements in an ICD and do not capture them in either terminal specification. Reference the interface control document in each terminal specification. This avoids double-booking the requirements with a potential for the requirements in two different places becoming different over time. The ICD remains in the program requirements tree for the life of the program.
b. **Interim ICD**. Use the ICD as a vehicle for reaching agreement on interface requirements and designs. All requirements in the ICD are placed in each terminal specification after agreement is reached on the meaning of the requirements for the two items. After approval of the ICD and assurance that the content of the two terminal specifications matches those requirements, the ICD is dropped from the requirements tree. The purpose of the ICD in this case is only to assist in the evolution of the needed interface definition. When that definition is complete, the ICD has fulfilled its purpose.

5.4 *Requirements analysis*

This manual recognizes the requirements categories listed in Figure 5-2 as an indentured breakdown. Tools are discussed below to help analysts define requirements in all of these categories. Product requirements are requirements that appear in product specifications Sections 3, 4, and 5. Generally, in systems and hardware, allocated functional requirements will have been translated into performance requirements as a prerequisite to inclusion in a specification; so, they only act as the seeds from which the performance requirements are derived. In computer software, functional requirements are often included in the specification under that heading and performance requirements are the quantification of those functional requirements. The term non-functional requirements is applied in software to group all of the indicated requirement categories. In system and hardware specifications the only top level categories generally are performance requirements and constraints.

Product Requirements
- Functional Requirements
- Non-Functional Requirements
  - Performance Requirements
  - Constraints
    - Interface Requirements
    - Specialty Engineering Requirements
    - Environmental Requirements
      - Natural Environmental Requirements
      - Cooperative Systems Imposed Environment Requirements
      - Non-cooperative Systems Imposed Environment Requirements
      - Hostile Systems Imposed Environment Requirements
      - Induced Environmental Effects Requirements
      - Environmental Impact Limiting Requirements
- Verification Requirements
  - Test Requirements
  - Analysis Requirements

Process or Programmatic Requirements
- Manufacturing Requirements
- Logistics Support Requirements
- Quality Assurance Requirements
- Test Planning Requirements
- Program Management Requirements

Figure 5-2 Requirements categories.

Verification requirements appear in Section 4 of a product specification and define what is expected by way of verifying that the product design synthesized from the requirements in Sections 3 and 5 of the specification does in fact satisfy those requirements. These are commonly test or analysis requirements but may also define demonstration and inspection requirements.

Process or programmatic requirements appear in statements of work, plans, and procedures and influence how programs are implemented, controlled, and accomplished. Most of the product requirements categories also apply here, but it is more convenient to categorize them as noted in Figure 5-2. Some of the same methods or tools used to gain insight into product requirements are appropriate to processes as well. The developing process is a system that gives birth to the product system. Process flow analysis is the company preferred approach to programmatic requirements analysis, more often called program planning.

The phrase "derived requirements" is sometimes used by people as if it were a distinct requirement category. It is not a separate category in this manual. Derived requirements, of whatever category, are those requirements not defined by the customer but exposed by the company through a requirements analysis process using the customer provided requirements as a basis. In the purest sense, all requirements except the ultimate requirement, the customer need statement, are derived requirements. Another term often seeming to have a categorization relationship is "driving requirement." This is a requirement, of whatever category, which has great leverage in determining the character of the system.

The preferred requirements analysis process for the company involves structured techniques that encourage insights into objects about which to write requirements. All requirements fall into one of two categories: performance requirements and constraints. The functional decomposition and allocation technique described in paragraph 3.4 exposes rudimentary performance requirements in the form of allocated functions or functional requirements. All other requirements fall into the constraints category and there are three kinds of those: interface, environmental, and specialty engineering.

On unprecedented systems the functional analysis and allocation process should take place before it is decided what shall accomplish the indicated functionality. During very early program phases involving unprecedented systems, it may be necessary to complete the functional analysis to some depth and expanding those function statements into fully phrased functional requirements with quantified values before accomplishing allocation in order to avoid a leap to point designs prior to a thorough analysis of the problem space. On precedented systems or at the lower levels of unprecedented systems, it may be desirable to speed up the allocation process by allocating functions directly to architecture followed by performance requirements definition based on those allocated functions.

Design constraints (interfaces, environmental, and specialty engineer) commonly cannot be fully defined until the corresponding architecture item has been defined as a result of the allocation of needed system functionality. Therefore, the definition of constraints may continue through item concept development into early preliminary design. In no case shall detailed design be authorized with an incomplete set of requirements unless a rapid prototyping development environment has been selected for the item characterized by cyclical development, building, and testing.

Constraints are boundary conditions that the design team must remain within while satisfying the item performance requirements. One should seek to minimize the constraints to only those that absolutely must be satisfied because all requirements constrain the design solution space and reduce the options available to the design team. It is possible to over constrain the solution space such that it is actually a physical impossibility to synthesize the requirements.

5.4.1 Team formation

Requirements analysis is the first activity of a team; so, it is only proceeded by actions taken to form the team. The PIT and PBT are formed by the program manager selecting team leaders and directing that those leaders coordinate with functional management to acquire the necessary personnel to accomplish program work. The team leaders must select and install appropriate lead persons and provide them any needed program-peculiar information. This group of leaders then completes team staffing, training, and acquisition of team physical resources.

The PIT is responsible for identification of the architecture and items within the architecture which will be assigned to teams. The PIT names the initial team leaders who are then responsible for completing team formation by acquiring qualified personnel skilled in areas needed to accomplish team work. The PIT is responsible for assessing PDT to training needs and arranging for that training as well as assisting PDT to acquire needed resources.

5.4.2 Performance requirements analysis

Performance requirements tell what the item or system must do and how well it must do it. The allocated function statements provide the primitive requirement attributes upon which performance requirements analysis can be performed. This involves determining a proper way to phrase the one or more full requirements derived from the allocated function and determining appropriate values. Valuation may require development and employment of models or simulations involving multiple requirements values exercised together through various values to determine a best mix of values.

5.4.3 Interface requirements analysis

Interface requirements are written about the intersections on N-square diagrams or lines on schematic block diagrams. The intersections or lines provide the attributes about which interface requirements are written. A given intersection or line may require one or more requirements statements to fully define the needed characteristics. The characteristics that are included in the requirements are a function of the technology of the interface: fluids, electrical signals, or physical attachment, for example.

The most difficult aspect of interface requirements is the fact that two parties are responsible for them, corresponding to the two items at the terminals of the interface. Those two parties must interact intensely to ensure they understand each other and evaluate each other's requirements for compatibility with their understanding. A PDT with responsibility for both items is responsible for integration of these requirements. Where different PDT or contractors are on the two terminals, the PIT is responsible for requirements integration.

5.4.4 Environmental requirements analysis

Environmental requirements analysis is accomplished on three levels: system, end item, and component. The system environmental requirements are generally taken from a customer standard on the kind of natural environment anticipated (space, underwater, land surface, for example). These requirements may have to be supplemented by hostile requirements driven by customer-supplied threat analysis data. Additional environmental effects may apply between the product system and non-cooperating systems in the form of electromagnetic interference.

The system environmental requirements must be mapped to the processes defined in a system process diagram that depicts a physical model of system operation and maintenance. The system environmental requirements will influence these process steps differently as a function of facilities within which they occur or different environments in different mission or maintenance phases. When this activity encompasses the complete life cycle of the product system, it is called a service use profile and it clearly defines the environmental influences and stresses applied to each end item as a function of the processes each end item is exposed to. So, the next step in end item environmental requirements analysis is to map the end items in the architecture to the life cycle processes.

It is then possible to determine the intersections between process environments and architecture applications to processes and therefore the end item environmental inputs. A given end item may map to multiple process steps each with different environmental effects in a specific environmental parameter. The analyst must integrate these inputs to derive a single value or range of values for each environmental parameter in accordance with some mathematical rule such as worst case, averaging, or root mean square.

Component environmental requirements either are equivalent to the corresponding end item requirements or have to be adjusted as a function of different environmental influences within the same end item as based on the space within the end item, in which they are located. The precise location may be influenced by induced environmental effects caused by energy sources within the end item interacting with the natural environment in addition to end item natural environmental requirements. End item zoning into equivalent environmental spaces is the preferred method for characterizing component environmental requirements. When it is determined within which end item space a component is to be located, the zone environmental requirements are simply copied for the component environmental requirements.

If the environments corresponding to the spaces within which components may be installed are inconsistent with the capabilities of one or more components, it may be necessary to alter the available environments through the introduction of environmental controls as part of the system. The existence of such a system may influence the relocation of some items to take advantage of a single environmental control system. This will require a balancing technique based on alternative benefits and detriments.

Adverse system impact on its natural environment is to be minimized. This is accomplished by identifying system energy sources and hazardous materials used in the system and tracing their contact path with the natural environment and finding ways to minimize adverse impact of those paths. The word minimize is used to recognize that the company may at some time be involved in systems with a military kind where the very destructive nature of such systems cannot be prevented from adversely influencing the natural environment during a time of hostile application of the system. These systems commonly exist in a state of readiness for many years and operate for war fighting only briefly. So, their peace time existence should, to the maximum extent possible consistent with their necessary war fighting capabilities, avoid adverse natural environmental impact.

5.4.5 Specialty engineering requirements analysis

There are many specialty engineering disciplines and each applies a particular requirements identification approach. These methods commonly implement one or both of the following two techniques:

a. Build a mathematical model that defines the requirement value for each item in the product architecture. Examples of this approach are reliability math models, weights models, and design-to-cost models. In the latter two cases, the sum of the values for any one level and branch in the architecture must equal the value for the parent item (unless margins are included). The former example follows the same pattern with a different mathematical rule. The requirement statement in each case is essentially the same for all items with different values taken from the model.
b. Identify one or more applicable documents that contain standard sets of requirements appropriate for the item.

All product systems will not require the same mix of specialty engineering disciplines. The program SOW and plans will identify the disciplines that must be brought to bear on the program. In paragraph 3.10 you will find listed a comprehensive list of specialty disciplines of which a program list will be a subset. These are all special views of the product system and its requirements which may require the attention of an engineer specially well versed in related skills.

5.4.6 Requirements validation

Validation is the process of assuring that identified requirements can be satisfied with available technology and resources, and within the laws of science. Validation actions for specific requirements are accomplished through development evaluation testing and analysis. Since validation takes budget and schedule, all requirements need not be validated. The PIT is responsible for identification of the driving requirements that must be validated.

5.4.7 Verification front end planning

Concurrently with the definition of requirements for product items, the PIT and PDTs must identify ways that these requirements will be verified. Each specification shall include a verification methods matrix telling what method will be used to produce evidence that the product satisfies its requirements. Corresponding verification requirements will be included in specification Section 4 and mapped to Section 3 requirements in the methods matrix.

Each requirement that will be verified by test and its corresponding verification requirements will feed into the integrated test planning process where a least cost effective test process will be designed. Each requirement that will be verified by analysis will be tagged to the responsible team and specialty discipline that must accomplish the verification work. Demonstration and inspection responsibilities will similarly be established.

5.4.8 Requirements relationships

There are several relationships between requirements and between requirements and other data that are helpful to retain and maintain but are not normally included in specifications. These are source information, rationale, and traceability.

5.4.8.1 Sourcing

A requirement source is the stimulus for the analyst first becoming aware that the requirement is needed for the item in question. The source will commonly be the analysis process itself through flowdown of higher tier requirements or allocation of functionality. But at the system level especially, insight into needed requirements will come from conversations and meetings with customer and user personnel or customer or user documents describing related needs.

On a program that does not apply a computer database for all requirements, it is possible to place a table in Section 6 (Notes) that captures source data for selected requirements. On a program using a computer database, a field should be included in the database for this information. The database application may not have a field specifically for source information, but a remarks field could be used for this and possibly other purposes.

5.4.8.2 Rationale

Rationale information captures reasons why the requirement was included and a basis for the required value or context. This information is very useful when the requirements are challenged. The rationale may not hold up to scrutiny, leading to deletion of the requirement, but it may also provide a sound basis for retention. In either case the rationale data has provided a useful service.

5.4.8.3 Traceability

The requirements contained within program specifications will be traceable through the hierarchy defined by the program specification tree overlay of the system architecture diagram. Since these items are also coordinated with the WBS overlay, the requirements are all traceable to the WBS structure. The development of WBS items is covered on the program schedule; therefore, requirements development is hooked to the schedule. The traceability information essentially consists of parent-child pairs captured in a requirements tool database. All requirements and synthesis decisions are captured in the program information system with information that links these decisions to the subject area and formal meeting within which the decision was derived.

Traceability helps to ensure that we have achieved a condition of completeness as well as minimization. If particular requirements are not traceable it may mean that requirements are missing or that we have extraneous, unnecessary requirements. In any case of failure to trace, we must ask ourselves if the requirement in question is really necessary and what the consequences would be if it were eliminated.

5.4.9 *Requirements flowdown and use of cloning*

On programs with an existing library of specifications in electronic media (database, word processing documents, or both), that library can be used as a basis for generating other requirements documents in a very cost effective fashion using either flowdown or cloning techniques. Generally, these techniques will not be as effective in defining the minimum necessary set of requirements for an item, but they will generally yield an acceptable specification more quickly than structured development.

In flowdown, an existing document is used as the basis for all of the documents in the immediately lower tier, the children. As shown in Figure 5-3, there are two effective approaches to flowdown. In method 1, each requirement in the parent specification is moved to all child specifications being developed and an appropriate value defined through an allocation technique by the responsible specialist. In method 2, the analyst cycles all parent requirements for one child before moving on to the next child specification.

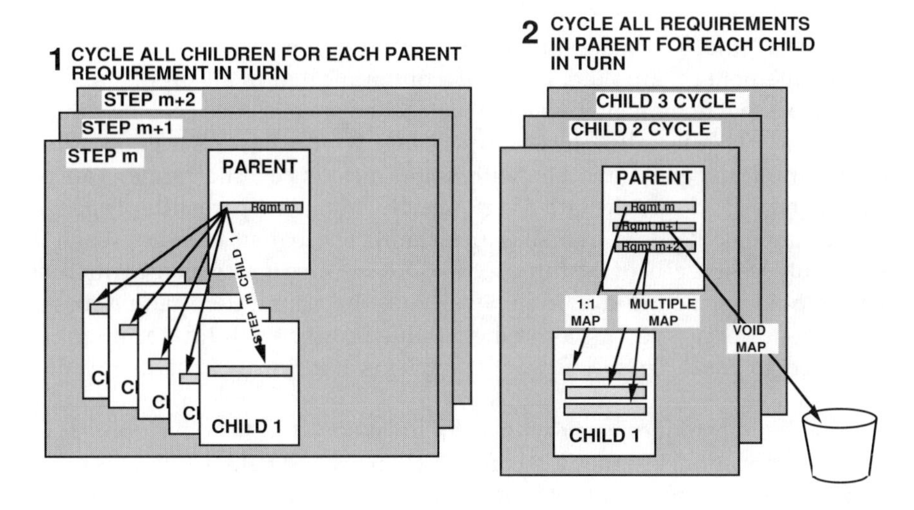

Figure 5-3 Requirements flowdown strategy.

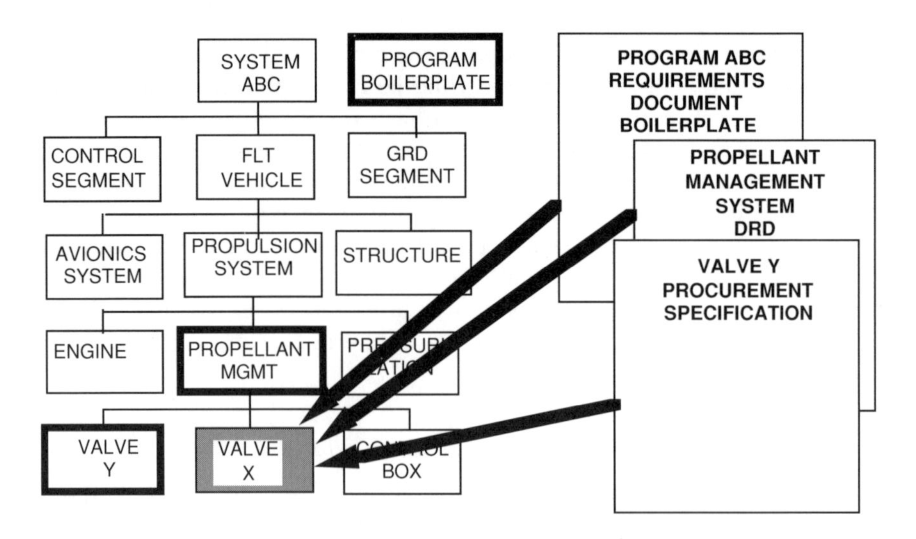

Figure 5-4 Cloning alternatives.

Three cloning techniques are recognized as a means to gain insight into appropriate requirements for an item. These are illustrated in Figure 5-4. All of these cloning techniques depend on a source document for the initial requirements. In a word processing environment, that document is copied into a word processing file and edited for the differences.

The program can prepare one or more specification standards, often called boilerplates. This can be done in two steps where the EIT maintains a set of company standards with very little content but a standard format. A program may then copy the company standard to a program standard file and include the results of the environmental requirements analysis for components, applicable documents listing and tailoring, qualification data boilerplate, and delivery boilerplate. Standards are most useful for procurement specifications to make sure that key information needed in the successful management of suppliers is clearly communicated to all suppliers.

An existing document much like the needed specification may be copied and edited. This is called like-item cloning. If you have one good valve specification and need another with some differences, the former can be copied and edited into the latter very rapidly. Finally, a parent item specification may also be used in a similar fashion. The parent item approach is commonly least useful of the three techniques because it may have a different paragraphing structure due to its level. But, the parent item requirements offer a good discipline for requirement value flowdown. If this technique is expanded to multiple specifications, it yields the flowdown strategy discussed above.

Cloning techniques carry with them great speed in the development of a new specification, but they also offer challenges and risks. Cloning is of no value for performance and interface requirements. It has value in specialty engineering requirements, environmental requirements, and applicable documents. There is a danger that content of the source document that does not relate to the new item is adopted simply by not being deleted. So, the analyst must be careful to not only add the peculiar data needed but also to delete the inappropriate content of the source document.

5.4.10 Use of specification standards

This manual encourages EIT development of specification standards for all specification types the company is required to create. These documents should be provided on-line in a standards library such that programs may download them to program storage, tailor them for program-peculiar content, and make them available to teams. The standards approach is most useful after the program has reached a degree of maturity with a sizable specification library that was created through application of an effective structured requirements analysis process. Bad standards will beget bad specifications; so bad standards rapidly prepared are worse than no standards.

Standards are useful where a word processing environment is used but can also be used where requirements databases are employed if the infrastructure is properly prepared. In the first suggested implementation, the database requires a text loader that parses the input document into paragraphs and loads database records accordingly. The standard can be loaded followed by editing or the editing can be done first. By loading it into the database it is possible to then hook up traceability that is helpful in ensuring proper content. In the second implementation, the database must have cloning capability built in such that the requirements for one document may be copied to a new document either en masse or on an individual basis.

5.4.11 Applicable documents, law, and regulations

Two kinds of requirements documents exist, program or product peculiar and general specifications and standards. Many customers prefer that certain standards be recognized in the development of their product. These standards are introduced into the program by referencing them in program specifications. Law and regulations may be similarly introduced, though these are sometimes applied through the contract rather than via specifications.

5.4.11.1 Applicable documents analysis

Applicable documents are documents referenced in specifications, plans, statements of work, and other documents that are said to apply to the purposes defined in the referencing document. Generally this action causes the complete contents, or some specific portion thereof, to have the same force of impact as the contents of the document itself. This is an efficient way to define sets of standard requirements so as to avoid re-doing the work on each program necessary to produce the content and provides standard approaches found to be useful over time. Applicable document requirements do not form a third kind of product or process requirements. Actually, the applicable documents are referenced in performance requirements and constraints.

Reference to these documents can also increase program cost while not providing any product advantage where the referenced document content includes both needed and unnecessary content and the unnecessary content is not excluded. All applicable documents should be screened for appropriateness and for unnecessary content using an organized approach. The PIT is responsible for implementing this organized approach on a program. Figure C-23 provides a company standard process diagram for this activity.

This approach shall include a conscious effort to minimize the appeal to applicable documents by including the requirements through direct content where the amount of material is not great and by questioning the real need for each document. Those applicable documents that do survive this screening must then be reviewed for tailoring needs. All of these documents fall into one of two categories: product and process documents. The product documents prescribe product requirements, not all of which may be needed for a given product. The design specialists must review them for need and the results reviewed by the PIT for acceptance. The content of all process documents must be studied by company process experts to determine the conflicts between those documents and internal company standards.

The company prefers to follow internal standards wherever possible in order to encourage the advantages of repetition. So, all external documents that are referenced in program documents must be adjusted to agree with internal standards to the maximum degree possible. This adjustment process is called tailoring. It is accomplished by adding content, deleting content, and by editing content. This editing can be done in one of two ways: in-context and legalistically. In-context tailoring consists of marking up a copy of the document, whether by redlining a paper copy of the document or electronically marking up a computer version of the document. Legalistic tailoring is captured by making a numbered list of changes with each change telling how to modify document content with phrases like: "Change paragraph 3.4.5 to read as follow: ...."; "Change 32 psi in paragraph 3.5.8 to 42 psi."; and "Delete paragraphs 3.7 through 3.15 and Appendix B."

Legalistic tailoring is acceptable where there are only a relatively few changes to be made such that the effect of the tailoring is easy to understand. Where the changes are detailed and expansive, in-context tailoring provides the user a much better understanding of the tailoring effect than the legalistic approach. While it may be necessary for contractual purposes to cite the tailoring formally in the legalistic form, in cases where the effects of the tailoring are hard to understand, it is useful to create an in-context version for internal use. One must make absolutely sure that this copy matches the formal definition perfectly.

Each applicable document listed in a system or program level document shall be subjected to a tailoring analysis using the following pattern:

a. Assign a responsible functional department and program person (document principal) most qualified to determine the relationship between content and the current company process definition.

b. The document principal must review content and prepare tailoring that brings about agreement between the document and internal company standard. If there is content that cannot be changed that involves a significant difference in the way the company functions, that program and customer must be assessed the cost of doing business in a different fashion, or company practices must be changed to permit it including providing any needed resources to implement the change.
c. The tailoring applied at the system level must be flowed down to suppliers in exactly the same way required by the customer unless there are compelling reasons for differences that will not entail a conflict with customer requirements. Suppliers may wish to use their internal practices for the same reason this company does. Where possible, they should be allowed to do so.

It is company policy that over time those applicable documents that most frequently appear in our contracts with customers will be subjected to a generic tailoring effort such that standard tailoring that perfectly matches our internal practices is available for all future work. Appendix A offers that kind of standard tailoring for the principal driving DoD customer standard for the work defined in this document, MIL-STD-499A. Tailoring for other standards are included in Appendix A when the company is exposed to those standards through requests for proposal.

5.4.11.2 Law and regulatory effects

Generally, a system created for use within the U.S. will be influenced by one or more laws or regulations with which the system must be compliant. Systems destined for use in other countries may be affected by the laws of those other countries. These laws and regulations may affect acceptable product operation, product development processes, or any combination of these factors. It is extremely important to understand these effects as early in the development process as possible since they can have a tremendous impact on system utility. If the customer will break a law each time the system is operated, the customer will be hard pressed to gain benefit from the system. Late realization of these impacts will invariably cause a significant redesign either at the expense of the customer or developer as a function of the type of contract and assignment of fault in the omission or serious shortfall in customer value.

The program, through the conscious efforts of both the customer and company, must make every effort to uncover and understand the impact of any laws or regulations that might affect the development or operation of the system as part of the earliest work accomplished.

It is possible that when these constraints are fully identified that the system may be over constrained to the end that no product system can be developed that is fully compliant with both the system requirements and the laws and regulations applied to it. In such a situation, it is necessary to actively seek out relief in a way that permits an acceptable solution space for product and process development before program budget is expended on particular attempts at design solution. One effective technique is structured constraint deconfliction, which is an organized way to challenge an over constrained condition. In this technique, every constraint on the solution space is clearly defined in terms of its effect on the solution space, the name or the organization of the principal stakeholder with decision-making authority for that constraint, and the effects, on both product solution space and the area or activity being protected by the constraint, of a roll back on that constraint to some less stringent position.

These effects are mapped to alternative design solutions and the weakest constraint positions are identified that have the greatest expansive effect on the solution space. Efforts are made, working with the identified stakeholders, to bring about an understanding, fully documented and formally approved, that provides the needed degree of relief.

5.4.12 Requirements margins, budgets, and safety factors

Margins, budgets, and safety factors are applied to selected requirements types. They provide an effective means to manage instability in requirement values. Care must be exercised in their use because engineers sometimes apply their own margins, resulting in increased cost and unnecessary capability.

5.4.12.1 Margin management

The program may elect to employ margin management in the development of selected requirements and corresponding design solutions. A margin is a portion of a requirement value set aside for use in controlling technical risk in managing product development. The designer must satisfy a more difficult requirement than necessary to satisfy minimum capabilities. Designers, who find it impossible or very difficult to meet their requirements, may appeal to PDT and/or PIT leadership for relief and access to the margin portion assigned to their item in that margin account or, alternatively, a margin derived from other teams or margin accounts. The margins provide solution space for escaping potential program risk when they materialize. They provide one means for mitigating program risks.

A requirement managed through a margin account is characterized as follows:

REQUIREMENT VALUE = TARGET VALUE + MARGIN

The requirement value is what appears in the specification. The designer must satisfy the target value in his/her design. The margin value offers protection for potential problems that may arise in the development process.

Margin accounts, when used, must be selected with care. They should be selected based on a historical difficulty in meeting certain requirements. Weight is a fairly universal parameter suitable for margin selection because of an almost universal difficulty in satisfying it. Requirements linked to program risks, and those involved in new technology developments, are other good candidates.

The use of margins must be coordinated with PDT education to avoid excessive margin inclusion. It is a reality that many engineers include margin in their own design work. If this is done in combination with a formal margin program, it will result in an unnecessary cost impact. PIT must be attuned to the potential for independent margin applications in managing a formal margin program.

If the margin accounts are properly selected, they will have been used up in the process of managing the program. In the process, the margins will have served as a way of making minor adjustments in requirements values during the development to account for things unforeseen in the original requirements definition process. Margin values not used in the development of the product will, of course, appear as excessive capability that could increase the cost of the product over what would have been the case without margins.

The design engineer is responsible for initiating an appeal to margin access based on problems meeting the current target values. The first appeal should be to the responsible PDT. If the problem can be resolved within that scope by simply reallocating margin values within that account and within the range of authority of that PDT, that ends the cycle. More complex solutions involving exchange between teams and margin accounts must be resolved at the PIT level.

5.4.12.2 Requirements budgets

Whether margin management is employed or not, a program may choose to apply a requirements budgeting approach for selected problem areas. A requirements budget is established across two or more disciplines and product requirements types for the purpose of controlling the use of a particular resource or capacity in satisfying system needs. For example, an RF gain budget may be established for a complete RF link between a transmitter and receiver. It will assign gain/loss values for receive and transmit antennas, receiver, transmitter, intervening space, and any wiring or switching devices in the circuit. Similarly, a budget may be established for aircraft guidance errors involving allocations for guidance platform accuracy, platform mounting accuracy, airframe asymmetry and alignment accuracy, and attitude or orientation sensing accuracy.

PIT is responsible for electing to implement and manage a budget program. A budget principal will be selected in each discipline or team where the responsibility for the component parameters reside. A budget lead person will be selected and that person shall be responsible for periodic budget briefings to the PIT. When it is clear that the design solution is adequate to satisfy the needs expressed in the budget program, the budget team will be disbanded.

5.4.12.3 Safety margins

Safety margins are applied to reduce the extent and cost of testing that must be applied to designs. The theory is that sufficient over-design mitigates against the possibility that failure will occur when the design is stressed only to the basic requirement value. Commonly, safety margins are set at one and one half times the real required value.

5.4.13 Requirements integration and synthesis

Requirements flow into specifications from many specialists and as a result the aggregate set for an item may contain contradictions and inconsistencies. The principal engineer for the specification development must apply integration skills during the development of the document to ensure that the final document is free of these problems because they will drive unnecessary cost and narrow the solution space, making solution more difficult for the team.

Unbalanced requirements pairs is one source of conflict, and reliability and maintainability requirements offer a good example of this condition. An item with a high reliability requirement and a low remove and replace or mean time to repair is using up availability resources that could be better applied to other items. If an item will seldom fail, it will not need to be repaired as often and it can be allocated a higher mean time to repair figure without adversely affecting system availability. The opposite case can lead to problems as well where a high mean time to repair has been allocated to an item with an allocated low reliability. The way these situations develop is that two different specialists are working their independent system models and making what appear to them to be sound judgments. The result is a suboptimum condition that must be detected by the principal engineer and corrected.

Unbalance between the requirements and the solution space can also occur. An item to which a low design to cost figure was allocated that faces very difficult technology problems is an inconsistency that will lead to an overrun condition.

The requirements set must be checked for omissions and unnecessary requirements. The former is admittedly hard to find, but a template for the kinds of requirements normally needed (a specification standard) is one way to catch obvious omissions. Where cloning from some existing specification has been applied, the document should be checked for unneeded carryover from the source document. Special attention should be given to trying to detect requirements that are not design free. This can more easily be done by someone very familiar with the detailed design process that will have to respond to the requirements.

If traceability has been included in the requirements database, that can be checked for item requirements that fail to trace to a higher tier. Such requirements should be scrutinized for continued retention in the document.

Verification planning, especially test plans, should be checked for inclusion of the needed verification events defined in the document. Logistics, quality, and production specialists, as well as the test specialist, on the development team should be specifically asked if the requirements satisfy their needs.

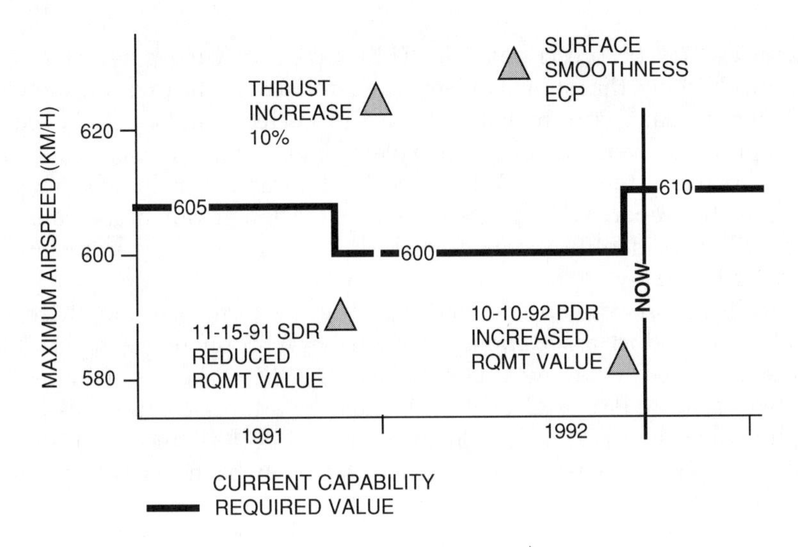

Figure 5-5 Typical technical performance measurement chart.

5.4.14 Technical performance measurement and risk management

Figure 5-5 illustrates a typical TPM chart and the accompanying text briefly describes the use of this data. We are concerned here with the way the list of parameters managed under the TPM program is selected and coordinated with C/SCS criteria and WBS.

The PIT shall determine a list of requirements that shall be tracked under a TPM program. This list may be in perfect alignment with a list of customer selected items or not but must include every item on the customer list. The criteria for requirements selected are

a. At least one parameter shall be selected in each product-oriented WBS.
b. At least one parameter shall be selected for each PDT.
c. As a goal (tempered by the size, degree of precedent, and complexity of the program), no more than 25 TPM items shall be selected for the whole program.
d. Each major program risk shall have at least one TPM established that gives information useful in understanding the current state of that risk.
e. Select key, driving requirements, the condition of which communicates maximum knowledge about overall program development success or difficulty. Good management requires hard choices about what to focus on. Choose those items that communicate forcefully the current state of the development program.

PIT, in consultation with Program Office and the customer, shall determine the TPM reporting frequency. Different internal and customer reporting frequencies may be prescribed. The TPM report for each parameter shall include a chart like that shown in Figure 3-6 updated to the present, showing the historical data, current value, future predicted trends, and keyed to related program events that help to explain changes that have occurred and are anticipated in the future. Where the TPM parameter is linked to a program risk, the report will update the current risk assessment as well. The report will also identify planned work that will affect the parameter adversely or favorably.

C/SCS reports define cost and schedule performance and these are traceable to product associations through the WBS references. Each TPM must also be tied to the most closely related WBS. Where difficulties are reported via the TPM or C/SCS, they must be addressed in terms of the complete picture of cost, schedule, and product performance. Solutions to problems exposed as cost or schedule shortfalls may be most easily resolved through product performance changes and vice versa.

The same process and techniques described for key product requirements value tracking can also be applied to the company's development and manufacturing process. This application is commonly referred to as management metrics. Key, top level, process parameters are selected that give clear signals about the state of the overall process. Values for these parameters are tracked in time and trends monitored against goals. Where performance is not measuring up to goals, action is taken to understand why this is the case and to correct the ultimate cause for the problem.

5.4.15 Supporting models configuration management

All program-peculiar mathematical or logical models and simulations used as a basis for specialized requirements analysis must be configuration controlled such that it is clearly known what version of the item/system requirements they relate to. Each specialist using such a tool has the obligation to maintain this record and these records should be checked by the PIT from time to time. Functional management and the EIT are responsible for maintaining a definition of general tools applied to programs.

5.4.16 Requirements documentation and baselines

Program specifications shall capture the results of the requirements analysis process for product items. These specifications shall be prepared in accordance with the standards required by the contract for content, style, format, types, and forms. The program specification tree shall be used to define what items shall require specifications. The following three classes of requirements documentation are identified:

a. **Customer specifications**. Approved and controlled by the customer at some point in their life cycle. These may include system and item specifications as well as interface control documents. These documents are commonly defined in the contract as CDRL items. Different customers may refer to these specifications below the system and segment or element level as configuration or end item specifications.

b. **In-house requirements documents**. Documents that capture the essential requirements for items that will be designed and manufactured within the company whose requirements are not otherwise covered in customer or procurement specifications. Generally, parts, materials, and processes fall into this class as well as some component and subsystem requirements documents. These documents will normally be DAL items.
c. **Procurement specifications**. These documents capture the requirements for items that will be developed and manufactured by a qualified supplier. They should be provided to the supplier along with a SOW which defines the process requirements corresponding to the technical product requirements contained in the procurement specification. Suppliers may cooperate with the company in the development of these specifications but their content shall reflect the minimum program requirements imposed by the company's customer. These documents will normally be DAL items and could, if prepared by the supplier under contract, be SDRL items.

The program specification tree shall be defined in terms of an overlay of the approved system architecture. This will include a top level specification tree diagram or architecture diagram annotation and a comprehensive specification tree database, listing, or table that gives specification numbers, architecture ID, specification type and form, responsible team or supplier, status, and planned and actual release dates.

A program may use a word processing or computer database approach to developing specifications. In the former, the specifications are created in word processors. In the latter case, the content is loaded into a computer database and the computer database generates the specifications in accordance with a required format or template. Master paper copies, created by either of these two methods, shall be signed by the appropriate PDT leader or PIT leader signifying approval. These master paper copies will be routed to the agent responsible for retaining the masters along with a computer word processing copy verified as equivalent, and when both are received by the agent, that agent is authorized to load the electronic copy into an on-line specification library for access throughout the program space via networked computer workstations.

If the program uses an integrated requirements database tool with specification generation capability, the specification content of the database shall be considered the master working copy and must be kept in perfect agreement with the word processor equivalent and master paper copy. Changes shall not be introduced into the word processor version independently. Changes must be introduced into the database content and specification partially or completely regenerated to maintain coordination of the different media.

Traceability of requirements should be accomplished on all programs, especially where the program uses an integrated computer database tool with that capability. Traceability must be accomplished and maintained when a customer specifically requires it. In other cases, traceability may be rejected if it is concluded by engineering management that maintenance of the data in word processor or spreadsheet form will detract from more urgent system engineering tasks that could not be accomplished in combination with traceability maintenance. In this event, program management must treat failure to accomplish traceability as a program risk and implement an abatement technique to prevent potential adverse program effects from materializing. All programs are encouraged to maintain traceability, but management must balance the risk of not doing it against other risks in reaching a final conclusion. The problem with traceability maintenance in the word processor mode is that some degree of double-booking of data is required and it may very well become divergent in the two sources. Thus, it requires considerable work to ensure coordination of the data and this energy may be more usefully spent in other areas at some risk.

At prescribed points in time defined in the Program Plan or Integrated Master Plan, the requirements defined in current specifications shall be baselined, meaning that they shall be locked up and not changed except in accordance with specific rules defined by the configuration management function. It is possible that more than one baseline shall be established, each at a different time, depending on customer and program phasing and their use of major decision-making points.

When a team has established a set of requirements for their future development and design work and that set of requirements has been published and approved by the proper authority, the team is authorized to proceed with design work (either conceptual, preliminary, or detailed depending on program phase). Upon completion of a requirements document that has been specifically identified in the specification tree for PIT review and approval, the PIT shall conduct an integration review. Program management will participate where the requirements document is a customer specification or is an interface control document with the customer or an associate item.

5.4.17 Requirements security and maintenance

A requirements baseline is a particular list of requirements for a specific list of items associated with a program event. The requirements defined in a baseline shall be protected against unauthorized change in accordance with provisions put in place by the PIT. Changes to a baseline will only be accomplished by a designated person on the PIT or responsible PDT. At any one time the complete set of program product requirements may include requirements protected under a baseline and those open to development prior to approval for a particular baseline. Generally, requirements baselines are established from top downward progressively.

Requirements in a development state may be changed with no controls by the responsible party. Changes encouraged by the customer shall be quickly evaluated for cost, schedule, and performance impacts and the customer informed of those impacts. Requirements changes that drive changes to a contract will not be accepted without appropriate contracts direction approved by the program manager.

5.4.18 Requirements tools

A requirements database is encouraged as a means to capture requirements that will appear in product specifications. The reason for this is that in the same database, from which the specifications may be printed, one can maintain traceability, rationale, and source data without inefficient multiple key strokes using different tools. An effective requirements database offers the most cost effective way to develop, retain, and maintain all of this data while also providing for printing all specifications from the same information source.

If tools are used for both system/hardware and software entities, the program should try to select tools that are interoperable for traceability across the interface.

If database tools are not available, the program must apply simple word processing and determine the relative risk of separately maintaining traceability data or not.

Figure 5-6 offers the preferred program tool environment. In the upper left hand corner the requirements analysis process conducted by humans results in requirements that flow into either a database or directly into word processed documents depending on the tools provided to the program. If a database is used, it generates a specification in the standard word processing application when the document is ready for review and approval. A word processor copy of the specification ready for review (no matter which method was used to originate it) is placed in an on-line buffer for read-only access by those who must review it.

The specification principal engineer is responsible for notifying reviewers of the availability of the document and the review window when their review must be complete. The principal then sets up a team meeting for formal review and approval of the specification. If the document was generated by database, the database is directly projected in the review; otherwise the text version is projected in the meeting room and made available via the buffer to those participating in the meeting remotely. The principal reaches agreements with meeting attendees and the decision maker about content changing the actual content of the document real time either in the database or text version, and upon meeting close, the document is ready for signature on a prepared signature page. A final paper master is run and routed to the release process. The final release action is to copy the document from the buffer into the formal specification library, file the paper copy, and erase the buffer copy.

Revisions are handled in the same manner except where a text processing environment is used without a database the released copy in the specification library is used as the basis for the revisions. When approved it is copied into the library as noted before. Since all program personnel may

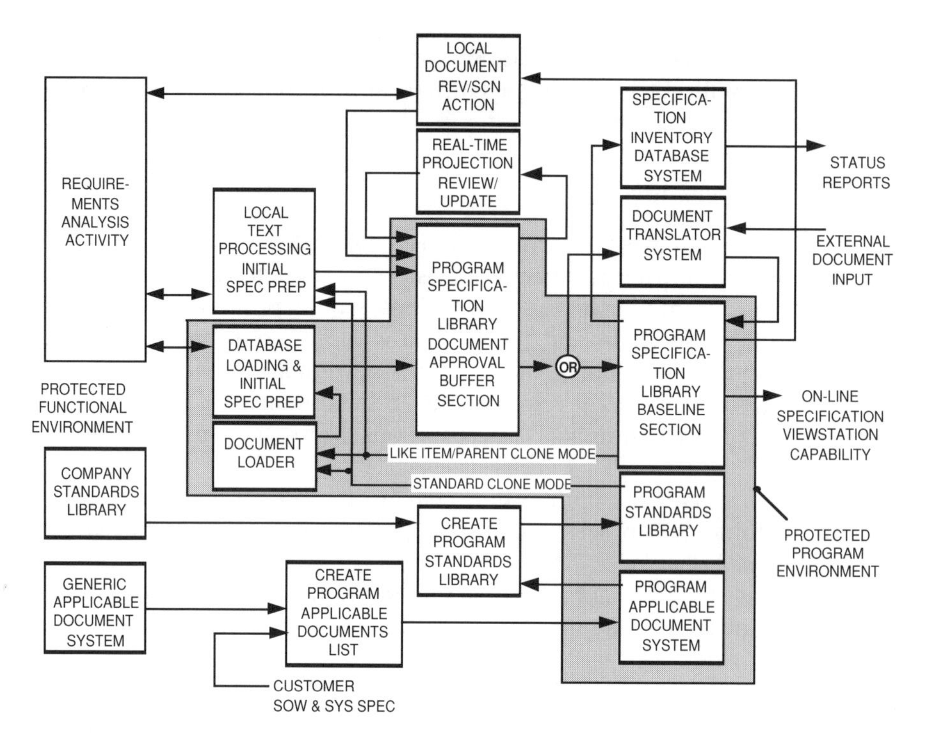

Figure 5-6 Requirements tool environment.

access the specification library from their workstation, there is no need to run paper copies and distribute them. Also, out-of-date copies of paper documents should not confuse engineering work.

The company maintains a generic applicable documents database with selected documents in electronic media. On a particular program this library is available for generating the program applicable documents listing and library with program-peculiar tailoring. The generic document list includes company preferred tailoring for documents commonly called by frequent customers that cause perfect alignment between the documents and internal practices encouraging program application of the company standard practices, resulting in continuous improvement of personnel skills and experience with that standard process.

5.4.19 Programmatic requirements analysis

Requirements for program processes shall be defined concurrently with product requirements. These requirements may include development of manufacturing, test and evaluation (verification), quality, deployment, operations, logistic support, training, and disposal. Each PDT must include specialties from these fields as appropriate for the development program. Team members must concurrently identify both product requirements and process requirements. Once these requirements are approved by PDT, PIT, or program management, the team shall accomplish the concurrent design of the product and related processes. The PIT shall integrate the work of the several PDT optimizing at the system level across both product and process design. Team inputs to system level quality and manufacturing processes are integrated by the PIT, for example.

Many of the programmatic requirements will be stated in terms of reference to a customer standard in the form of an applicable document. These documents must be reviewed as noted above for agreement with company standards and tailored wherever possible for agreement with company methods, plant features, resources, and capabilities.

5.5 Design and integration

Design work is accomplished by synthesizing predetermined requirements. Synthesis is the process of transforming a prescribed set of requirements and scientific and engineering knowledge into a solution to the problem posed by the requirements. It is a design activity involving the creative thought processes of skilled engineers supported concurrently by many specialists. Concurrent support means that the design engineer joins actively with the specialists in conversation and joint work to establish a design solution that satisfies the concerns of all of the specialists' views of the requirements. It also means that both the product design and the development, manufacturing, logistics, and operational process are concurrently designed such that they are together optimized at the system level. All items in the product will fall into one of five categories and the synthesis process appropriate to these categories is described in subordinate paragraphs.

5.5.1 Synthesis as a function of product type

The process appropriate to the development of product designs is determined by the item's supplier status. Items for which the company accepts development responsibility will obviously require a more comprehensive process than other items. Products to be supplied by others will require company review and approval of predetermined planning and work products focused on ensuring delivery to the company of the needed product.

5.5.1.1 New company development items

The design for items that require new development or significant modification involving development work by the company shall be synthesized by an assigned concurrent team in response to a previously defined and approved set of requirements shown to be traceable to parent item or system requirements. The team personnel and initial leadership shall be selected by the PIT. The team should have participated in the item requirements analysis and transition upon PIT approval of the requirements to a design mode. Otherwise, they must take some time to understand the requirements before beginning the preliminary design process.

Given an understanding of the requirements, which form the problem definition, the team shall develop alternative concepts for satisfying these requirements and trade the benefits and liabilities of these candidate solutions. This process may produce an obvious preferred solution but may require a formal trade study to focus the best decision. In either case, the team must ensure that the candidates embody the item requirements across the spectrum of specialist views represented on the team and in the item requirements. The best vehicle for team discussion is a meeting where everyone's views are respected and encouraged, if not required.

The meeting area should include a war wall decorated by sketches and information useful in encouraging understanding of the requirements and alternative solutions. This area should also provide for computer projection from team and program data stored on networked computers. This space should also be connected for telephone service to permit conference calls with speaker phone effect for exchanging ideas with persons not able to attend the meeting. The purpose of each meeting should be spelled out at the beginning and agreed upon by participants. Each meeting should allow the widest possible input from team members without allowing the strongest personalities to overwhelm the more timid persons, who will have ideas every bit as good as the more vocal.

Once the team has identified all of the viable alternative concepts it can or chooses to, the team must ensure common understanding of these alternatives and evaluate them against the requirements and a predetermined criteria based on cost and other factors defining alternative goodness. All candidates must satisfy the requirements or they must be deleted from the list of viable candidates. The selection of a preferred candidate should be made after careful evaluation of the relative merits and deficiencies of the candidates against the criteria. This whole process could take only a couple of hours in a simple case or stretch into several weeks in a very difficult selection process, but it must be accomplished within the time made available or the PIT advised of the difficulties encountered that will extend the selection process.

The lead design person must accept the team lead in conceptualizing product alternatives while other team personnel take the lead in conceptualizing process alternatives as follows:

a. Manufacturing process. The lead manufacturing person must develop alternative manufacturing approaches suitable for alternative product concepts.

b. Quality process. The lead quality engineer must consider how appropriate inspection actions can be accomplished in context with the alternative manufacturing processes as appropriate to the alternative product candidates.

c. Test and evaluation. The lead test engineer must conceptualize needed testing for product and manufacturing process alternatives.

d. Material acquisition. The lead procurement or material person must evaluate material acquisition impacts of the product alternatives.

e. Operations process. Where the product will be operated by humans in a very intimate way, operations specialists may be required to develop alternative views of operational concepts and match them with alternative product candidates.

f. Logistics process. The lead logistics engineer must devise alternative process flows supporting field use of the product and understand the ways these flows will differ for the different alternative product candidates.

The concurrent development process entails the communication of the ideas resulting from this work by specialized personnel and the sharing of information about the resulting aggregate whole that will result from the integration of these different product and process possibilities. The team seeks to join their aggregate mental capability into a single effective force for the purpose of selecting the best possible design concept for both product and related processes. This sharing process must include effective discussion between these specialists with all in attendance to benefit from the opportunities for synergism offered by the whole team. It will not be as effective for a team leader to have pair-wise discussions with team members. The total is greater than the sum of the parts in this case. The sharing and selection process will also be speeded up by sketches that illustrate ideas expressed in words by the specialists.

Once a preferred alternative concept has been selected based on the evidence available to the team, the team must then focus on preliminary design to validate that the requirements can be satisfied within the available technology and that a particular concept will be effective in satisfying the requirements. This will entail preparation of product layout drawings and sketches and similar descriptions of mating processes. This work will require support by assigned analysis and specialty engineering personnel to assure that the concept will satisfy requirements. The preliminary design must be reviewed and approved by the PIT (or a higher tier PDT) before the team is permitted to pursue detailed design of the product and processes.

5.5.1.2 Associate items

Some items that are part of the system or that interface from outside systems may be under the development responsibility of associate contractors with no direct control possible by the company. The control is exercised through a letter of agreement signed by both companies and the common customer which defines how business will be accomplished across the interface. The PIT is responsible for all development work defined by these interfaces using an ICD to manage the development of the interface definition and for communication of the results to the affected PDT.

5.5.1.3 *Supplier items*

All supplier items shall be procured in accordance with an engineering drawing package or procurement specification and a SOW or purchase order as a function of the nature and complexity of the item and the degree of development and testing that the supplier must accomplish. The PDT responsible for the parent item shall be responsible for supporting the procurement process.

5.5.1.4 *Customer supplied items*

The customer may require that specific resources already owned be used in the system. These items should be clearly identified and mutually agreed upon by company and customer personnel. Responsibility for these items will be assigned to a PDT by the PIT and the assigned PDT shall be responsible for integrating those resources into the system. If the item is of great significance, the PIT may accept the responsibility or assign it to a separate PDT. Wherever possible, the existing design of these items will be respected. Where this will entail considerable development cost, the responsible team must evaluate life cycle effects of making changes vs. adapting to the existing design. Customer input should be energetically solicited in this process and no final decision reached until customer agreement is forthcoming.

5.5.1.5 *Non-development items*

All items within a new or modified system that do not require any development action are referred to as non-development items. This commonly includes customer supplied items although these items may require some modification. Where possible, every reasonable effort will be made to preclude modification of existing items such that they may remain in this category consistent with satisfying system requirements.

5.5.1.6 *Materials and parts control*

Programs shall make every effort to use materials and parts from a standard company list approved, where necessary, by the customer. New materials and parts not currently on the list shall be evaluated by company materials and parts personnel before being added to the list. These materials and parts will be acquired from suppliers that have a proven history of supplying trouble-free materials. An effort shall be made to minimize the number of different fastener types (screws, bolts, nuts, etc.) that the customer will have to deal with in maintaining the product.

5.5.1.7 *Computer resources*

Computer hardware resources required in a system shall be sized to provide memory and throughput margins agreed upon by the customer and program management. Program-peculiar computer software shall be developed in accordance with a disciplined process defined in the company Software Development Plan (Manual 08).

5.5.2 *Progressive control rigor*

The degree of control and rigor used in the systems engineering process must be adjusted during the program life cycle. In the very earliest phases creativity must be allowed to flourish at the expense of rigor. In this period there is not a great investment in prior work; so this work policy is consistent with good economic policy. As a program matures and the investment in formal information products increases, this investment must be protected against frivolous changes and this requires a degree of control that is inappropriate in earlier phases.

5.5.3 *Product development team activity*

Section 4 addresses team organization and responsibilities. This section discusses the process employed by the PIT and PDT to accomplish their responsibilities. The team leaders must provide and encourage a suitable environment for conducting integrated product development as noted below. PDT are assigned development responsibility for specific items in a system architecture. These assignments are made by PIT.

5.5.3.1 PDT requirements definition and synthesis

The first task of the PDT after it is formed (see paragraph 5.4.1) is to accomplish a requirements analysis for the top level item (if not previously accomplished by the PIT) and then to expand this analysis downward through the architecture for which the team is responsible. The team must select a method for accomplishing a structured decomposition process for their portion of the system within which their requirements analysis work will be accomplished. The team must also select a development environment appropriate to the product entity. A particular program may elect to limit team selection of certain company approved decomposition models and development environments or, at least, approve them through their PIT. Figure C-7 illustrates the product team requirements analysis process.

Concurrently with the identification of product requirements, the team will also develop process requirements for related production, logistics support, material and procurement, deployment and operations, and quality processes. The product requirements will appear in a specification for the team item. The process requirements are captured in program plans, schedules, cost determinations, and program directives. The product and process requirements should be released together as a package. In most cases, the item assigned to a team will be sufficiently complex that a tree of requirements documents will have to be released to define its architecture. These requirements should be determined in a top-down fashion within the team's area of responsibility.

5.5.3.2 PDT design development

Figure C-8 illustrates the product team design development process, following the preferred program phasing development environment approach, partitioned into preliminary and detailed design phases. Preliminary design is intended to validate that it is possible to develop a design complying with the requirements and to define a design concept that will be pursued in detailed design. The preliminary concept should be reviewed prior to authorization to undertake detailed design in order to control risk. The reviewer should determine that the team is aware of the item requirements, that they have responded to them, and that it is feasible to comply with them.

Detailed design will result in a detailed definition of the design suitable for use by production to define manufacturing and assembly steps and for quality identification of inspection steps. The product design process will be accomplished concurrently with related process design development for manufacturing, logistics support, material and procurement, deployment and operations, and quality process design. The complete package of process and process designs should be released together.

The reviewer for both preliminary and detailed design is the PIT Leader for the top level team responsibility, for any items defined as end or configuration items, and by exception for all PDT items. The PDT Leader is the reviewing authority for all other team items.

5.5.3.3 Computer software design

Many of the company products will include, even consist completely of, computer software. While the development of computer software follows essentially the same process described in this manual (requirements definition, design, and verification), it is sufficiently specialized that the company has published a computer software development manual and that manual should be consulted for detailed coverage of this area.

Commonly, PDT will be assigned at a sufficiently high level that they have responsibility for entities that include both hardware and software. In this case, the team may choose to apply one development environment to the hardware aspects and another to the software subject to approval within the team and by PIT.

5.5.4 Program integration team activity

The PIT is the program technical authority reporting to the Program Manager. It is for all system level technical activity and for integrating the work products of the several PDT.

5.5.4.1 System requirements analysis

Figure C-4 illustrates the PIT requirements analysis process used to define system level requirements down through the level where PDT are assigned responsibility. This is a concurrent process with participation by production, logistics, quality, material and procurement, and operations personnel to develop coordinated system level product and process requirements. The PIT has overall program authority for the requirements analysis process. It may restrict the company approved development environments and decomposition models available to the PDTs.

5.5.4.2 Team identification and chartering

The PIT reviews the expanding system architecture, developed through functional decomposition and structured through concept synthesis, and identifies PDT responsibilities. The team selects initial team leaders and assists the new PDT Leader to assemble an effective team and acquire needed resources. The PIT provides any needed PDT training. The PIT should provide the PDT with a set of requirements for the top level product item for which they are responsible. Alternatively, the PIT may choose to assign the PDT that initial responsibility. The PIT seeks to perfectly align PDT assignments with the product architecture such that teams can be unambiguously assigned budget, schedule, and product performance responsibilities that coordinate with WBS, SOW, and specification content. This also results in the simplest possible PDT interface accountability pattern, providing for ease of PIT interface development and management.

5.5.4.3 Interface integration

As illustrated in Figure C-5, the PIT is responsible for ensuring that all teams are working with the same interface definition at the level teams are assigned. The PIT accomplishes this by preparing a system schematic block diagram or N-square diagram in graphical, listing, or both media. Team interface development work is monitored for compatibility across team boundaries. The team is responsible for external interface development and works through an interface working group with associates to manage that development.

5.5.4.4 Specialty assessment and system optimization

The PIT includes specialized engineers who integrate PDT work in their specialty areas and interact at the PIT level to optimize across these disciplines at the system level. PIT personnel specifically observe PDT performance looking for symptoms of sub-optimization. These include: ease or difficulty in accomplishing team responsibilities relative to other teams, very complex design concepts, very intense interfaces between team items, inclusion of subsystems with very limited functionality, and design features with unbalanced allocations in terms of cost, reliability, weight, and other parameters.

5.5.4.5 Risk analysis

While risk identification and abatement is a responsibility of management throughout the program organization, management of the overall risk program is a PIT responsibility. This includes identifying a list of current risks, assigning responsibility for risk abatement, organizing and facilitating periodic risk reviews as part of regular program meetings, and tracking progress. Risks may flow from all of the areas identified on Figure C-5 and are intertwined with program technology issues as noted on that figure. The PIT identifies requirements that will be managed through technical performance measurement as a means to control potential risk and follows cost/schedule control system reports to identify potential cost and schedule risks.

5.5.4.6 *Transitioning critical technologies*

The program-specific plan will identify specific critical technologies needed for that program. This section of the SEM/SEMP explains how the program shall create the list of critical technologies, manage the acquisition of needed technologies, and bring about a condition of balance between program requirements, available technology, and program risk. Figure 5-6 illustrates a process that shall be used on program to manage technology. It is the risk management subset of the activity illustrated in Figure C-5 covering the system integration process. These activities suggested by the blocks on Figure 5-7 (which is an excerpt from Figure C-5) are explained in subordinate paragraphs.

5.5.4.6.1 *Technology management (108152).* The company maintains a generic available technologies list corresponding to the company product line. This list is imported to the program by the PIT as its initial list. The list is augmented during the development based on any new technologies uncovered in searches and demonstrated by the program. The PIT maintains this list for reference by PDT and the PIT in conducting technology assessments of PDT designs.

5.5.4.6.2 *Technology availability assessment (108153).* Each design concept that PDT develops must have defined with it a list of technologies that will be required to support that concept. This list will be formally reviewed with the design concept and its requirements at the item requirements review and/or preliminary design review. Designers should be encouraged to appeal to the most mature and proven technology that will enable a viable design concept compliant with identified requirements. The technology list is also reviewed concurrently by the PIT in this assessment activity. The PIT reviews the design concept or preliminary design for suitability of the planned technology and any design features that cannot be supported by the indicated technologies and compares the technologies required with the list of available technologies provided from the technology management activity.

If either of these assessments produce a mismatch, the responsible development team is notified of the conflict and a risk is entered into the risk list. Otherwise, the design concept is approved from the technology perspective. The PIT may reach a third conclusion where the technology required is on the available technologies list, but it feels that an alternative technology also on the list would result in less risk or greater customer value. This point is coordinated with the responsible team.

5.5.4.6.3 *Technology search (108158).* When the technology assessment process concludes that there is a mismatch between needed technology and available technology and the responsible PDT concludes that no currently identified technology is adequate to satisfy necessary customer requirements, the PIT is responsible for conducting a search for a technology that will satisfy the requirements. If an existing technology is found, it is

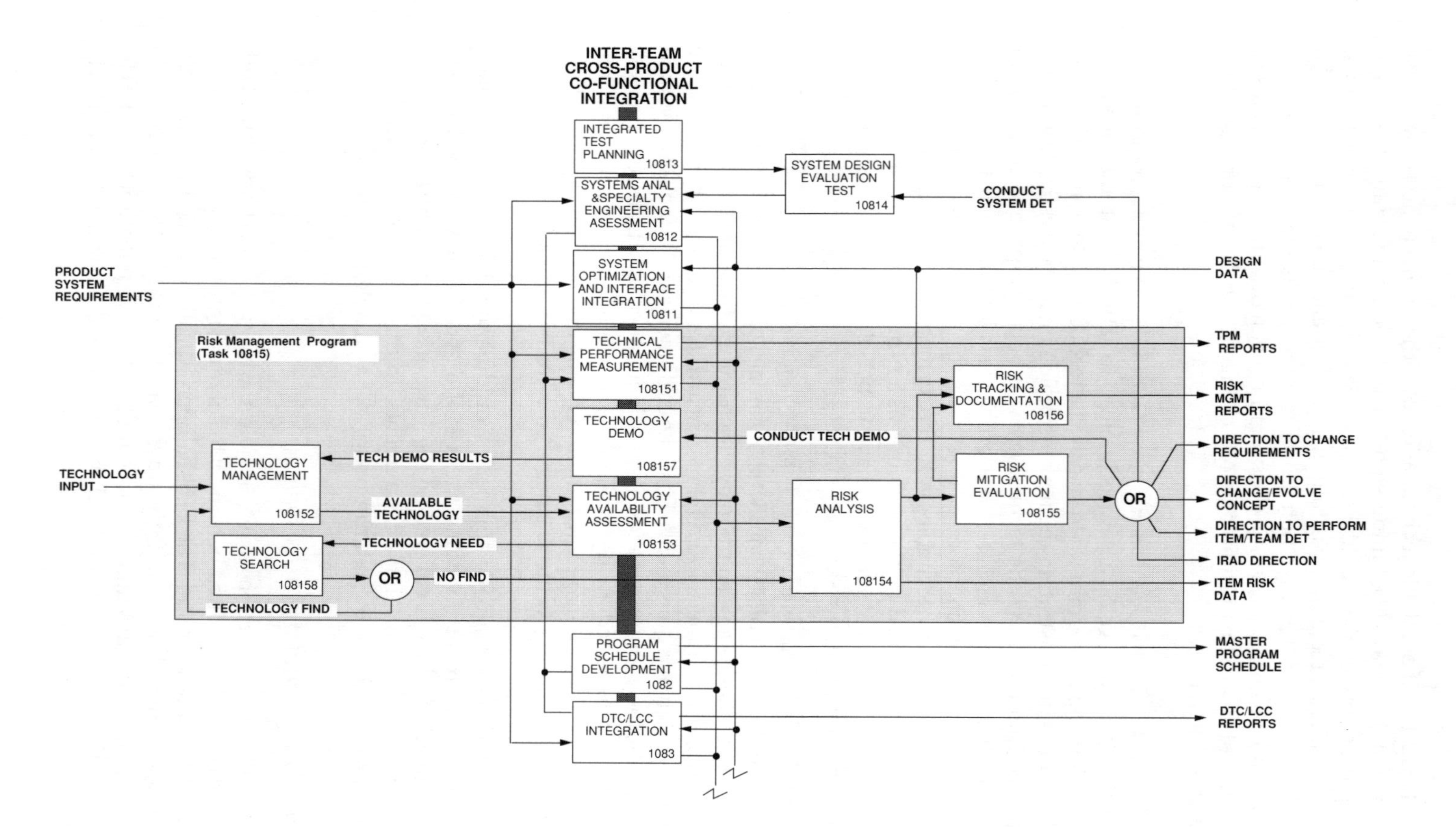

Figure 5-7 Technology development environment.

added to the list of available technologies after the means of accruing access to that technology have been understood. Acquisition may entail a personnel training program, capital acquisition, or outlay of funds to purchase rights.

If no available technology is found, a risk is identified that must then be mitigated in one of the ways noted on Figure 5-7. This may entail demonstrating a new technology within the program, changing the requirements to permit use of an existing technology, or a change in the design concept to use existing technologies.

5.5.4.6.4 Technology demonstration (108157). When a program concludes that it must develop a new technology to satisfy a customer need, the PIT will develop a program to do so called a technology demonstration. The mechanism of demonstrating the new technology may be as simple as creating a description of how the technology will be accomplished or entail an extensive study, new science, or test article development and testing. The result may be a new technology that satisfies the program need or a failure to demonstrate the new technology. In the former case, the technology may be proprietary or owned by the customer as a function of the contract. In the latter case, another alternative will have to be selected to solve the original technology conflict and risk.

5.5.4.7 Pre-planned product improvements

In the event of a technology conflict that can be solved in the near term by a requirements change agreed upon by the customer, there are two approaches to encourage continued movement toward final compliance with the customer's original requirements. First, the customer could establish a planned deviation of the requirement for a specific number of articles until a new technology is expected to be available that would permit satisfying the requirements. The original requirement remains in the system specification, but a less demanding value is carried in a formal deviation for a specific number of articles.

A second programmatic approach would be to establish a pre-planned product improvement involving the use of a specific new technology, the development of which is covered on the same or separate contract, to permit compliance with the original requirements. In this case, the specification is changed to encourage the interim use of the available technology.

5.5.5 Design and analysis tools

Each engineering function is responsible for acquiring and applying with skill tools that allow trained personnel to solve design and analysis problems common to that discipline. The EIT shall periodically review the tools available in each discipline and study ways that these tools may be effectively used together and ways that the tools inhibit effective integration work. The results of this study will be folded into functional department tool growth paths.

Each functional department will make available its standard tool set to all programs within any constraints imposed by the number of licenses legally held by the company.

5.6 Test and evaluation development

Test activities shall be planned and developed concurrently with the development of product requirements and designs. Test requirements will be determined during product requirements definition and integrated test planning work accomplished during the design process. Test personnel will actively participate on the PDTs and the PIT.

5.6.1 Test applications and categories

Testing is an effective means to validate and verify requirements. It produces documented evidence useful in reaching understandable conclusions about the true characteristics of items tested and their degree of correlation with related requirements. Validation is a process of proving that it is possible to satisfy requirements within the current state-of-the-art with available technology. Verification is a process of proving that a particular design solution does in fact satisfy predetermined requirements. Both validation and verification can be accomplished through testing or analysis.

Testing is appropriate when it can achieve the validation or verification goals more rapidly, more believably, or at less cost than analysis. Three kinds of testing are recognized and described in subordinate paragraphs.

5.6.1.1 Development evaluation testing

Development evaluation testing (DET) is accomplished during preliminary design to support the responsible PDT in defining a valid design approach compliant with item requirements. It is a requirements validation mechanism that shows whether it is possible to satisfy the requirements using particular technologies and design concepts. Each PDT is responsible for developing a list of requested DET and these lists are integrated by PIT into an integrated DET plan requiring minimum resources while satisfying all valid test requests. Where analysis can be used to satisfy a test request, PIT may redirect PDT work to that effect.

The DET process is located on Figure C-8 and expanded on Figure C-16. Note that the program must develop the test article, the documentation, and the test apparatus as well. These actions may entail special procurements or internal fabrication to acquire the necessary resources.

Tests that can be developed and controlled from within a PDR should be managed and accomplished in that fashion. Where tests entail work or responsibility by two or more PDT, they should be controlled by PIT. The responsible agent must report test results for use by the responsible PDT(s).

5.6.1.2 Qualification and system testing

Items that comprise a system must be qualified for the application. This means that we must prove that their design is adequate for the environment and interfaces to which they will be exposed. Qualification can be demonstrated through testing, analysis, demonstration, or inspection as appropriate to the item. Where an item has been previously employed in a similar application, it may be qualified by analysis if it can be shown in that analysis that the new application is no more demanding than the prior application. This is often called qualification by similarity, but involves an analysis to establish the degree of similarity. In all of these cases, evidence is developed in the form of reports that supports, or fails to support, the premise that the item meets the requirements. This evidence must be collected into the program information library and referenced in the verification compliance matrix for use at the program FCA and SVR.

PIT will review system architecture and identify them as new, modified, or existing as a prerequisite to defining qualification methodology. On unprecedented systems, many of the items will require qualification through testing. On systems comprised of largely precedented items, many items can be qualified by similarity with previous applications. The PIT must develop and maintain a verification compliance matrix that defines the methods to be used to qualify the items. Those items that have been qualified by previous applications should be reviewed at an internal qualification evaluation review (QER) conducted by PIT to verify adequacy of the logic supporting the conclusion that formal verification is not needed. Those items that survive this review will be excluded from the qualification testing process. This review should be held during the early detailed design period to permit adequate time to plan an effective qualification test if the logic is insupportable.

The PDT responsible for an item is responsible for developing the qualification requirements for an item in the form of appropriate content of the development specification Section 4 linked to the content of Section 3. The team is also responsible for development of individual qualification test planning and procedures subject to PIT integration into a least cost qualification plan. Figure C-13 illustrates a generic qualification test development process that expands the one block on Figure C-8. Figure C-14 illustrates the generic qualification test implementation process.

PIT must coordinate all qualification testing to ensure that it will, in the aggregate, satisfy system verification needs in combination with other verification (analysis, demonstration, and inspection) sources. Test reports will be produced by the test agent (PDT, PIT, or vendor) and be reviewed by PIT for verification adequacy before inclusion in the program information library.

5.6.1.3 Acceptance testing

Acceptance testing is accomplished on every article produced or some sample thereof to verify that the product meets all product requirements. The acceptance test requirements are contained in the item product specification Section 4. These requirements are selected to sense a minimum number of top level conditions that are indicative of correct and flawed responses to specific stimuli. The responsible PDT must produce the test requirements, plan, and procedure as illustrated in Figure C-15. PIT will review these data, signaling readiness to accomplish tests upon availability of the item and test apparatus.

The test procedure will be proofed on the first article and monitored by PIT. Upon successful test of the first article, the procedure, with some possible changes, will be authorized for use on all articles or on a sampling basis as defined in the integrated test plan.

5.6.2 *Integrated test planning*

Testing is costly so the program must minimize testing to control program cost. This may be achieved by minimizing the number of special test articles, combining tests, and accomplishing tests at higher system levels where possible.

The program should develop an integrated test plan accounting for all development, qualification, and acceptance test planning. The PIT is responsible for this document but must accept and review all PDT test requests in the process of creating and maintaining the document.

DETs are requested by, and requirements for those tests are defined by, a PDT or prescribed by the PIT. The requesting agent must clearly identify the ultimate purpose of the test and provide rationale for its selection as well as consequences if not selected. Selected DET must have procedures developed that are responsive to the test requirements and purpose and resources acquired defined in those procedures.

Qualification test requirements are contained in Sections 4 of all of the program specifications. These requirements must be responded to by one of three kinds of organizations: the company, a supplier, or the customer or one of its other associate contractors. All test requirements shall be partitioned with respect to responsibility, and those for which the company accepts responsibility shall be used as the basis for the portion of the program integrated test plan related to qualification testing. The PIT will ensure that suppliers are clearly informed of their responsibilities for qualification testing through statements of work, procurement specification, and supplier data requirements lists content.

The senior PIT test and evaluation person is responsible for integration of the test requirements contained in specifications, and identified in the verification methods matrix with a test method, into a least-cost effective program to accomplish the verification actions defined in the specifications. Each verification requirement shall be traceable to the verification event defined in the integrated test plan, to the test procedure that will guide the event, and the test report that gives test results.

Acceptance test requirements are contained in product (or part two) specifications, if prepared on the program. In this event, the acceptance test planning process is almost identical to the qualification test planning process. All test verification method requirements are accounted for in a minimized set of acceptance tests. Where part two specifications are not prepared, acceptance testing requirements will be determined by responsible PDTs and integrated by PIT under a special acceptance test task team.

5.6.3 Test article development

Special test articles are designed by the PDT responsible for the item represented by the test article. These articles are developed to support the product design process by providing engineering with opportunities to explore alternative design features, validate requirements, and gain confidence that credible design solutions have been defined.

All special test articles shall be configuration controlled by the responsible team such that their configuration is always known with respect to the current and past product baselines. Special test articles need not be maintained current with the current product baseline, but must never be used for test purposes where the evidence obtained is compromised by the wrong configuration. Where the article is not in the current product baseline configuration and there is any question of a compromise driven by configuration mismatch, the responsible team will perform a change impact analysis to determine the cost and schedule effects for updating the configuration to the current product configuration. The responsible team, and PIT, will jointly review the relative risks before proceeding with modification or testing in the absence of changes.

5.6.4 Technology demonstrations

Technology demonstrations are implemented as a part of a larger program or independently to develop technology and provide confidence that a particular technology is available. It is commonly accomplished through testing applied to a test article embodying the new technology. The tests are focused on producing evidence that the related technology is mature.

Technology demonstrations are most often driven by programs and when this is the case they are controlled through the responsible program PIT or PDT as appropriate to the demonstration scope. Technology demonstrations focused on independent research and development are controlled by the EIT or a functional department named by the EIT.

5.7 System analysis development

System analysis work accomplished for development evaluation, qualification, or acceptance purposes is accomplished by many different specialists and does not have the advantage of a single integrating influence common to all test activity where all program personnel are drawn from a single test and evaluation functional department. The PIT must, therefore, provide the equivalent integrating influence for the program for analyses without the aid of a universal company analysis strategy and functional home for that strategy. All development evaluation analyses must be defined in the original program plan and budget and schedule allocate these tasks consistently with the need dates of the users of this information.

The need dates for analytical results must account for the speed with which the analysis in question may be accomplished with the tools available.

In cases where the available tools will not support the need schedule, new tools must be found or a less demanding schedule elected. The difficulty with acceptance of a known mismatch in analysis capability and need dates is that the design process will proceed on risk without the analysis results, using more than enough margin to cover the final analytical results. When the results are available and they are less demanding than the assumed results, the excess margins tend to remain in the product and the analysis cost was essentially wasted. When the analysis results conclude that even the assumed margins are insufficient, the design will have to be changed with possible impacts in many other areas resulting in cost and schedule problems and possible performance problems where the needed capability can be satisfied only with extreme difficulty.

Qualification analysis requirements are defined by the union of all requirements in all development (part one) specifications identified in the verification methods matrix for verification by analysis and the corresponding verification requirements in Section 4 of the development specifications. The PIT shall identify the responsible principal engineer for verification of each of these requirements, include that information in a verification compliance matrix, and track the development of analytical evidence in the form of released analysis reports.

Analysis is less common than test for acceptance purposes. Where the program prepares part two specifications, the acceptance analysis verification is accomplished essentially the same as qualification analysis verification. If part two specifications are not prepared, PIT shall determine if any analyses are required to supplement acceptance testing.

In addition to analysis actions that are necessary in their own right, some analysis activities may be required in association with testing to determine what the results of the testing means. This may be true for all three applications of analysis discussed above but is more common the case for development analysis.

5.8 Requirements verification

Verification is the process of proving that requirements have been satisfied in the design solution and is accomplished by qualification or acceptance testing (for development and product requirements, respectively) and analysis.

5.8.1 Verification methods

Four specific methods are identified for verifying compliance with development and product requirements: test, analysis, demonstration, and inspection. These methods are explained in subordinate paragraphs.

5.8.1.1 Analysis

Special analyses shall be conducted to verify compliance of requirements where testing is not feasible or can be avoided without incurring risk. Testing is generally preferred to analysis to prove compliance. Analysis is generally preferred to testing as a function of cost. The precise mix of test and analysis for validation and verification on a program involves a balance between these factors. Analyses for specific items shall be performed by the responsible PDT. Analyses that span the boundaries established for two or more PDT shall be performed by PIT. PIT is responsible for managing the overall program analysis process for the purpose of validating and verifying requirements.

5.8.1.2 Test

A test involves manipulation, commonly in some fairly complex way, of a test article by some combination of human test engineers and a test apparatus. Testing shall be applied to product development for one of three purposes when it is determined to be the most cost effective way of accomplishing the corresponding ends. Subordinate paragraphs describe each of these three purposes and provide guidance in each case. Refer to the process diagrams referenced in Appendix C for a representation of the covered processes.

Every test shall be characterized by a goal, test requirements, a test plan, and/or procedure. The goal defines the overall objectives. The test requirements provide detailed definition of what must be accomplished and may be covered in Section 4 of a specification or result from grouping together requirements from several sources in the interest of economy. Test plans offer guidance on the use of company resources to accomplish desired test results defined in test goals and requirements. Test procedures define in detail what has to be done to satisfy test goals and requirements. All test goals, requirements, plans, and procedures may be grouped together into an integrated program test plan.

5.8.1.3 Other verification methods

Requirements may also be verified by demonstration or inspection. Verification by demonstration is appropriate where human activities are involved. An activity is accomplished and in the process it is shown that the requirement has been satisfied. Inspection is a process of observation of a reality and comparison of results with a standard without the use of complex instruments.

Where items have been used in previous system applications with similar environmental influences and requirements, it may be possible through analysis to show that further testing is not required for the new application. All items falling into this category shall be evaluated in a design equivalency review. Those that pass this review shall not require new verification testing or analysis for the application. Those that fail to pass this review shall be subjected to appropriate testing or analysis actions in accordance with the new program requirements.

5.8.1.4 Verification management

Each specification shall include a verification methods cross-reference matrix prepared by the responsible team. For each requirement in the specification, this matrix shall tell what method or combination of methods shall be used to verify the requirement and show the paragraph number of the verification requirement in Section 4 that prescribes how the verification action shall be characterized.

The PIT shall be responsible for all system verification planning. The PIT shall review all lower tier verification requirements and integrate them into a program verification plan involving the optimum balance of test and analysis actions needed to control known product design risks. All test actions shall be coordinated by the PIT and addressed in a program integrated test plan that minimizes program resources to achieve verification goals. All requests for special test articles shall be approved by PIT.

PIT shall maintain a verification compliance matrix that lists all requirements for all items requiring verification coordinated with corresponding verification requirements, verification method, verification event identification (specific test or analysis action), and verification management status information.

5.9 System test and evaluation

A program shall define and carry out a test and evaluation program for the purposes noted in paragraph 5.6 in accordance with an integrated test plan, and test procedures prepared for each test event. This program is the responsibility of the PIT. The PIT may assign responsibility to a PDT to accomplish one or more tests but shall coordinate results with other testing. Refer to the company Test and Evaluation Manual for program implementation details.

5.10 Quality assurance support

Quality assurance personnel must participate in the development of the product requirements and design solution, but this is a bi-directional relationship. Concurrent development requires that product and process be concurrently developed. The quality process requirements must be developed during the time product requirements are being developed and quality process design developed while the product design is developed. Throughout these concurrent processes, responsible teams must maintain compatibility between the product and process definition. Refer to the company Quality Assurance Practices Manual for quality assurance details.

5.11 Production support

Production facilitization, tooling, materials, and processes requirements and plans or designs must be concurrently developed or adapted as the product design process matures from requirements analysis through detailed design such that the physical reality of production is fully compatible with product design features. Refer to the company Manufacturing Planning Manual for details from the manufacturing perspective.

Through first article production, the PIT and/or appropriate PDT shall be responsible for all production support work through cooperative activity with the responsible facility-oriented Product Manufacturing Team (PMT). Subsequent to first article acceptance testing, the previously responsible PDT shall be disbanded and the PMT will accept that responsibility supported by the PIT directly.

5.12 Computer software coding

For company purposes, software coding and acceptance testing are defined as production processes. Software development work ends with identification of the requirements and associated architectural definition which defines the design including languages and computing equipment to be used. The responsible PDT will act as the production agent. On a program or team following the waterfall, phased environment production will follow approval of the design at CDR. On a program or team employing the spiral, rapid prototyping environment, some coding will occur in a cyclical pattern of design, build, test.

5.13 Customer readiness for product delivery

On minor programs or programs involving continuous production of similar products, simple production and delivery may be adequate with no special preparation of the customer for accepting products into their normal pattern of use. On major programs it may be necessary to take special actions to prepare the customer for delivery.

5.13.1 Law and regulation considerations

Products and services regulated by government prescriptions pose special problems that must be clearly understood in the development of the product or service such that it will be not only suitable for the desired functionality but legal to apply to the intended purpose. Within the United States this may entail federal, state, county, regional, district, and city laws and/or regulations. These may not all be clearly identified in the contract (if a contract is involved); so, it falls to the company to be knowledgeable about the law with respect to its business and customer bases.

Some products and services may require some form of certification prior to use in certain areas or conditions. These requirements should be woven into the program schedule and acquired in a timely way with respect to deployment or delivery.

5.13.2 Real estate acquisition

The product may require customer acquisition of real estate not held at the time development begins. The contract should make clear who is responsible for acquisition of that real estate. If the customer is responsible, the program ensures that it is suitable for product use either directly or subsequent to development.

5.13.3 Construction

Construction of real estate improvements such as roads, buildings, and utilities is not a normal part of the company's product line, but if the contract requires company responsibility for these improvements, they must be covered in the planning and suitable contractors selected to provide them at the optimum time relative to their use.

5.13.4 Base activation

Prior to delivery of the product to its primary operating location, that location must be surveyed to assure compatibility with the product. At the appropriate times relative to product development and delivery, any improvements must be accomplished, their capability verified, and any certifications required by law or regulation obtained. Personnel must be acquired and trained to operate the product.

5.13.5 Deployment and delivery

The first delivery of product may be to a system development test and evaluation site where it may be operated in accordance with the integrated test plan to verify that the system requirements have been fully satisfied. Unfavorable test results will result in design changes, deviations, waivers, or requirements changes to ensure consistency between program requirements and product capabilities.

Proven product elements must be moved from production sites to prepared locations by suitable means and initially set up for customer/user use. These deliveries will continue as defined in the contract and approved logistics planning data so as to satisfy an initial operating capability and to provide planned logistics support capabilities. Some company-provided training and operational testing may be scheduled to occur during this initial period.

5.14 Post-delivery support

Technical personnel provided to help the customer/user reach an acceptable level of competency with the delivered system must work in a difficult environment with two masters. The company and program must realize that it is in the company's best long term interest for these persons to focus on serving the customer rather than protecting the company image. This will sometimes result in temporary embarrassment and unfavorable customer attitudes about the product. But, efforts to prevent the customer from full knowledge of their product will have a very detrimental effect with longer term consequences. The best approach to taking bad tasting medicine is to swallow quickly.

Technical field personnel must be encouraged to support the customer's needs as their primary responsibility and must not be penalized in terms of promotion and pay matters as a result of having done so even where some temporary difficulty for the company results. Where problems are exposed, the company must lead the movement to quickly understand and correct any problems found to interfere with customer success with the product. Naturally, problems that involve capabilities for which the system was not designed must be corrected in ways that properly compensate the company for development and implementation of the needed changes.

5.15 System modifications

The company may be called upon to re-engineer or modify product systems previously produced and delivered by the company or by other companies. The fundamental difference in the work in these two cases is a function of the amount of information the company will have easy access to on the previous application.

5.15.1 Modification development responsibilities

The need for modifications may become known during the development process after a design baseline has been established or much later when the product is transitioned to production or even has been delivered. The responsibility for development of a modification is dependent upon where the program is in its life cycle when the need for changes is first understood and undertaken. If a modification is undertaken while the responsible development PDT is still functioning, that PDT should be made responsible for the change, given that the change can be completely contained within that team's area of responsibility. If the PDT has been removed from the program at the time the change is undertaken, the PIT may accept the responsibility directly or establish one or more new modification teams.

5.15.2 Modification impact analysis

The responsible development agent must first understand the system impacts from the anticipated change. If the development information base is still available, it may be used to identify impacts in a very effective way. Functional or process flow, architecture, and interface diagrams may be marked up to show impacts and the consequences of those impacts traced into product and process effects. Requirements traceability data may be used to understand the potential reach of changes affecting system requirements. Verification traceability data can be used to observe the impact on testing and analysis work previously accomplished and to suggest new testing and analysis work that will be required.

If the product has been delivered or deployed in part or in total at the time the change is contemplated, the logistics consequences of the modifications must be clearly understood. All of the items that will be affected, including spares, support equipment, and technical data, must be identified and the specific changes needed for those items developed. Change impact analysis must be extended to computer software and specific changes defined.

5.16 Design for disposal

All of the resources composing every system will eventually have to be disposed of. The materials used in the manufacture of systems determine the degree of difficulty the customer will have in disposing of the system. If it contains hazardous materials, those materials will require special disposal actions with potentially high costs.

During system development the design process must identify all energy sources and hazardous materials. Efforts must be undertaken, consistent with contractual requirements and budget, to minimize the danger to society from these sources at such time that the system is disposed of. The specialty discipline most appropriate to accomplish this work is system safety. During reviews, the reviewers must ensure that these hazards are identified and their effects mitigated.

5.17 Material control

Throughout the period covered under contract, the company must maintain control of all material involved in the development, production, and test of the product. This includes periods of temporary storage. Material shall be protected from damage, confusion with other material, or degradation while under company control. Needed material shall be provided to the production process in a fully functional condition in accordance with production planning. Unusable material shall be clearly identified with respect to the nature, timing, and cause of the problem and disposed of so as not to re-enter the production process. It shall be determined, when flawed material is identified, under whose control the flaw developed and conditions that resulted in the problem corrected to preclude future occurrences. The customer will be informed about these incidents in accordance with contract requirements.

Material identified as unsatisfactory will be disposed of in one of the following ways:

a. Rework to conform and confirm that it satisfies requirements,
b. Accept the material for use based on a formal exception,
c. Re-identified for an alternate application for which it is satisfactory, or
d. Reject it for use of any kind and dispose of it in accordance with detailed program or company material management instructions.

Detailed material management practices are contained in the Material Management Practices Manual (MMPM), company manual 07.

5.18 Production process definition and control

The production process shall be developed concurrently with product requirements and designs as well as the inspection and acceptance test process for mutual compatibility. The process shall be defined in computer or paper descriptions with clear and effective version control of the descriptions. Product and process revisions or changes will be cross-coordinated to ensure continued compatibility. The PIT shall review and approve all production process releases, original or changes thereto. Company manufacturing activities will be guided by the Manufacturing Practices Manual (MPM), company manual 05.

5.19 Inspection and acceptance testing process control

The inspection and acceptance test process shall be developed concurrently with product requirements and design as well as the production process. This process shall be documented and versions controlled. The PIT shall review and approve all revisions or changes. Revisions of the inspection and acceptance test process will be cross-coordinated with production process and the product configuration. Company quality assurance activity will be guided by the Quality Assurance Practices Manual (QAPM), company manual 04.

5.20 Logistics and operational support

Company responsibility for post-delivery support for logistics, maintenance, spares, servicing, and operational activity shall be defined in contract coverage. Where these services are required, they shall be performed in accordance with logistics and operational procedures prepared for consistency with product design features. If the customer is to be responsible for post-delivery support, they may require delivery of logistics and operation data and material. Company integrated logistics activity will be guided by the Integrated Logistics Support Manual (ILSM), company manual 06.

section six

Specialty engineering integration (MIL-STD-499A Part III)

6.1 Specialty engineering basis

Specialty engineering disciplines evolved because it was not possible for design engineers to master both their design discipline and the necessary depth of the many related facets of the design knowledge base. Design engineers in hardware and software disciplines are broadly qualified in a particular design discipline with skill and experience in the creative synthesis of requirements into possible design solutions compliant with those requirements. Since they generally have not mastered the many specialized disciplines addressed in this section, they must depend on specialty engineers to support and participate in the design process from those many perspectives.

The problems and products with which the company deals are too complex, customer cost and schedule requirements too confining, and the consequences of failure too serious to tolerate independent designer synthesis of design solutions followed by specialty engineering assessment and consequent redesign work. This serial process wrongfully evolved in many companies as a way to protect design engineers from interference by the growing number of specialists during the creative design process. This approach has been discredited and replaced by an appreciation for the need for effective teamwork between the design and specialty communities during the creative design process. This partnership is called concurrent engineering. It entails close cooperation in the development of requirements followed by close cooperation in the development of compliant designs.

6.2 System and specialty engineering discipline summaries

The specific analytical disciplines that will be brought to bear on a given program are a function of the nature of the product system. The purposes for applying these analysis disciplines are generically described as follows:

a. Define appropriate requirements for given items of the system from a specialized perspective. Interact concurrently with other specialists to identify and to resolve any specialty engineering requirements conflicts.
b. Help the designer understand the meaning of the requirements and provide examples of how these requirements have been successfully satisfied previously.
c. Support the design engineer by accomplishing a specialized analysis to determine how to satisfy a specific requirement. This may involve a computerized stress analysis or thermal design analysis, for example, that the design engineer is not qualified to perform.
d. Determine if the current design solution, or a set of alternative solutions, satisfies the requirements. This an assessment activity that should form a part of the requirements verification evidence documentation.

The SOW will define what analyses are required on a program and the program schedule must show when these analytical activities must be brought to bear to support planned schedules. PIT and PDT management are responsible for ensuring that analysis work is accomplished to satisfy their needs in a timely way. In each of these areas required on a particular program, the work of the specialty is coordinated across the product line by a PIT member from that specialty discipline. Specialists are assigned to a PDT as needed and accomplish related work, under the supervision of the team leader, using standard specialty techniques defined in the company specialty standard and enforced by the PIT specialty engineer. There follows a brief description of the principal activities of each of these disciplines.

6.2.1 *Reliability*

Reliability is a measure of the probability that an item or system will continue to function for a specific duration and under prescribed conditions. It is measured in terms of the probability of failure or the mean time between failures in hours. The engineer allocates the system reliability figure to lower tier items in the architecture forming a reliability model. The design team fashions a design that satisfies this allocated figure which is verified by assessing the reliability of the parts and computing the resultant reliability figure for the item as a function of the way the parts are connected and used. Part and component reliability figures are commonly extracted from reference documents listing proven reliability figures for specific kinds of components.

6.2.2 *Maintainability*

Maintainability is a probabilistic statement of the time it will require to repair a failure. This can be stated in terms of the remove and replace time, total repair time, or other parameters. The maintainability engineer allocates system level repair time to items in the system and tracks design team performance in responding to these allocation. As design alternatives are evaluated, the maintainability engineer looks for features that will encourage or deter maintenance actions.

6.2.3 Testability, integrated diagnostics, and built-in test (BIT)

A system will generally require some means of determining whether it is in an operable condition under certain circumstances. This specialty determines an optimum way to accomplish this end through identification of an integrated view of product and support equipment features and capabilities that together will assure effective testing.

6.2.4 Availability

Availability is a measure of the probability that the system will be available for use at any point in time. It is measured in terms of a particular combination of the system reliability and maintainability.

6.2.5 Logistics support analysis

The logistics specialty engineering activity seeks to identify features that will result in optimum supportability in terms of maintenance (testing, servicing, handling, etc.); spares provisioning; training; and technical publications.

6.2.6 Operability

Operability in concert with human engineering seeks to optimize the ease of operation of the system and the effectiveness of the system. A process diagram is used as a basis for evaluating the steps needed to operate the system. Human activities needed to operate the system in each of these steps are studied and operability features offered to the development team.

6.2.7 Transportability, mobility, and portability

Some elements of the system may require some degree of physical mobility. This specialty discipline determines the character of the needed mobility and defines ways to satisfy these needs.

6.2.8 Human engineering

The human engineer seeks to ensure that design features reflect human capabilities with respect to recognition of critical conditions and ease of actions that must be taken to operate and maintain system items.

6.2.9 System safety and human hazards

The safety engineer identifies safety requirements based on customer needs and cooperates with design and analysis personnel to understand and identify safety hazards to life, health, and property value. The principal approach is to build a model of operation in cooperation with the maintainability and logistics personnel of the system operations and support process and to examine this process for conditions that can cause hazards to develop. The product is evaluated for ways to prevent these conditions from ever developing or ways to control the risks when they do occur. A hazard list is prepared and ways are found to eliminate or control each hazard.

6.2.10 Environmental analysis

Environmental analysis is accomplished to determine to what environments the product system shall be exposed, to characterize those environments with precision, and to identify product characteristics needed to survive in those environments. Environmental aspects involved always include the natural environment but may also include a hostile element activated by persons or groups intent on reducing the effectiveness of system capabilities and a non-cooperative element entailing other systems that may unintentionally interfere with system operation. In the larger sense the environment includes everything that is not in the system; so even cooperative systems that purposefully interface with the system are technically part of the environment. The latter are normally handled as system interfaces by the PIT.

6.2.11 Environmental impact analysis

The system must operate within a prescribed environmental definition. The system and the environment will interact in certain ways and the goal is to minimize the adverse impact of the system on its environment. This is accomplished by understanding the interface between the system and the environment in terms of all materials and energy that are exchanged across this interface. Each of these interfaces is studied for ways to reduce environmental impact. Environmental laws and regulations are studied for compliance issues.

6.2.12 System security

The system security discipline seeks to identify risks to the system and identify ways that the system can avoid these risks. The threats to the system must first be listed and then each one of these threats must be dealt with in the design to preclude their occurrence from having an adverse effect on system operation.

6.2.13 Producibility

It is possible to design a product such that it is either very easy to manufacture (or produce) or it is very difficult to produce. Producibility seeks to bring about the former condition by close and cooperative work with the design engineers to encouraging the use of processes already clearly characterized and with simple design features. This may entail the use of robotics or human touch labor with different product characteristic needs.

6.2.14 Life cycle cost

Life cycle cost (LCC) is defined as the total cost of a system over its life cycle from development through disposal. LCC is a technique to determine and track during development the total cost over the complete life of a system. This includes the non-recurring cost of development and deployment, the recurring cost of manufacturing, testing, and training, the operations, maintenance, logistics support cost during its useful life, and the disposal cost of the system at life's end. This total cost may be allocated to system elements and used as a target for development. Design to cost is a component of LCC. LCC is tracked by the PIT based on PDT inputs and PIT evaluation of those inputs.

6.2.15 Design to cost

Design to cost (DTC) is an organized way to allocate non-recurring development cost (an element of LCC) to system elements to control the total system cost. DTC is a technique to encourage cost-conscious behavior in the development team toward the end that the product development cost targets are met. DTC is applied like any other allocable quantitative requirement such as reliability or weight. A system development cost number is first identified and this cost is then allocated down through the hierarchy based on anticipated development difficulty. The design team identifies a target figure recognizing a margin to protect themselves from exceeding the required value and tracks estimated cost as a function of the design choices made. DTC is used as a selection parameter for alternative design solutions. The PIT makes the original DTC allocations, integrates the current team estimates, and tracks this parameter in time. While the DTC allocations are managed by PIT, the responsible PDTs are accountable for satisfying them.

6.2.16 System cost/effectiveness analysis

System cost/effectiveness analyses will be employed to support the development of life cycle-balanced products and processes and to support risk management activities. It is both a very important specialty engineering analysis discipline and an integral part of the program decision-making and control apparatus. A system level Measures of Effectiveness (MOE) hierarchy shall be defined for the system as a basis for computing cost and effectiveness parameters for alternative solutions that must be traded one against the other. All analyses shall be performed in the context of the responsibilities of the PDT and PIT as a function of the scope of the analysis. There being a limited number of specialists skilled in these analyses, personnel will be normally assigned to the PIT and temporarily attached to a PDT if that kind of analysis must be accomplished within the area of responsibility of the PDT.

6.2.17 Electromagnetic compatibility

On systems that include sensitive electronic circuits that could be upset by strong electromagnetic fields or that are capable of generating such fields that could interfere with other systems, the system will be studied for sources of interference and any that are identified will be corrected to within the levels prescribed by law, program requirements, or sound engineering judgment or recognized standards.

6.2.18 Radio frequency management

Systems which include radio frequency emitters, such as radio or radar transmitters, must have their frequency assignments coordinated with available spectrum assignments and one or more controlling organizations.

6.2.19 Electrostatic discharge

Systems operating in the atmosphere are susceptible to a build up of electrostatic charge that, if allowed to reach a high potential relative the surrounding charge, can have a detrimental effect on sensitive on-board electrical equipment.

6.2.20 Mass properties

The mass properties engineer is responsible for ensuring that the design falls within weight and center of gravity (CG) constraints established for the product. The principal method involves allocation of available weight to system elements and monitoring the design process to see that responsible teams and designers remain true to their allocations. A weights table is established that lists all of the system elements and their weights with subtotals and grand total. Weight margins may be established to protect the project from weight growth problems and provide for management of difficult weight issues as the design matures.

The mass properties engineer must also compute the CG of elements where this is a critical parameter. In maintenance situations this data may be required not only for a whole end item (hoisting and lifting, for example) but for various conditions where the item is incomplete as in assembly and disassembly operations.

6.2.21 Materials and processes

Materials and processes seeks to standardize on a minimum number of materials qualified for the product application and a standard series of manufacturing processes appropriate to manipulate those materials. Some materials and processes may require company specifications, written by materials and processes, to characterize them where adequate definition does not exist elsewhere. Designers must select materials from the standard list provided by materials and processes. The standard materials list is the responsibility of PIT. Any materials and processes specifications will be initially prepared by the PDT that first requires its use and thereafter reviewed, approved, and maintained by PIT.

6.2.22 Parts engineering

The role of parts engineering is to assure the use of parts qualified for the application and to standardize on the fewest possible number of different parts. This is accomplished by development of a standard parts list from which designers may select parts. Any suggestions for additions to the list by designers are reviewed to determine if a suitable part has already been identified or an existing listing can be applied to the new application. Some parts may require a company parts specification written by parts engineering. The parts list is the responsibility of PIT.

6.2.23 Contamination control

Contamination control seeks to limit contamination of the product during manufacture by any foreign material generally defined in terms of particles larger than a certain size. Control is exercised by requirements for processing within areas qualified for a prescribed level of cleanliness and special transportation and handling process instructions.

6.2.24 Guidance analysis

If the system includes an element that must move from one point to another with a degree of position precision, this discipline provides requirements that encourage the needed accuracy and evaluates design features to assure that those requirements will be satisfied.

6.2.25 Structural dynamics and stress analysis

This discipline determines the needed strength of structures under static and dynamics conditions under all system conditions. Computer tools are used to model the structure and support structural design personnel in selection of materials and design concepts.

6.2.26 Aerodynamics

If the system involves movement of an element through the atmosphere at speed, it may require an aerodynamics analysis or wind tunnel test to assure that its shape minimizes drag and offers adequate lift to balance weight under all conditions of flight.

6.2.27 Temperature and thermal analysis

Heat sources and sinks are identified and the resultant temperature of items in time is determined. Involved in positioning and mounting of items for thermal control and elements involved in altering the environment within which items are located.

6.2.28 Quality engineering

Quality engineering skills shall be applied concurrently during the development of product requirements and designs to ensure compatibility between the product and the quality assurance process. Quality assurance procedures, methods, and positioning within the manufacturing stream will be coordinated with these product features.

6.2.29 Disposal analysis

During the development of a system, the eventual disposal of the system shall be considered in accordance with tasks defined in the SOW. Features that encourage safe and low cost disposal will be included in system characteristics.

6.2.30 Deployment planning analysis

If the program involves products that must be moved into use within a customer environment, this process of creating the initial operating capability will be subjected to analysis to determine optimum methods and techniques. Results of this analysis will be applied to requirements and designs as appropriate.

6.2.31 Survivability and vulnerability

This may include nuclear, biological, and chemical analyses as a function of the threats posed by a hostile force. The effects of these agents are defined for the benefit of design teams and design alternatives reviewed for compliance with recognized effective solutions to the problems posed by the agents.

6.2.32 Value engineering

Value engineering is a structured method for finding ways to improve a product after it has entered a production status. It involves the employment of cross-functional teams in the performance of trade studies focused on selection of the most cost effective solution to a production problem. This process may be married to a pre-planned product improvement program for the purpose of determining the precise way that the pre-planned improvements will be implemented. Value engineering, when implemented, will be managed by PIT. Each PDT or PMT will be responsible for value engineering of the products for which they are lead.

6.3 Generic specialty engineering process description

The specialty engineering process entails three steps repeated by each discipline within the context of their knowledge base and toolset. The concurrent process demands that these three steps be applied at the right times with respect to the on-going design process and that there be effective communications between specialty practitioners and design engineers. The three steps are applied in the sequence listed.

6.3.1 Specialty requirements definition

During a requirements identification period preceding the design work, the specialty community must define appropriate requirements for given items of the system from a specialized perspective. The specialty engineers apply their requirements analysis techniques and toolsets to define needed characteristics from the perspective of their discipline. Two common requirements techniques are

a. An appeal to authority in the form of specialty standards imposed by the customer or accepted as industry standards. The specialty engineer reviews the selected or prescribed standards and determines any parts of them that should not apply for specific reasons. Tailoring is offered to exclude those requirements. These requirements are commonly stated in terms of required compliance with the standard.
b. Mathematical models are commonly used to assign numerical values for each item in the system architecture. The model computes a system value based on the component values assigned. Commonly these models may also be used as an allocation or flowdown aid to partition a parent item value into values for subordinate elements. These models also will identify conflicts where specific values are inconsistent. Requirements are stated simply as complying with the value extracted from the model.

Specialty engineers must also interact concurrently with other specialists to identify and to resolve any specialty engineering requirements conflicts. This work must be orchestrated or coordinated by the PIT working across specialty engineering disciplines at the system level and within the PDT by those responsible for integration at the team level.

This requirements work must include both the product requirements identification activity and the process requirements activity. The intent is to develop a set of product and process requirements that are consistent and system optimized.

6.3.2 Education and design support

Each specialty engineer must help the designer understand the meaning of the requirements and provide examples of how these requirements have been successfully satisfied previously. This may require accomplishing a specialized analysis to determine how to satisfy a specific requirement, for example, a computerized stress analysis or thermal design analysis associated with a particular design concept that the design engineer is not qualified to perform. The emphasis must be on assuring that the right design is developed on the first pass. We cannot wait until the design assessment process to discover that the design is deficient in one or more specialty areas. This activity is at the very heart of the concurrent design process.

6.3.3 Design assessment

As the design matures, each specialty engineer must formally determine if the current design solution, or a set of alternative solutions, fully satisfies the requirements. This assessment activity should form a part of the requirements verification evidence documentation in the form of analysis reports keyed to specific requirements.

6.4 Specialty engineering tools

Table 6-1 lists all of the specialty disciplines that may be involved in the company's product line and identifies principal tools used. The continued development of these tools is the responsibility of functional management. Their effective application to programs is the responsibility of program management through the PDT and PIT.

6.5 PIT integration work

The PIT has overall system development responsibility. This includes the several specific tasks described in subordinate paragraphs as well as the system level responsibility.

6.5.1 Technical management and audits

The PIT shall manage the overall technical development process for the Program Manager/Director. This includes defining development teams, assigning work to teams, auditing team products for overall compatibility, formally reviewing and approving team products as a precursor to continued work, and making decisions affecting the work of all teams.

6.5.2 Cross-team interfaces

A key responsibility of the PIT is for all interfaces that cross the system boundary to external systems and the environment as well as all interfaces

Table 6-1 Specialty Engineering Disciplines and Tools

Specialty discipline	Specialty toolbox
Reliability	Reliability model
Maintainability	Maintainability model
Availability	Availability equation
Survivability and vulnerability	Atomic, bacteriological, and chemical effects
Electromagnetic compatibility	EMI effects
Radio frequency management	Spectrum allocation definition
Electrostatic discharge	Atmospheric model
Human engineering	Human capabilities definition
System safety	Hazard analysis
Health hazards	Hazard analysis
Environmental impact	Weather and geological models
System security	Threat analysis
Producibility	Manufacturing queuing models
Supportability and ILS	Maintenance models and queuing simulation
Operability	Simulation
Testability, integrated diagnostics, and BIT	
Transportability, mobility, and portability	
Mass properties	Weights model
Materials and processes	Approved M&P list
Parts engineering	Approved parts list
Contamination control	Contaminant size definition
Guidance analysis	Simulation
Structural dynamics	
Stress analysis	
Aerodynamics	Wind tunnel
Temperature analysis and thermal control	Thermal model
Life cycle cost and design to cost	Cost model

that have one terminal in each of two different team items. All other interfaces will have both terminals in one team's items and are the full responsibility of the corresponding team. The PIT shall monitor team interface development work and assure that compatibility is maintained. Where interface documentation is required between associates, the PIT shall be responsible for that documentation and coordination of the results through the development teams.

6.5.3 System optimization

The PIT is responsible for assuring that the overall system achieves an optimum solution not only with respect to the product requirements but with respect to process needs as well. PIT will continually or periodically review the team requirements and solutions for sub-optimized situations. These can often be detected in the form of unbalanced team ease of performance. One team may have a very easy time of satisfying its requirements while other teams are very burdened. These differences may be relieved by rebalancing the driving forces. PIT should also be alert for complexity, the handmaiden of sub-optimization. Complex solutions can sometimes be simplified through a different allocation of functionality resulting in simpler interfaces and simpler team solutions.

6.5.4 System level process integration

Each PDT must develop process inputs related to their team items. Commonly, however, teams cannot, by themselves, complete the related work because it is affected by the inputs from other teams as well. PIT must integrate the effects of these team process inputs in at least the following areas: manufacturing process, inspection process, procurement process, logistic support process, system deployment and site activation process, and operational processes.

6.6 Concurrent development team activity

Teaming is motivated by a need to pool the collective knowledge of many specialists because we are specialized in the interest of mastering a larger knowledge base than any one person is capable. A second teaming motivation is to organize a large problem into a series of smaller related problems that are compatible with management span of control limits. The fact that we must decompose large problems into a series of smaller ones imposes a need for excellent communications between the teams and people staffing those teams.

These teams must be staffed with qualified specialists in fields supportive of the team goals. Once formed, the team membership must be protected from unnecessary personnel changes because good communication within the team depends on forming bonds between the members. Frequent personnel changes break these bonds and inhibit effective collective work.

6.6.1 Team formation, leadership, and staffing

The PIT and PBT shall be formed and leaders selected by the Program Manager/Director. All PDT shall be formed and initial leaders selected by the PIT, subject to Program Manager/Director approval, with specific product responsibilities as overlays on the evolving system architecture. All development work will be accomplished within these organizational structures under the leadership of the designated team leaders. The PIT is a system level PDT thus has similar responsibilities but at the system level. The PIT and PDT leaders are all at the same level on the Program Staff but the PIT shall have review and approval authority on all PDT information products including lower tier requirements, design definitions, test and analysis data, and system process inputs.

The PDT will be staffed in accordance with the need for specialty views driven by the technologies and techniques appropriate to the product element about which the team is formed. The team shall, wherever possible, be physically collocated to take full advantage of a close working relationship entailing easy, direct voice communication, person-to-person interaction, and efficient control and management. Where physical collocation is not possible, remote team members will be linked into the team process by good telephone, FAX (if needed), and networked computer communications.

6.6.2 *Team requirements work*

Upon formation, the team will be provided with a set of requirements, which may or may not be fully formatted into a formal specification, an area of responsibility for system architecture, a brief statement of work telling what must be accomplished, and responsibility for specific master schedule milestones relative to the area of architecture assigned. The team must accomplish work to expand the requirements definition from the top level, for which they are given an initial set of requirements, down through all of the lower tier items for which requirements are needed as defined on the specification tree overlay of the architecture diagram. The team will apply an organized requirements analysis process to concurrently develop appropriate requirements for product items and corresponding processes.

The team must determine item architecture expansion from the top level item assigned using one of the approved decomposition techniques listed in paragraph 3.4.1. As items are identified through allocation by the team of exposed functionality coordinated with PIT, the team expands the allocated functionality into performance requirements.

The team is responsible for development of interface requirements jointly with other teams sharing interfaces with their item. This process is audited by the PIT. Top level team item environmental requirements developed by PIT are allocated to lower tier items using some combination of direct assignment of top level environmental requirements, system service use profile expansion, and end item zoning. Lower tier specialty engineering requirements will be flowed down from the top level specialty requirements by the assigned team specialists.

The PDT is responsible for establishing traceability of all team requirements to the top level team requirements. The PIT is responsible for traceability between the top level team document and system requirements. Verification requirements will be developed concurrently with product requirements and be coordinated with PIT for integrated test plan development.

The team must minimize the set of applicable documents called as compliance documents in the lower tier specifications to those required at the system level and ensure that any tailoring in those documents perfectly matches the tailoring used at the system level. If there are other standards that must be used, the team must accomplish tailoring for the application. If the document is called by other teams, PIT must coordinate any tailoring differences.

The PDT must reach agreement with PIT on the top level requirements before work is accomplished to expand those requirements. Subsequent work may require changes to these requirements based on sound rationale, but there must be an initial agreement on the top level requirements. Lower tier requirements must be reviewed by the PDT leader and, by exception, by the PIT. Concept development work may proceed in parallel with requirements work for the purpose of validating that the requirements can be met with planned technologies, but no detailed design shall be accomplished on any item until the requirements for that item have been approved.

6.6.3 Procurement items

Upon approval of the team requirements (by PIT or PDT leadership, as appropriate to the item level), the team shall begin design development and or procurement, as appropriate. The team must coordinate its make-buy decisions with the PIT make-buy architecture overlay. Any item that is to be purchased shall have developed for it one of the following combinations of documents based on the indicated logic:

a. Procurement specification & SOW — Item requiring development by the supplier.
b. Engineering drawing package — Item that will be built to print.
c. Purchase order — Off-the-shelf item already in production by the supplier.

6.6.4 Team design process

The work of teams during the design development process can be partitioned into the several phases commonly recognized in this process. The first design phase is concept exploration where the team determines how to satisfy the allocated functionality. Commonly this is accomplished by identification of two or more alternative approaches followed by a formal decision-making process to identify the relative merits of the alternatives. Preliminary and detailed design phases incrementally improve upon the detailed knowledge about the design. Finally, the team must cooperate in the requirements verification process.

6.6.4.1 Concept exploration

The concurrent development process for design shall engage all team members in identification of viable alternative design concept solutions. Where the alternatives permit a relatively simple selection logic, judgment decisions shall be arrived at and the selection rationale documented. Where the selection logic is complex involving difficult questions, the team will apply the trade study approach to define selection criteria, value candidates in these criteria, and make a selection based on the best overall solution. Each alternative will be evaluated by each specialty and the decision-maker appraised of these many conclusions. Team meetings will be extensively used to realize synergism between the several specialized team members. But, the team must be very careful to study these alternatives expeditiously and come to a sound but timely decision on the best approach. The team leader is responsible for encouraging the selection process so as to support item schedules.

6.6.4.2 *Preliminary design*

Once a concept has been selected, the team must move quickly to develop a preliminary design responsive to all item requirements. In so doing, the principal designer(s) needs the assistance of all specialty engineering personnel to ensure that the design satisfies each set of concerns. Specialty checklists are discouraged as a means to communicate specialty requirements to the principal designer(s). The specialty engineers should be an immediate team resource to the principal designer(s) for direct input on the current direction in which the team is moving. The specialist must be sensitive to the depth of understanding of their specialty on the part of the designer(s) and provide concurrent help as needed whether asked for or not. The goal should be to blend the aggregate knowledge of the team into the equivalent of one all-knowing engineer for the duration of time it takes to create the preferred design solution.

The preliminary design solution must be reviewed by the PDT leader or PIT team as a function of item scope and importance to system success. This approval process must address requirements compliance, interface compatibility, design risk, and technology maturity. Once approved, the team must carry on into detailed design to develop detailed engineering drawings and related process designs coordinated with system level concerns by PIT.

Any special test articles and unusual analyses that will be needed in support of detailed design must be identified during the preliminary process and introduced into the program plan and schedule if not already included. Addition of significant cost and schedule impacts will require substantial justification but must be addressed if needed. Alternative ways to resolve the problem perceived should be evaluated before full commitment is made to substantial changes in the program.

6.6.4.3 *Detailed design*

The team's detailed design responsibilities include both product design and process design. The process parallels that described for preliminary design but requires detailed solutions to every product and process design problem encountered. The solutions to all of these problems will, during the course of the work, improve team knowledge about the solution, possibly forcing alternative approaches not conceived as part of the original plan defined in preliminary design. In these cases, the team must coordinate them through PIT with other potentially affected teams. These changes should be minimized through sound preliminary design development, but some will be necessary because of individual human and process imperfection and unforeseen circumstances. The test of the team's effectiveness in doing its requirements and preliminary design work is measured in the number of avoidable design changes subsequent to detailed design release.

The intimate team interaction characterizing preliminary design must continue through detailed design. If the depth and breadth of the team architecture responsibility warrants, it may be necessary to provide the team with sub-teams responsible for major elements in the team's responsibility. If this is done, then the senior PDT must take on the PIT role for the junior PDT reviewing and approving their work.

If special test articles, laboratories, or mock-ups are required, the team must develop an article design and acquire the device. It must be run and results collected in support of a predetermined decision-making process focused on alternative design solutions. The team must be very careful to control the device configuration such that it is always known what the correlation is between the data produced from the device and the current or planned product configuration.

6.6.4.4 Requirements verification

For each item that requires a specification, as shown in the specification tree overlay of the architecture diagram, the responsible team shall develop an item verification matrix that tells by what method it shall be determined that the product design satisfies each requirement. This matrix must be reviewed and approved by PIT verification and then expanded to define precisely how that method will be accomplished. The methods are test, analysis, demonstration, and inspection. Test methods will be coordinated with integrated test planning to ensure that adequate resources will be made available in a timely way.

The team is responsible for accomplishing any analyses needed to verify product compliance. Most integrated testing will be accomplished or coordinated by the PIT, but simple tests involving simple test apparatus may be accomplished by PDTs.

6.7 Concurrent development aids

Teams require specific facilities and capabilities to encourage success in integrated product development. These include good communications, specific facilities features, information access, and an effective meeting environment.

6.7.1 Good communications

The most fundamental communications provision is to place people physically close together as a function of the intensity of the human interfaces they must maintain. The most intense human interchange is going to occur between persons on the same team. The system development methodology described in this plan is specifically designed to evolve a system definition to take advantage of this kind of arrangement. Place team members close together to encourage good, immediate verbal and visual communications.

Programs must also be provided with excellent electrical communications capabilities in the form of telephonic and fax equipment. Electronic mail should also be provided throughout the program space through the computer network serving the program.

6.7.2 Facilities

Program facilities must be arranged to encourage physical collocation by teams. Each team space must be provided with a wall expanse or meeting room space within which the team may congregate for meetings. This space should be served by speaker phone capability and computer projection from the network serving the program. White boards should also be provided for sketching transient ideas.

6.7.3 Information access

It is imperative that teams share their information not only when it is perfected but during the development of that information. Information must be shared in process. The program should be served by an information database of some kind on a computer network.

6.7.4 PDT integration meetings

The principal means to achieve the goals of the integrated product development is realized in the team meeting. These meetings must be conducted in an open atmosphere encouraging active participation of all members within a minimum framework assuring order. The team leader is the default chairman of these meetings, but the leader should employ different members as facilitators who encourage adherence to the agreed agenda and seek out opinions from those reluctant to give voice to their views.

The two greatest dangers in PDT meetings are (1) the premature formation of a team attitude supportive of "group think", leading to defense of faulty positions held by the team and (2) paralysis of the decision process driven by excessive interest in perfection. The team leader must consciously guard against group think by encouraging specific persons with a demonstrated skill for argumentation to question what appear to be unanimously approved conclusions and by periodically telling meeting attendees of the dangers of group think.

While it is important for teams to thoroughly ventilate ideas and consider all reasonable alternatives, it is also important that the team reach conclusions in a timely way. Team leaders must be capable of decisive leadership encouraging thoughtful consideration of alternatives while driving to firm decisions in a timely way based on the best available knowledge at the time.

section seven

Notes

7.1 Acronyms

In all cases where an acronym is used in this document, it is included parenthetically after the expanded version the first time that term is used in the text. Table 7-1 also lists all acronyms used in this document and explains those not obvious.

Table 7-1 Acronym Explanations

Acronym	Meaning	Explanation
ABD	Architecture Block Diagram	A hierarchical diagram formed of blocks and connecting lines that depicts the structure of the physical model of the system solution.
CDR	Critical Design Review	Major review that reviews final design and authorizes building articles for test purposes.
CDRL	Contract Data Requirements List	A list of data deliverables for a contract.
CIT	Corporate Integration Team	Cross-division team responsible for balancing corporate resources across divisions.
C/SCS	Cost/Schedule Control System	A system to define, track, and report cost and schedule requirements.
C/SCSC	Cost/Schedule Control System Criteria	A system of criteria for systems used to track cost on programs and report cost and schedule performance. Contractors doing business with DoD must use systems that comply with this criteria.
CPI	Company Private Information	Information that can be properly withheld from a customer.
DAL	Data Accession List	A list of all program documents not on the CDRL that can be acquired by the customer for a separate cost.

Table 7-1 (continued) Acronym Explanations

Acronym	Meaning	Explanation
DIG	Development Information Grid	An information structure installed on a computer network server for use in capturing PDT information through PDR.
DoD	Department of Defense	
DRB	Development Review Board	A board chaired by the PIT to review PDT and PIT work products.
EIT	Enterprise Integration Team	Team responsible for integration across all programs in the company.
ILS	Integrated Logistics Support	A coordinated approach to the planning and implementation of logistics services.
IMP	Integrated Master Plan	Master plan used in the U.S. Air Force integrated management system approach to system acquisition.
IMS	Integrated Master Schedule	Master schedule approach used in the U.S. Air Force Integrated Management System.
NASA	National Aeronautics and Space Administration	
PBT	Program Business Team	A team of administrative, finance, and contracts personnel supporting the program manager.
PIT	Program Integration Team	The cross-functional team responsible for system level development.
PDR	Preliminary Design Review	Major review that reviews design concepts and gives authority for detailed design.
PDT	Product Development Team	A team of cross-functional specialists selected to develop a particular element of a system.
RFP	Request For Proposal	Formal customer document requesting submission of response to the business opportunity described therein.
SBD	Schematic Block Diagram	A block diagram depicting the interfaces that must exist between elements of the system shown on the ABD.
SDR	System Design Review	A formal DoD review prior to preliminary design work.
SDRL	Supplier Data Requirements List	A list of data deliverables for a subcontract.

Table 7-1 (continued) Acronym Explanations

Acronym	Meaning	Explanation
SEM	System Engineering Manual	Functional manual describing/ defining the company system engineering process.
SEMP	System Engineering Management Plan	Plan commonly required on DoD contracts to describe the contractor's system engineering process.
SEDS	System Engineering Detailed Schedule	Schedule document called for in MIL-STD-499B that may continue to be used.
SEMS	System Engineering Master Schedule	Schedule document called for in MIL-STD-499B that may continue to be used.
SOW	Statement of Work	A document that identifies all work required on a program.
SRR	System Requirements Review	Major DoD program review to review system requirements.
TPM	Technical Performance Measurement	A quantitative reporting technique for key requirements that shows parameter history as well as future predictions.
WBS	Work Breakdown Structure	A finance-oriented overlay of the system architecture intended to track cost in terms of specific items in the system.

7.2 Special terms

7.2.1 Validation

Validation is a process of assuring that a given requirement can be satisfied within the laws of science and available technology. It is commonly accomplished through analysis and testing to demonstrate that one or more design concepts will satisfy the requirement and otherwise be acceptable on the program.

7.2.2 Verification

Verification is a process of proving that the design solution complies with requirements. It is commonly accomplished through analysis, testing, demonstration, and inspection.

COMPANY XYZ

FUNCTIONAL
SYSTEM ENGINEERING MANUAL
(SEM)

AND

GENERIC PROGRAM
SYSTEMS ENGINEERING
MANAGEMENT PLAN
(SEMP)

APPENDIX A

TRACEABILITY TO EXTERNAL STANDARDS

Appendix A

Traceability to external standards

A.1 Overview

Paragraph A.2 explains the historical relationship between this document and MIL-STD-499A and associated data item description (DID). Subsequent paragraphs identify specific external systems engineering standards and provide in each case a traceability matrix between them and this SEM/SEMP plus tailoring information for the external standard to cause compliance of this document with the external standard.

Paragraph A.3 covers MIL-STD-499A. Paragraph A.3.1 and Table A-1 provide a traceability matrix between MIL-STD-499A and the contents of this SEM/SEMP to demonstrate the degree of compliance between the two documents for DoD contracts. Paragraph A.3.2 contains the company standard tailoring that must be introduced into the statement of work (SOW) in any contract where MIL-STD-499A is called for compliance.

Paragraph A.4 covers data item description DI-S-3618 commonly called on contracts requiring compliance with MIL-STD-499A. Paragraph A.4.2 and Table A-2 offer a traceability matrix between the DID requirements and the content of this document. Paragraph A.4.2 contains the tailoring necessary for the customer's contract data requirements list (CDRL) that calls for delivery of a SEMP prepared in accordance with the indicated DID.

Paragraph A.5 covers ISO 9001. Paragraph A.5.1 provides a traceability matrix between 9001 and this SEM/SEMP. Paragraph A.5.2 notes that no tailoring is provided. ISO 9001 primarily requires that a company have a written process description and that they follow it. Since the company has a written SEM/SEMP and company policy requires that it be followed, no specific tailoring need is foreseen.

If future contracts require adherence to other external systems engineering standards, they are to be added to this appendix with traceability and tailoring information required to show SEM/SEMP compliance with those standards.

A.2 External systems engineering standards

MIL-STD-499A may or may not apply contractually to a given program as a function of the customer requirements. The DoD standard was used as a basis for the company SEM/SEMP because it is the only comprehensive standard available and at least part of the company's customer base includes potential military procurement and development. If this SEM/SEMP is applied on a program with a MIL-STD-499A compliance requirement, the proposal, and subsequent contract, should make abundantly clear how the company intends to tailor the standard. This appendix presents that standard tailoring to cause alignment between this company SEM/SEMP and MIL-STD-499A. This tailoring should be included in the customer's SOW during the proposal period.

On a program where MIL-STD-499A is required and is listed for compliance in the statement of work, its status should be changed to reference or guidance and replaced by this SEM/SEMP as the compliance document. It is uncommon to include tailoring for a reference or guidance document, but the tailoring should be included in the SOW because it defines a relationship between the government standard and the internally controlled SEM/SEMP. Customer concern for lack of control over future changes to this document can be addressed by an understanding, contractually formulated or otherwise, that so long as the internal document does not deviate from the agreed to tailoring of MIL-STD-499A, it is acceptable. Alternatively, the customer can be asked to certify this SEM/SEMP as compliant with the tailored MIL-STD-499A, in which case MIL-STD-499A may appear as a compliance document with the understanding that compliance will be attained through this SEM/SEMP.

If the program has no obligation to comply with MIL-STD-499A nor any other system engineering standard, this document may be applied to the program without conflict. If a customer calls for some other system engineering standard, then the proposal team must perform a tailoring analysis on that other standard to bring about a condition of maximum equality between that standard and this SEM/SEMP. As lessons learned from such experiences, the resulting tailoring should be entered into this appendix to preclude subsequent repeat of this work.

The purpose of these tailoring exercises is to enable company repetition in its system engineering process encouraging continuous improvement in the company's capability. So, this tailoring process should not be approached as an exercise in company arrogance. There is a much higher purpose behind company tailoring interests. All company customers will benefit from a constantly improving company system engineering capability, and company personnel will benefit from increasing professional stature within the company and among their peers in industry.

A.3 MIL-STD-499A

A.3.1 MIL-STD-499A traceability

The SEM/SEMP includes the three parts required by MIL-STD-499A paragraph 5.1, System Engineering Management Plan (SEMP). The contents of paragraph 5, Detailed Requirements, must be assessed at the complete document and program level; so, Section 5 is not included in the detailed traceability table. Table A-1 provides a traceability matrix between the content of

MIL-STD-499A paragraph 4, General Requirements, and Appendix A, Task Statements, and the contents of this document. Table A-1 satisfies the requirements defined in paragraph B.5, Practice Documentation Quality Assessment, for the purpose of establishing an external benchmark score of 100 with respect to this standard.

A.3.2 Company tailoring for MIL-STD-499A

The rationale for the changes included below in the form of legalistically phrased items is to preserve the generic nature of the company SEM/SEMP and a single systems engineering process for all programs. Inclusion of program-peculiar information in the SEM/SEMP will force the preparation of a program-peculiar SEM/SEMP for each program and make it difficult to preserve a single systems engineering process needed to encourage repetition and continuous company improvement in a common process. Include these statements in MIL-STD-499A tailoring included in the contract statement of work.

1. Change paragraph 5.1.1 to read as follows:
   "5.1.1 Contractual provisions. The contractor may apply to the program a generic SEMP prepared for use on all of its programs provided that its content is traceable to this standard. Program peculiar tailoring of that generic SEMP may be included in a program directives manual that is consistent with MIL-STD-499A tailoring included in the program statement of work. Only those items which are basic to the satisfaction of program objectives and the applicable portions of this Standard will be placed on contract."
2. Add paragraph 1.4. (c) as follows:
   "This Standard does not specifically require inclusion of any program peculiar data in the program SEMP. Where the contractor chooses to apply a generic SEMP to the program, program peculiar data may be included in a program peculiar plan or separate CDRL item."

A.4 DID DI-S-3618

This data item description is commonly called in a CDRL for a SEMP defining the format and content for a SEMP on a program that must comply with MIL-STD-499A. Paragraph A.4.1 provides traceability between the DID formatting and content definition and this SEM/SEMP. Paragraph A.4.2 provides tailoring that should be included in the CDRL that calls this DID to ensure that the company SEM/SEMP will satisfy customer expectations for the document.

A.4.1 DID traceability

Table A-2 lists all paragraphs in the DID DI-S-3618 DD Form 1664 block 10, Preparation Instructions, and traces their appearance in this document to a section/paragraph number.

Table A-1 Traceability — MIL-STD-499A to Company SEM/SEMP

499A Paragraph	Title	SEM/SEMP Paragraph	Title
4.a	Technical Objectives	1.5	Technical Objectives
4.b	Baselines	4.11.5	Baseline Management
4.c	Technology	5.5.4.6	Transitioning Critical Technologies
4.d	Realistic System Values	5.2.2	Unprecedented Systems
4.e	Design Simplicity	4.9	Development Controls
4.f	Design Completeness	4.9	Development Controls
4.g	Documentation	4.11.5	Baseline Management
4.h	Engineering Decision Studies	4.9	Development Controls
4.i	Cost Estimates	4.7	Program Work Definition and Authorization
4.j	Technical Task and Work Breakdown Structure Compatibility	4.7	Program Work Definition and Authorization
4.k	Consistency and Correlation of Requirements	5.4	Requirements Analysis
4.l	Technical Performance Measurement	5.4.14	Technical Performance Measurement
4.m	Interface Design Compatibility	5.11.7	Product Interface Management
4.n	Engineering Specialty Integration	6	Specialty Engineering Integration and Concurrent Development
4.o	Engineering Decision Traceability	4.9.3	Decision Database
4.p	Historical Data	4.9.3	Decision Database

4.q	Responsiveness to Change	4.11.6	Change Control
4.r	Compatibility With Related Activities	4.2.3	Program Organizations
10.1	Technical Program Planning and Control	4	Technical Program Control
10.1.1	Development of Contract Work Breakdown Structure (WBS) and Specification Tree	4.6	Program Planning Documentation
10.1.2	Program Risk Analysis	4.10	Risk Management
10.1.3	System Test Planning	5.6	Test and Evaluation Development
10.1.4	Decision and Control Process	4.9	Development Controls
10.1.5	Technical Performance Measurement	5.4.14	Technical Performance Measurement
10.1.6	Technical Reviews	4.9.4	Design Reviews and Audits
10.1.7	Subcontractor/Vendor Reviews	4.14	Subcontractor Technical Controls
10.1.8	Work Authorization	4.7	Program Work Definition and Authorization
10.1.9	Documentation Control	4.11.5	Baseline Management
10.2	System Engineering Process	5	System Engineering Process
10.2.1	Mission Requirements Analysis	5.2.2	Unprecedented Systems
10.2.2	Functional Analysis	5.3	Functional Decomposition
10.2.3	Allocation	5.3.3	Functional Allocation
10.2.4	Synthesis	5.3.4	Architecture Synthesis
10.2.5	Logistic Engineering	6.2.5	Logistics Support Analysis
10.2.6	Life Cycle Cost Analysis	6.2.14	Life Cycle Cost Analysis
10.2.7	Optimization	5.5.4.4	Specialty Assessment and Optimization
10.2.8	Production Engineering Analysis	6.2.13	Producibility
10.2.9	Generation of Specifications	4.11.5.1	Product Requirements Baseline

A.4.2 Company tailoring for DID DI-S-3618

The rationale for the changes included below in the form of legalistically stated tailoring items is to preserve the generality of the company SEM/ SEMP. This program-peculiar data will be listed or provided in program-peculiar documentation where required. Include these statements in the CDRL calling for a SEMP that must be compliant with MIL-STD-499A.

1. Reverse the order of Parts 1 and 2 for consistency with MIL-STD-499A. The DID was originally prepared for MIL-STD-499.
2. In DID paragraph 2.d, the reference to key personnel does not require naming specific individuals, rather positions these personnel would fill.
3. In DID paragraph 2.f, delete the reference to review schedules. This information will be provided in program-peculiar schedules provided by other means.
4. Under DID paragraph 2.g, the SEMP need not provide a list of program-specific parameters for TPM, rather provide a process through which parameters are identified, tracked, and managed. The specific list may be acquired by the customer through a CDRL or the DAL.

A.5 ISO 9001

A.5.1 SEM/SEMP traceability to ISO 9001

Table A-3 lists the requirements paragraphs (section 4) of International Standards Organization standard ISO 9001 dated 1978(E) by paragraph number and name followed in each case by the SEM/SEMP paragraph that is responsive to that requirement to show compliance of the SEM/SEMP with ISO 9001. The title of ISO 9001 is *Quality systems — Model for quality assurance in design/development, production, and servicing*. Where the ISO 9001 paragraph is a title-only form, the SEM/SEMP paragraph number cited is NA for "not applicable". Some of the content of ISO 9001 is focused on areas of interest outside the scope of a SEM/SEMP, but they are included in this SEM/SEMP for completeness and noted as not applicable.

Table A-2 Traceability — DID to Company SEM/SEMP

Data Item Description		SEM/SEMP	
Paragraph	Title	Paragraph	Title
1	System Engineering	5	Systems Engineering Process
2	Technical Program Planning and Control	4	Technical Program Planning and Control
3	Engineering Integration	6	Specialty Engineering Integration and Concurrent Development

Many of the paragraphs in ISO-9001 contain only a paragraph heading and these are identified in Table A-3 with the symbol "NA", for not applicable, in the SEM/SEMP column. Other ISO-9001 content applies more appropriately to a program plan or company policy manual to which the SEM/SEMP is subordinate. In these cases, you will find the symbol HL in the SEM/SEMP column meaning that the reader must consult a higher level document to find the related material. A third kind of non-coverage exists in the form of coverage in another company document to which this document is not subordinate. These cases are identified by the symbol "OD", for other document, in the SEM/SEMP column and a reference to the manual where that information is found.

A.5.2 ISO 9001 Tailoring

None offered.

A.6 EIA standard 632

A.6.1 SEM/SEMP traceability to EIA standard 632

Table A-4 provides traceability information between Electronics Industry Association Standard 632 sections 4 and 5 (requirements sections) and this SEM/SEMP. This standard began life as MIL-STD-499B which was not approved because it was in conflict with an evolving DoD policy to appeal to commercial standards for manufacturing and management.

A.6.2 EIA Standard Tailoring and Adjustments.

As shown in Table A-4, this SEM/SEMP covers the requirements defined in EIA 632 and there is no need for tailoring of the standard. The company systems engineering process does, however, require functional analysis (using one of several methodologies) as a prerequisite to requirements analysis and the sequence of discussion of these two topics in the standard appears in conflict with this fundamental and universally accepted systems engineering sequence. When this SEM/SEMP is applied on a contract calling the EIA standard, this difference and the basis for the company position must be made clear to the customer. Under the company methodology, functional analysis is a technique for analyzing needed system and element functionality as a prerequisite to definition of system architecture and appropriate performance requirements for those architecture items. Requirements analysis expands the terse function statements into quantified performance requirements (before or after allocation) and identifies appropriate design constraints for the items identified through functional decomposition and allocation. Together with other techniques described in this document including computer software requirements analysis, these techniques form the company system requirements analysis process.

Table A-3 Traceability — ISO 9001 to Company SEM/SEMP

ISO 9001 Content		SEM/SEMP	
Paragraph	Title	Paragraph	Title
4	Quality system requirements	5.10	Quality Assurance Support
		6.2.28	Quality Engineering
4.1	Management responsibility	NA	
4.1.1	Quality policy	OD	Quality Assurance Practices Manual
4.1.2.1	Responsibility and authority	4.2.3	Program Organizations
		3.7	Management Style and Working Environment
4.1.2.2	Verification resources and personnel	5.8	Requirements Verification
4.1.2.3	Management representative	4.2	Company Organizational Structure
4.1.3	Management review	4.2	Company Organizational Structure
4.2	Quality system	OD	Quality Assurance Practices Manual
4.3	Contract review	5.2	System Definition
4.4	Design control	NA	
4.4.1	General	4.6	Program Planning Documentation
4.4.2	Design and development planning	4.7	Program Work Definition and Authorization
4.4.2.1	Activity assignment	4.7	Program Work Definition and Authorization
4.4.2.2	Organizational and technical interfaces	4.11.7	Product Interface Management
4.4.3	Design input	4.4.2	Process Inputs
4.4.4	Design output	4.4.1	Major Outputs, Deliverables, and Results
4.4.5	Design verification	5.8	Requirements Verification
4.4.6	Design changes	4.11	Configuration Management
4.5	Document control	NA	
4.5.1	Document approval and issue	4.12	Data Management
4.5.2	Document changes/modifications	4.12	Data Management
4.6	Purchasing	NA	
4.6.1	General	4.14	Supplier Technical Controls

4.6.2	Assessment of sub-contractors	4.14	Supplier Technical Controls
4.6.3	Purchasing data	4.12	Data Management
4.6.4	Verification of purchased product	5.8	Requirements Verification
4.7	Purchaser supplied product	5.17	Material Control
4.8	Product identification and traceability	4.11	Configuration Management
4.9	Process control	NA	
4.9.1	General	OD	Production Practices Manual
4.9.2	Special provisions	OD	Quality Assurance Practices Manual
4.10	Inspection and testing	OD	Quality Assurance Practices Manual
4.10.1	Receiving inspection and testing	OD	Quality Assurance Practices Manual
4.10.2	In-process inspection and testing	OD	Quality Assurance Practices Manual
4.10.3	Final inspection and testing	OD	Quality Assurance Practices Manual
4.10.4	Inspection and test records	OD	Quality Assurance Practices Manual
4.11	Inspection, measuring, and test equipment	OD	Quality Assurance Practices Manual
4.12	Inspection and test status	OD	Quality Assurance Practices Manual
4.13	Control of non-conforming product	OD	Quality Assurance Practices Manual
4.13.1	Non-conformity review and disposition	OD	Quality Assurance Practices Manual
4.14	Corrective action	OD	Quality Assurance Practices Manual
4.15	Handling, storage, packaging, and delivery	OD	Integrated Logistics Support Manual
4.15.1	General	OD	Integrated Logistics Support Manual
4.15.2	Handling	OD	Integrated Logistics Support Manual
4.15.3	Storage	OD	Integrated Logistics Support Manual
4.15.4	Packaging	OD	Integrated Logistics Support Manual
4.15.5	Delivery	OD	Integrated Logistics Support Manual
4.16	Quality records	OD	Quality Assurance Practices Manual
4.17	Internal quality status	OD	Quality Assurance Practices Manual
4.18	Training	4.2.1.2	Company Training Program
4.19	Servicing	OD	Integrated Logistics Support Manual
4.20	Statistical techniques	OD	Quality Assurance Practices Manual

Table A-4 SEM/SEMP Traceability — EIA Standard 632

EIA Standard Content Paragraph	Title	SEM/SEMP Paragraph	Title
5.2.3	Electromagnetic Compatibility and Radio Frequency Management	6.2.17	Electromagnetic Compatibility
		6.2.18	Radio Frequency Management
5.2.4	Human Factors	6.2.8	Human Engineering
5.2.5	System Safety and Health Hazard	6.2.9	System Safety and Human Hazards
5.2.6	System Security	6.2.12	System Security
5.2.7	Producibility	6.2.13	Producibility
5.2.8	Product Support	6.2.5	Logistics Support Analysis
5.2.9	Test and Evaluation	5.9	System Test and Evaluation
5.2.10	Integrated Diagnostics	6.2.3	Testibility, Integrated Diagnostics, and Built-In-Test (BIT)
5.2.11	Transportability	6.2.7	Transportability, Mobility, and Portability
5.2.12	Infrastructure Support	6.2.30	Deployment Planning Analysis
5.2.13	Other Functional Areas	NA	
5.3	Leveraged Options	5.2.3.3	Re-Engineering Existing Systems
5.4	Pervasive Development Considerations	NA	
5.4.1	Computer Resources	5.12	Computer Software Coding
5.4.2	Materials, Processes, and Parts Control	6.2.21	Materials and Processes
		6.2.22	Parts Engineering
5.4.3	Prototyping	3.2.2	Rapid Prototyping
5.4.4	Simulation	4.11.5.3	Product Representation Baseline
5.4.5	Digital Data	3.6	Program Information Environment
5.5	System and Cost Effectiveness	5.4	Requirements Analysis

5.5.1	Manufacturing Analysis and Assessment	5.18	Production Process Definition and Control
5.5.2	Verification Analysis and Assessment	5.8	Requirements Verification Planning
5.5.3	Deployment Analysis and Assessment	6.2.30	Deployment Planning Analysis
5.5.4	Operational Analysis and Assessment	5.20	Logistics and Operational Support
5.5.5	Supportability Analysis and Assessment	5.20	Logistics and Operational Support
5.5.6	Training Analysis and Assessment	5.20	Logistics and Operational Support
5.7	Technical Reviews	4.9.4	Design Reviews and Audits
5.7.1	Structuring Reviews	4.9.4	Design Reviews and Audits
5.7.2	Alternative Systems Review (ASR)	4.9.4.1.1	Alternative Systems Review (ASR)
5.7.3	System Requirements Review (SRR)	4.9.4.1.2	System Requirements Review (SRR)
5.7.4	System Functional Review (SFR)	4.9.4.1.3	System Functional Review (SFR)
5.7.5	Preliminary Design Review (PDR)	4.9.4.1.4	Preliminary Design Review (PDR)
5.7.6	Critical Design Review (CDR)	4.9.4.1.5	Critical Design Review (CDR)
5.7.7	System Verification Review (SVR)	4.9.4.1.7	System Verification Review (SVR)
5.7.8	Physical Configuration Audit (PCA)	4.9.4.1.8	Physical Configuration Audit (PCA)
5.7.9	Subsystem Reviews	4.9.4	Design Reviews and Audits
5.7.9.1	Software Specification Review (SSR)	4.9.4.2.1	Software Reviews
5.7.9.2	Preliminary Design Review (PDR)	4.9.4.1.4	Preliminary Design Review (PDR)
5.7.9.3	Critical Design Review (CDR)	4.9.4.1.5	Critical Design Review (CDR)
5.7.9.4	Test Readiness Review (TRR)	4.9.4.2.4	Test Readiness Review (TRR)
5.7.9.5	Functional Configuration Audit (FCA)	4.9.4.1.6	Functional Configuration Audit (FCA)
5.7.9.6	Physical Configuration Audit (PCA)	4.9.4.1.8	Physical Configuration Audit (PCA)
5.7.10	Functional Reviews	4.9.4.3	In-House Reviews
5.7.11	Interim System Reviews	4.9.4.3	In-House Reviews
5.7.12	Review Responsiblities	4.9.4	Design Reviews and Audits

COMPANY XYZ

FUNCTIONAL
SYSTEM ENGINEERING MANUAL
(SEM)

AND

GENERIC PROGRAM
SYSTEMS ENGINEERING
MANAGEMENT PLAN
(SEMP)

APPENDIX B

PROGRAM SYSTEMS ENGINEERING ASSESSMENT CRITERIA

Appendix B

Program systems engineering assessment criteria

B.1 Assessment process

The company employees a two-dimensional assessment system to monitor systems engineering performance and assure that all programs are applying an effective systems approach based on a specific criterion defined in this appendix that is traceable to the contents of this manual. The first dimension is for program compliance with internal practices defined in this document. The second dimension is for benchmarking internal practices against recognized standards for performance of systems engineering. The former metric is focused on individual programs while the latter is focused on the company as a whole since the enterprise employs a common systems approach on all programs. Paragraph B.2 covers the metrics used. Paragraph B.3 provides the internal assessment criteria and questions related to that criteria upon which the internal metric is based.

Figure B-1 illustrates the general process used to determine overall process metrics. For lower tier practices, the corresponding task is selected on a specific program and performance audited against internal practices for that task. In the case of the overall systems engineering process that this document is focused upon, the process definition is contained in this manual. A person is selected by the EIT to accomplish the audit for a specific program. That person conducts an audit based on a series of questions included in paragraph B.3.2 checklist. The result is a number in the range of 0 to 100. A program that faithfully implements this manual will receive a high score. A program that does not will suffer a low score.

Other specific metrics are also used to assess performance as well as the overall systems engineering metric discussed in this appendix. Those are defined by other means by the EIT and individual program PIT.

Programs are called upon to make improvements in implementation where low scores are detected. This will entail program study of the problem followed by program determination of a corrective course of action reviewed by the EIT. The EIT tracks program performance on all audit corrective action plans. Programs with low scores are also selected for audit more frequently than those that exhibit higher scores.

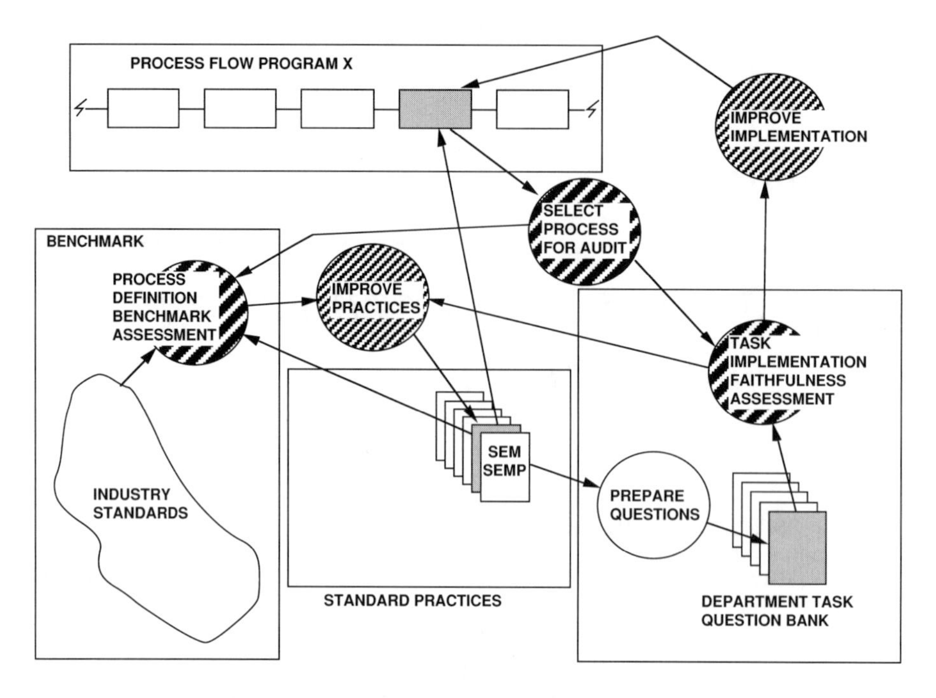

Figure B-1 Systems engineering audit process.

At the same time or at some other time this manual is checked against a recognized benchmark for compliance by the EIT or under their guidance. The most recent benchmarking score is used in establishing the combined systems engineering metric for a program. This metric takes the form of X,Y where X is the internal practices assessment figure and Y is the most recent benchmarking score. This combination tells how well the program is performing the systems engineering process with respect to our internal practices and how those internal practices stack up against a particular standard of systems engineering performance. The EIT is responsible for SEM/SEMP improvements suggested through benchmark audits.

B.2 Assessment metrics and tracking

The company uses two metrics to determine the adequacy of its systems engineering performance and to measure progress over time toward a goal of world class systems engineering. This goal is a moving target because there is continuing progress in industry in improving the process. The company must avoid institutionalizing any of its practices, rather it must constantly be seeking ways to improve them. The company maintains contact with customers, other companies, and engineering societies to ensure access to the latest process methods and standards. Periodically the company conducts an assessment of this manual with respect to an accepted industry standard to ensure internal identification of the most effective techniques available.

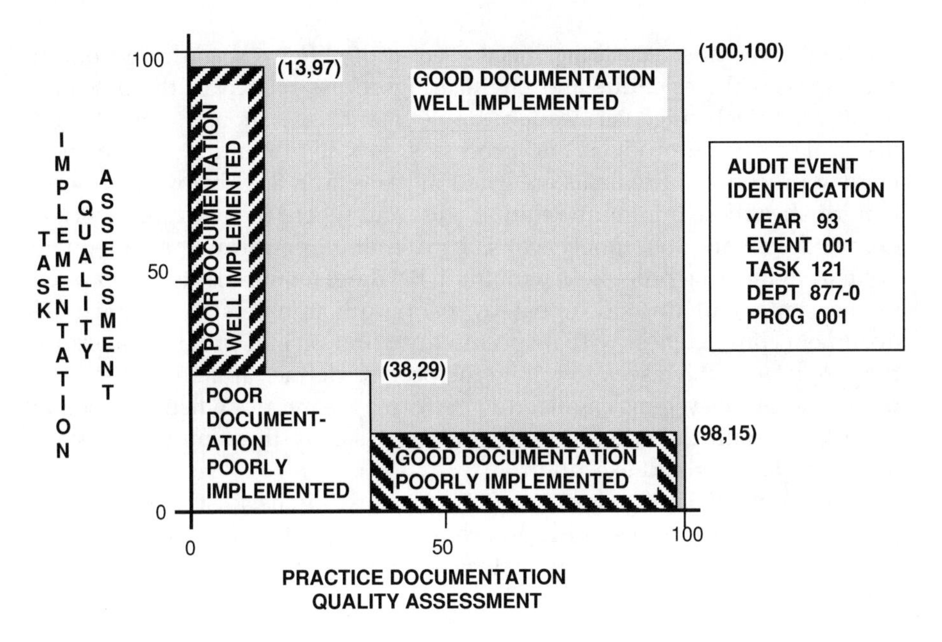

Figure B-2 Two dimensional assessment scoring examples.

Figure B-2 illustrates the two-dimensional nature of the overall systems engineering metric. The vertical axis corresponds to the internal practices metric covered in paragraph B.3. The horizontal dimension is for the benchmarking figure determined by comparison of this document with an external source reference as described in paragraph B.5.

The two metrics computed are M_y, the internal practices compliance metric (vertical axis of Figure B-2), and M_x, the practices quality metric (horizontal axis of Figure B-2). These should be charted in time in the Figure B-2 format with each x,y data point annotated for date and program. Commonly the M_x metric will remain fixed for a period of time subsequent to an internal manual update and benchmarking exercise while the M_y metric for a given program more frequently adjusts for improvements centered about the current practices. Over time, the metric M_x,M_y should move up and to the right toward the top right corner of the chart. If it remains stationary, it indicates a lack of process improvement.

An aggregate M_y metric for the whole company may be calculated and tracked by averaging the individual program M_y metrics, with or without scaling factors for relative program importance.

It is the responsibility of the EIT to respond to low M_x metrics uncovered through audits and it is the responsibility of a program PIT to respond to a low M_y metric with corrective action.

B.3 Minimum systems engineering criteria

Systems engineering is an interdisciplinary, or cross-functional, approach to evolve and verify an integrated balanced set of system product and process solutions that satisfy customer needs across the life cycle of the system. It encompasses the scientific and engineering efforts related to the development, manufacturing, verification, deployment, operations, support, and disposal of system products and processes. Twenty minimum system engineering process requirements are listed in paragraph B.3.1 below. A process that satisfies these minimum requirements and where the indicated work is executed well and in a timely way is by definition performing an adequate system engineering process. In paragraph B.4 these requirements are mapped to the content of this document by paragraph numbers to demonstrate document completeness with respect to this criterion and provide assurance that programs which seek to follow the process defined in this manual will receive accurate systems engineering performance metrics when assessed in accordance with the criteria. Programs must satisfy these minimum standards in planning and execution.

Paragraph B.3.2 provides a set of questions in checklist form that the person accomplishing a specific program systems engineering audit will use to derive the metric for the aggregate effect of the criteria listed in paragraph B.3.1.

B.3.1 The criteria

a. **The Customer Need**. A clear statement of system need shall be developed by or agreed to by the customer. The customer may be a single, clearly identified entity as in some DoD contracts or a collective entity defined by a marketing function on a commercial program. In the latter case, this criterion may be provided by siting a perceived customer need based on market research that must be satisfied.
b. **System Requirements Capture**. A set of system requirements shall be derived from the need to define the product mission, system performance goals/requirements/figures of merit, external interface definition, system environment definition, and top level specialty engineering constraints in the simplest format acceptable to the customer.
c. **Problem Space Decomposition**. An organized means shall be in place to decompose the large problem expressed by the customer's need into a more detailed understanding of the problem space and assignment or allocation of the functionality thus exposed to elements in the system architecture (or physical solution space) consisting of hardware, computer software, facilities, materials, and personnel.
d. **Cost Definition**. A means shall be in place to allocate and collect recurring and non-recurring cost to/for specific product entities and capture them within the company's, and possibly the customer's, cost accounting system. Cost performance shall be tracked as a means of assuring that cost target will be satisfied.
e. **Schedule Definition**. A means shall be in place to define planned work scheduling and to track work performance to that schedule.
f. **Development Responsibility**. An organized method shall be in place for clearly and comprehensively assigning responsibility for

development of the product system elements and their associated employment process to personnel and/or teams. Where the customer requires financial reporting, this method must be mutually consistent with the customer's definition of work breakdown (in support of effective cost collection) and the company development team concept applied on the program.

g. **Requirements Before Design**. A means shall be in place to define and enforce the definition of requirements for items prior to design work being permitted. It may be necessary to define an item concept prior to completion of some of the design constraints, but these must be in place and approved by appropriate authority before a commitment to detailed design is initiated.

h. **Requirements Capture and Publication**. A means shall be in place to capture all requirements, correlated with their architecture item, and to make them available for reference, discussion, and critical review by all program personnel. If required by the customer, specifications shall be published and maintained in a format compatible with customer documentation requirements. Programs may use requirements database content directly without paper publication if desired and not inconsistent with customer requirements.

i. **Traceability and Flowdown**. Traceability of requirements to sources, other requirements, and the development process shall be accomplished. Ideally, all requirements should trace to the ultimate requirement, the customer's need. This process shall ensure that higher tier requirements have properly influenced lower tier requirements, especially for procured items and that customer requirements have been appropriately flowed down to lower tier elements. This requirement encourages the use of a computer database tool for the purpose of capturing requirements and providing for traceability.

j. **Requirements Verification**. An organized means shall be in place for verifying that the final product design satisfies the requirements prescribed. This means shall provide for associating specific compliance evidence in the form of analysis and test reports to specific requirements. This requirement encourages the use of a computer database tool linked to the requirements database suggested in item "i" above.

k. **Interface Development**. An organized way shall be devised and put in place to develop compatible internal and external interfaces between elements depicted on the system architecture, to minimize the cross-organizational interface, and to clearly assign responsibility for the development of each interface. Interfaces between items under the control of the company and associates will be developed in accordance with a plan approved by all parties including the customer. Interfaces with items to be provided by suppliers shall be controlled by the same company agent responsible for the item being procured.

l. **Design Capture**. An organized way shall be put in place to capture and retain the results of development team product and process concept and design work so as to be easily accessible at any time by all program personnel and available for iterative improvement by responsible engineers and analysts.

m. **Management Oversight**. An organized method shall be put in place for gaining management insight into the results of development team work coupled with a means to compare these results with planned work and to direct the work of teams based on these insights and program goals and requirements.
n. **System Integration**. An organized means shall be in place to integrate the work of separate product development teams and to optimize at the system level across design, manufacturing, quality, test, operations, and logistics support to overcome problems resulting from suboptimization at the individual team level or individual discipline.
o. **Specialty Engineering Integration**. A means shall be put in place to perform product and process development work concurrently. The concurrent development process shall encourage cooperative development work between all relevant specialists so as to apply the right disciplines at the best possible time and period to achieve the most desirable results in the following areas unless in conflict with program requirements:
    1. This shall include a means to define what the system and its components must do in the customer's environment to maintain needed readiness and operational capability and to attain mission success.
    2. This method must provide for development and evaluation of the system maintenance and logistic support concepts and coordination of those concepts with the evolving system design concept and a mechanism for gaining insight into needed procedural and human task-oriented requirements.
    3. The means shall also provide for evaluation of the system and its components for reliable and safe operation.
    4. The concurrent development mechanism shall result in development of requirements and plans for testing, inspection, and production processes concurrently with product design requirements and processes.
p. **Design Rationale Capture**. A means shall be put in place to capture design rationale and traceability of the decision process that led to the preferred system concept. All requirements and the designs corresponding to those requirements shall be reviewed by responsible management personnel and approved. Relevant details of these reviews shall be captured for the purpose of communicating direction and for subsequent reference.
q. **Requirements Configuration Control**. A means shall be put in place to clearly define the current system requirements and concept baseline during the period leading up to PDR. Subsequent to PDR the requirements baseline shall be captured in a set of specifications and reports, or computer data equivalents.

r. **Design Configuration Control**. A means shall be put in place to positively control the configuration of the product design subsequent to the formal approval of that design and to control the configuration of all representations of that design in the form of models, simulations, mock-ups, test articles, test results, and analyses. Design configuration management should not begin before PDR and must begin no later than CDR.
s. **Product Configuration Control**. The configuration of physical products produced through procurement, assembly, and manufacturing shall be clearly defined and tracked in time as a function of changes. Each product entity shall be clearly linked to its design and requirements configuration definition.
t. **Continuity Restitution**. There shall be in place a means to detect that a discontinuity has occurred in planned program work and to reach sound decisions that will return the development process to a condition of control.
u. **Risk Management**. A means shall be in place to systematically identify potential cost, schedule, and performance risks and to mitigate these risks over the life of the program such that their effects can be positively responded to should they materialize during the period the program is in force.

B.3.2 Internal practices checklist

The final three pages of this appendix provide a systems engineering internal practices checklist, a copy of which may be used directly for program audit purposes. It includes a series of questions for each item in the paragraph B.3.1 criteria list. To use this checklist note the weighting factor (WT) and place that number in the SCORE column for that question if the answer is yes. Make no entry if the answer is no. Upon completion of the questions, add up the numbers in the SCORE column to yield S_t (score total), divide by 50 (ΣW_t, weighting factor total), and multiply the quotient by 100 to give the final metric value in the range 0 to 100. This metric M_v corresponds to the vertical axis of Figure B-2.

In the case of each question a simple answer by a program person should not be the only source of information. The assessor should gain access to tangible evidence of corresponding program performance in the form of a document, entered computer information, or observed behavior.

Depending on the program phase at the time of the assessment, the questions may have a slightly different flavor. In an early phase, there will be no evidence of accomplishment of later phase work but there should be evidence of corresponding planning. In later program phases, there will be no current activity on earlier work but there should be evidence of having completed the related work in the earlier phase.

B.4 System engineering process criteria traceability

Paragraph B.3.1 lists a minimum system engineering process criteria and Table B-1 maps the criteria to the content of this document for the purpose of demonstrating process coverage completeness with respect to the company's own systems engineering criteria. Refer to paragraph B.3.1 for process criteria names. Table B-2 maps this same criteria to the generic company program phases illustrated in Figure 3-2.

B.5 Practice documentation quality assessment

The external systems engineering assessment process begins with the selection of a standard for comparison. This standard is reviewed by the EIT and each specific requirement identified, ideally by a paragraph number. Each of these references is listed in tabular form and mapped to the content of this manual. Where the item maps to this manual a score in the range of 1 to 10 is assigned using the list offered below for guidance. If no corresponding practices content can be found, a score of 0 is recorded. Intermediate scores may be entered. Obviously, there is a degree of subjectivity in this scoring process and more than one person should contribute to the scoring within the boundaries established by time and budget.

0 There is no reference in company practices to this matter
1
2
3
4
5 Company practices address this area but improvements can be made
6
7
8
9
10 Company practices are fully compliant or judged to be better than the external standard

The aggregate count of these mapping events (ΣE_i) is divided by 10 times the total number of source document requirements ($10E_t$) and multiplied by 100 to derive a metric in the range of 0 to 100. This metric (M_x) corresponds to the horizontal axis of Figure B-2. The formula for this metric, therefore, is

$$M_x = 100\Sigma\, E_i/10E_t$$

Appendix A offers traceability between MIL-STD-499A and the content of this manual. It is believed that this manual rates a 100 score when measured against MIL-STD-499A, but the EIT need not necessarily select MIL-STD-499A as the external standard. Even if that standard is selected, a particular audit process may reach a conclusion of a less than perfect score.

Table B-1 Content to Process Criteria Map

SEM/SEMP	Generic process criteria																				
Paragraph	a	b	c	d	e	f	g	h	i	j	k	l	m	n	o	p	q	r	s	t	u
4.1																					
4.2						X							X								
4.3						X															
4.4																					
4.5																					
4.6	X	X		X																	
4.7				X	X	X							X								
4.8					X								X								
4.9												X	X			X					
4.10				X	X								X							X	X
4.11								X			X		X				X	X	X		
4.12																					
4.13																					
4.14																					
4.15											X										
5.1																					
5.2	X	X		X																	
5.3		X	X				X														
5.4							X		X		X				X	X					
5.5							X					X		X	X						
5.6																					
5.7																					
5.8										X											
5.9													X								
5.10													X								
5.11																					
5.12																					
5.13																					
5.14																					
5.15																					
5.16																					
5.17																					
5.18																					
5.19																					
5.20																					
6.1															X						
6.2				X											X						
6.3															X						
6.4															X						
6.5														X	X						
6.6												X		X	X						
6.7												X			X	X					

Table B-2 Generic Phase to Process Criteria Map

Program	Generic process criteria																				
Phase	a	b	c	d	e	f	g	h	i	j	k	l	m	n	o	p	q	r	s	t	u
1	X	X	X	X	X								X	X	X	X				X	X
2	X	X	X	X	X								X	X	X	X				X	X
3		X	X	X	X	X	X	X			X	X	X	X	X	X	X	X	X	X	X
4			X	X	X	X	X	X	X		X	X	X	X	X	X	X	X	X	X	X
5				X	X	X			X	X	X	X	X	X	X	X	X	X	X	X	X
6				X	X					X			X	X	X	X	X	X	X	X	X
7				X	X								X	X	X			X	X	X	X
8										X							X	X			
9		X	X	X	X								X	X	X						

This external audit should be accomplished or updated under any of the following circumstances:

a. Upon initial release of this manual.
b. When this manual undergoes a significant change.
c. When a different external standard has been selected.
d. If 3 years has passed since the most recent audit.

SYSTEMS ENGINEERING
INTERNAL PRACTICES CHECKLIST
PROGRAM AUDIT FORM

PROGRAM ______________________ DATE ________________

AUDITOR ______________________

QUESTION		WT	SCORE
A1	Does a customer need statement exist?	1	_____
A2	If the customer need statement exists, ask three people picked at random on the program and ask them to tell you what the customer need is. If two can show you a copy of the need, point to a document which contains it, or verbally give an interpretation of it that in your judgment is in agreement with the actual statement, score the question yes.	1	_____
B1	Does a system specification, or a specification for the highest level item for which the company program is responsible, exist?	2	_____
C1	Determine how the architecture of the system was determined or is being determined. If an organized decomposition process defined in this manual was used for those portions of the system that entail new development, score the question yes. Yes may be assigned if a very large percentage of the system is predetermined and an impact analysis was accomplished.	2	_____
D1	Does the program have a cost structure defined for all work with allocations made for specific cost centers?	1	_____
D2	Is cost tracked to the predetermined cost allocations and management effort effectively applied to encourage cost compliance?	1	_____
E1	Does a current program schedule exist?	1	_____
E2	Is work accomplished tracked to the schedule and management effort effectively applied to encourage schedule compliance?	1	_____
F1	Is all program work defined as in a statement of work?	1	_____

SYSTEMS ENGINEERING
INTERNAL PRACTICES CHECKLIST
PROGRAM AUDIT FORM (continued)

PROGRAM ______________________ DATE ________________

AUDITOR _______________________

	QUESTION	WT	SCORE
F2	Has all of the work been clearly allocated to specific organizational entities?	1	______
F3	Is the work allocated to specific organizational entities traceable to the cost and schedule requirements?	1	______
G1	Is there evidence that requirements are being or have been developed andapproved before detailed design is undertaken?	3	______
H1	Does a program specification tree exist?	1	______
H2	Review the program specification tree. Is there a clearly defined mechanism for identification of specification types, forms, and formats?	1	______
H3	Select one specification at random and determine if it follows the program standard for specifications. Does it comply?	1	______
I1	Does the program maintain requirements traceability data?	1	______
I2	Examine traceability data. Are you able to find inconsistencies in the datawithin a period of no more than 10 minutes?	1	______
J1	Do specifications contain a verification methods matrix?	1	______
J2	Does evidence exist of verification planning in the form of a verification compliance matrix, verification planning documentation, or integrated testplanning?	1	______
K1	Is there a clear definition of interface responsibilities on the program in the form of a schematic block diagram or N-square diagram with some means to correlate architecture responsibility with interface responsibility?	2	______

SYSTEMS ENGINEERING
INTERNAL PRACTICES CHECKLIST
PROGRAM AUDIT FORM (continued)

PROGRAM ______________________ DATE ________________

AUDITOR ________________________

	QUESTION	WT	SCORE
L1	Are the results of product and process design work captured in an enduring way that can easily be accessed by all program personnel?	2	_____
M1	Does the program have an effective way for management to gain insight into work performance and status?	2	_____
N1	Does the program have an effective means to integrate the product work ofprogram personnel and teams?	2	_____
N2	Does the program have an effective means to integrate the product and process work on the program to the end that manufacturing, quality, test, and logistics processes are compatible and optimum with respect to the product design?	2	_____
O1	Is there a written list defining what specialty engineering disciplines shall be applied on the program?	1	_____
O2	Are all specialty disciplines involved in the requirements definition process?	1	_____
O3	Is there evidence of designer-specialty engineer interaction to understand specialty engineering requirements?	1	_____
O4	Are the documented results of specialty engineering assessments of the design included within the verification evidence?	1	_____
P1	Can the program produce the rationale for design decisions and actions?	2	_____
Q1	Is there evidence of an ability to baseline the requirements at specific points in time and that these baselines are respected?	2	_____

SYSTEMS ENGINEERING
INTERNAL PRACTICES CHECKLIST
PROGRAM AUDIT FORM (continued)

PROGRAM ______________________ DATE ________________

AUDITOR _______________________

QUESTION		WT	SCORE
R1	Does the program actively control the configuration of product representations (models, test articles, simulations, etc.)?	2	_____
S1	Does the program have a means to clearly control the configuration of specific designs and their physical product entities?	2	_____
T1	If there has been any history of discontinuity in past program work, is there evidence of having expeditiously identified and corrected the problem?	2	_____
U1	Does the program have a means to identify cost, schedule, and technical risks, a way to assess their impact, assign mitigation responsibility, and track status?	3	_____
	TOTAL SCORE (ΣS_i)		_____
	TOTAL WEIGHT (ΣW_t)	50	

$M_y = 100\Sigma\, S_i/\Sigma W_t = 100\ (\quad)/(50) =$ ________

AUDIT REMARKS

COMPANY XYZ

FUNCTIONAL
SYSTEM ENGINEERING MANUAL
(SEM)

AND

GENERIC PROGRAM
SYSTEMS ENGINEERING
MANAGEMENT PLAN
(SEMP)

APPENDIX C

GENERIC PROCESS DIAGRAM

Appendix C

Process diagram legend

C.1 Process diagram purpose

The process flow diagram included in this appendix offers an overall system life cycle process definition for the enterprise. It defines the tasks and common sequence of tasks required to acquire business, accomplish system development, manufacture product, operate the product, and logistically support the product. Table C-1 lists all process diagrams included in this appendix.

Table C-1 Appendix C Process Diagram List

Figure	Title
C-1	Overview
C-2	Top level process diagram
C-3	Generic product and process development (Task 1)
C-4	System requirements analysis (Task 105)
C-5	System integration (Task 108)
C-6	Form, charter, and train team (Task 111)
C-7	PDT item concurrent requirements analysis (Task 113)
C-8	PDT concurrent item product and process design (Task 115)
C-9	Structured decomposition analysis (Task 10511)
C-10	Functional flow analysis (Task 105111)
C-11	Architecture synthesis (Task 106)
C-12	System level design reviews and audits (Task 109)
C-13	Item qualification test development (Task 115141)
C-14	Qualification testing (Task 4)
C-15	Item acceptance test development (Task 115142)
C-16	Item design evaluation testing (Task 115144)
C-17	Systems analysis and specialty engineering assessment (Task 10812)
C-18	Production (Task 3)
C-19	Operations (Task 8)
C-20	RFP-proposal cycle (Task D)
C-21	Prepare proposal (Task D7)
C-22	Create program infrastructure (Task D6)
C-23	Applicable documents assessment flow (Task DE)
C-24	Marketing and pre-contract studies (Task E)

C.2 Diagram structure and symbol use

The process diagram is constructed of simple blocks corresponding to tasks in the process interconnected by arrow-headed lines showing the common sequence of tasks. Some diagrams also include logical AND and OR symbols to join and separate flow in accordance with the corresponding logical constructs. A concurrent bond symbol, consisting of a heavy bar, is used in some diagrams to signify the intense human interaction necessary between the tasks so joined to achieve success in concurrent development work.

C.3 Task numbering

All tasks are identified by a task number in the lower right hand corner. These are unique character strings composed of one or more concatenated characters, each one of which is from the set: {0 9}**U**{A-Z less O}**U**{a-z less l}. Assignment of these strings follows a hierarchical pattern. When a given task, such as 134 is expanded to its next level of detail, the lower tier tasks are numbered in the set {1341, 1342, 1343, ... 134Z}. The task numbers assigned are the same ones listed in the TASK NUMBER column of Table D-1 where you will find a brief description of the task, a company document reference where the task is explained in sufficient detail to understand how it is performed, and the generic company phases in which you would expect to apply the task.

C.4 Task layering

Some diagrams illustrate more than one layer of the process. This is done by enclosing all sub-tasks within a boundary identified as the parent task number. In these cases, the parent task number may or may not be located in the lower right corner.

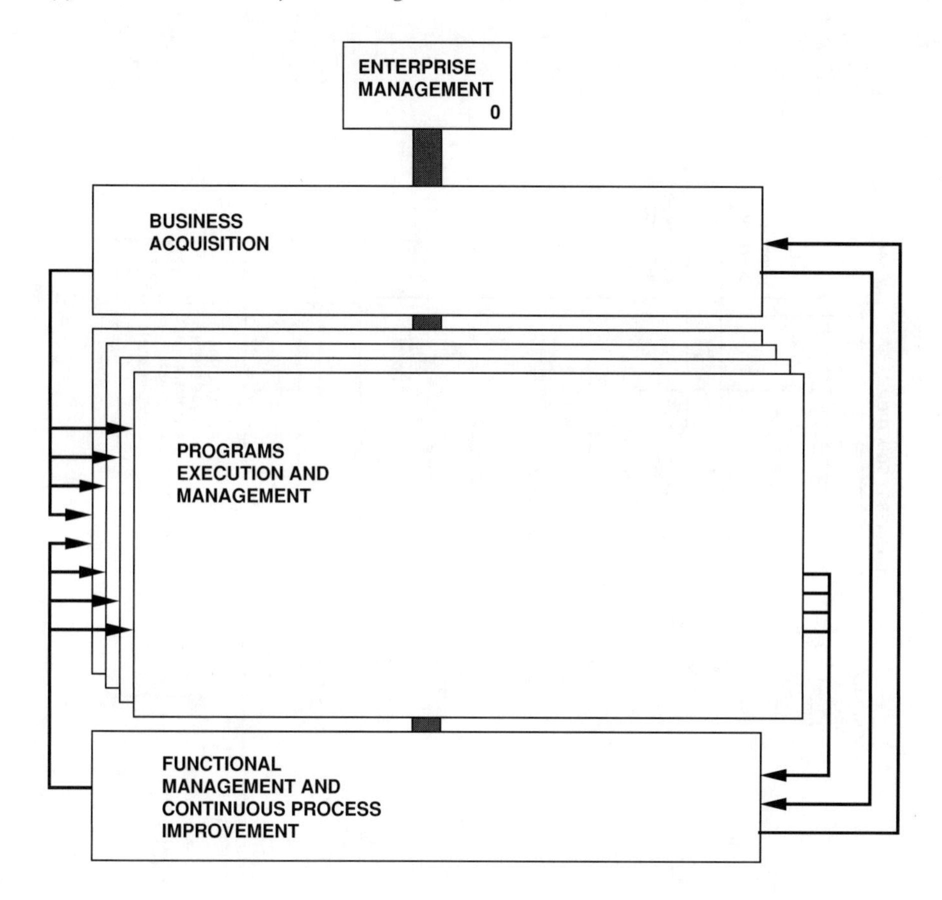

Figure C-1 Overview.

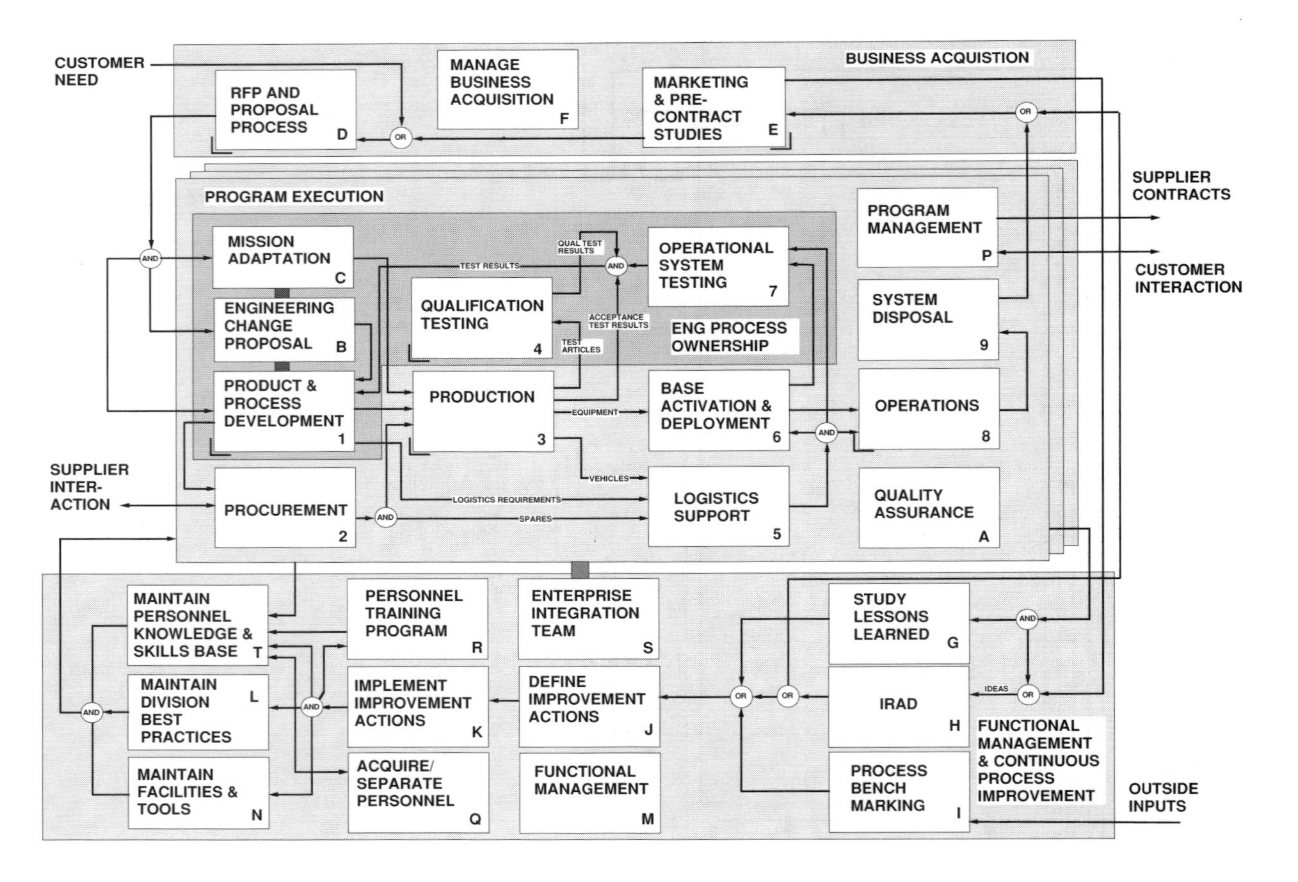

Figure C-2 Top level process diagram.

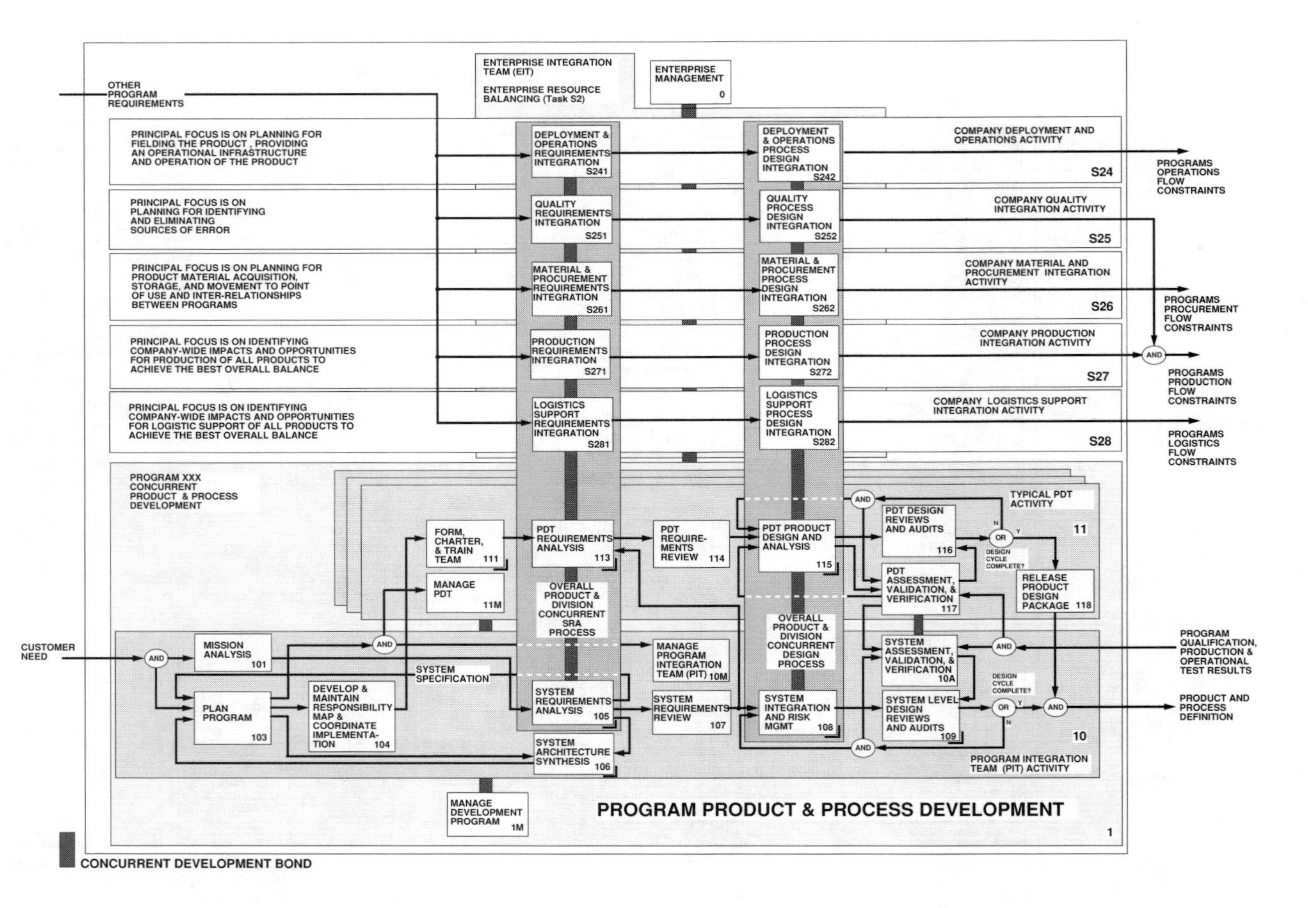

Figure C-3 Generic product and process development (Task 1).

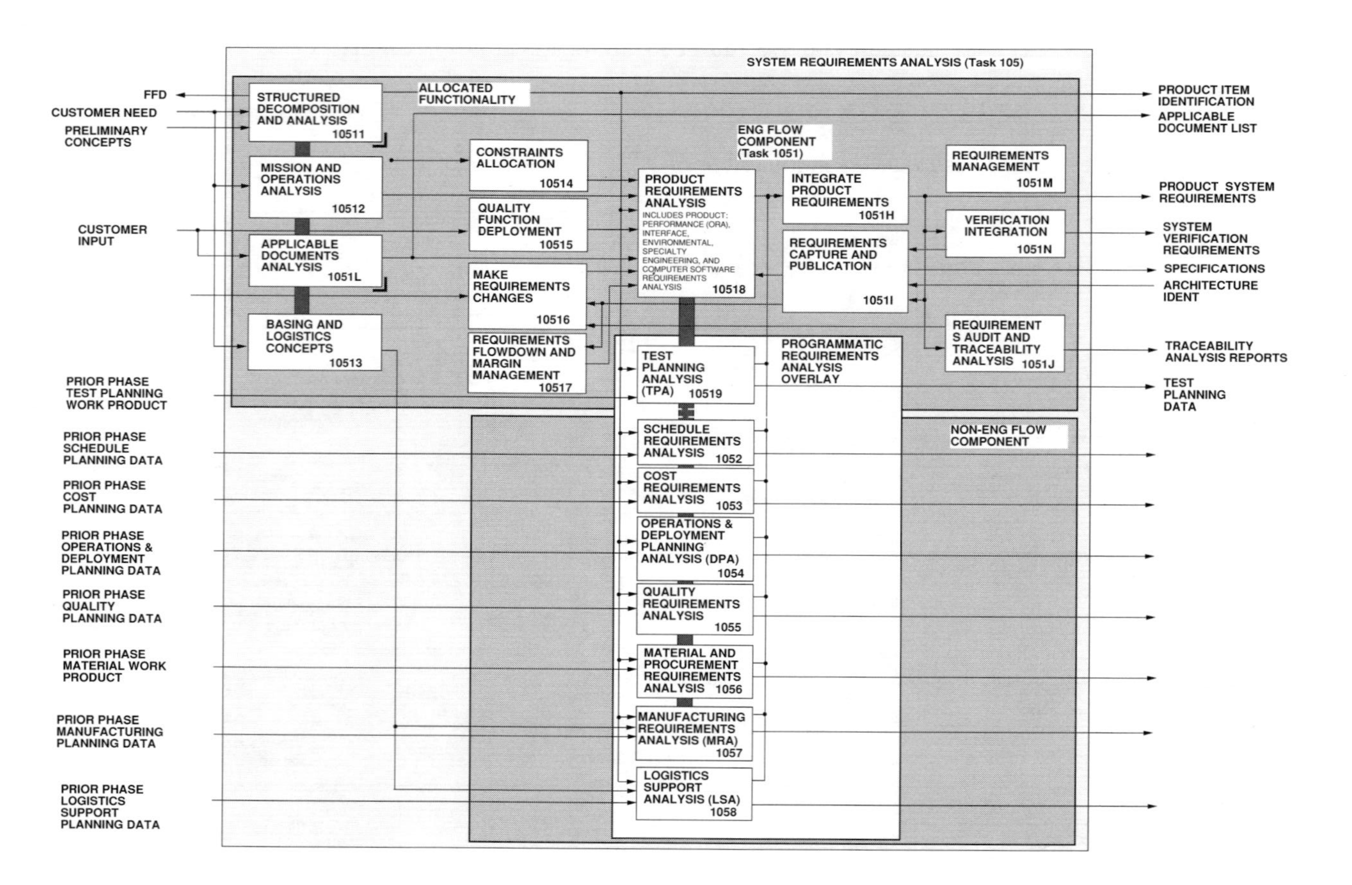

Figure C-4 System requirements analysis (Task 105).

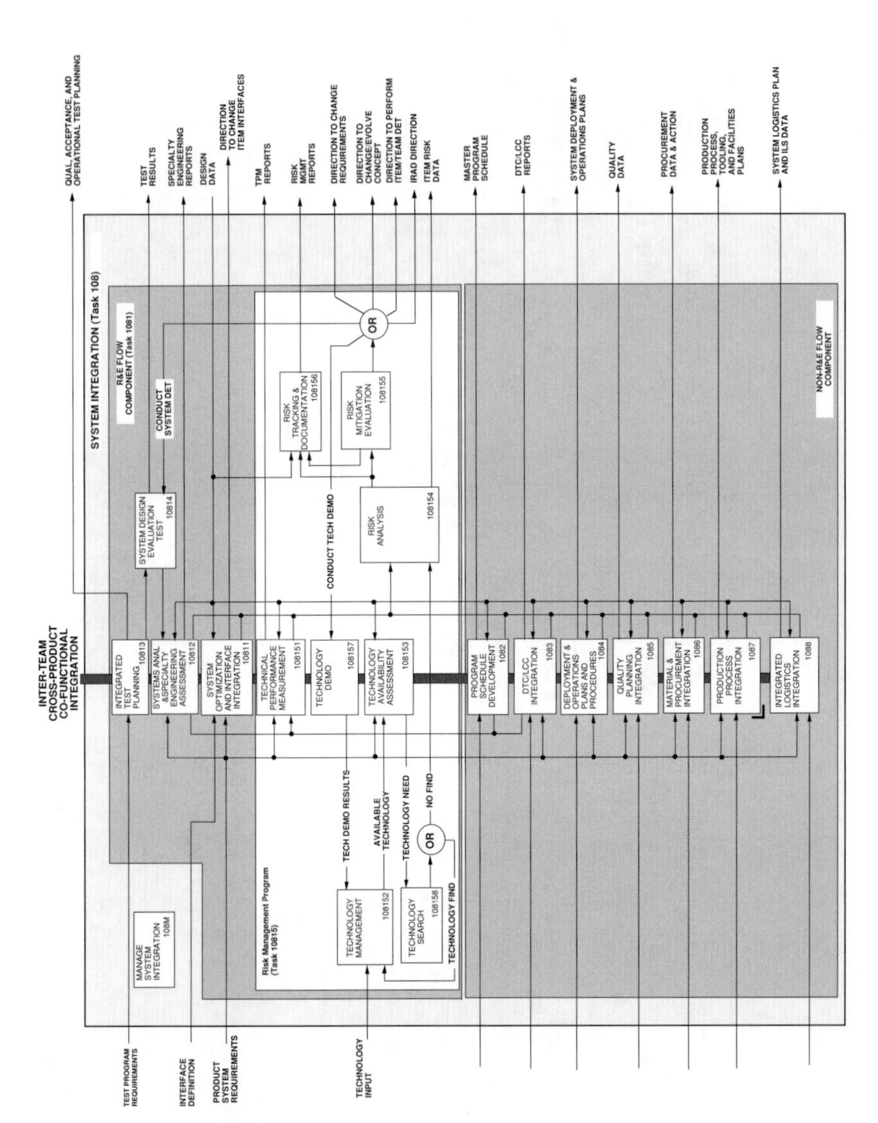

Figure C-5 System integration (Task 108).

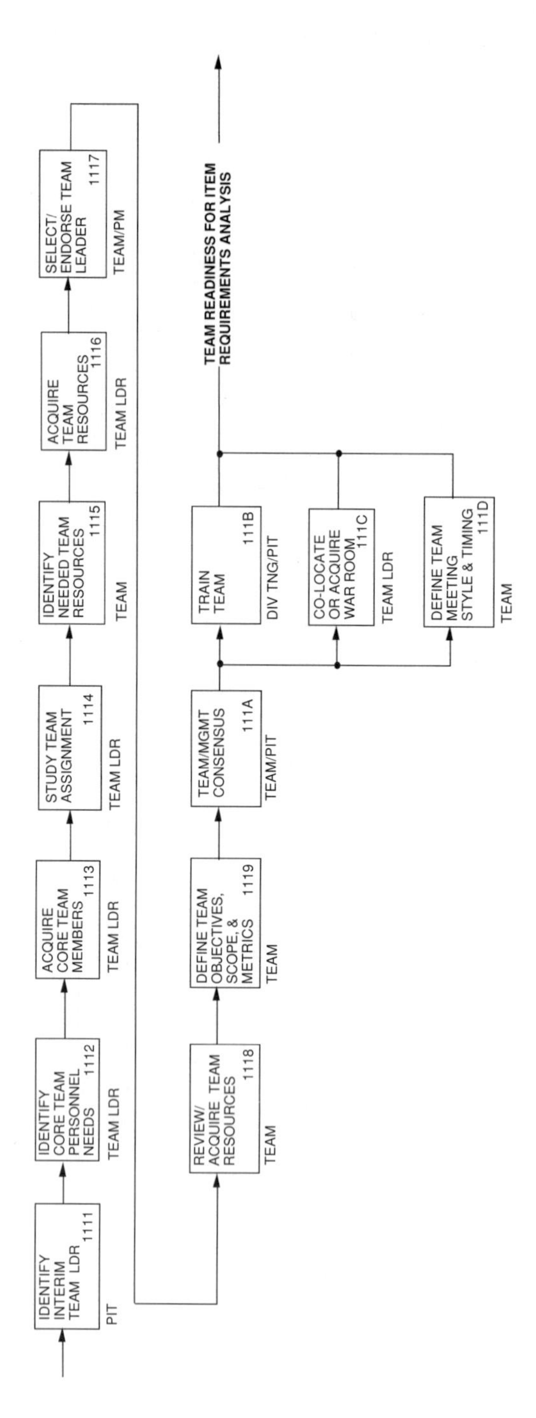

Figure C-6 Form, charter, and train team (Task 111).

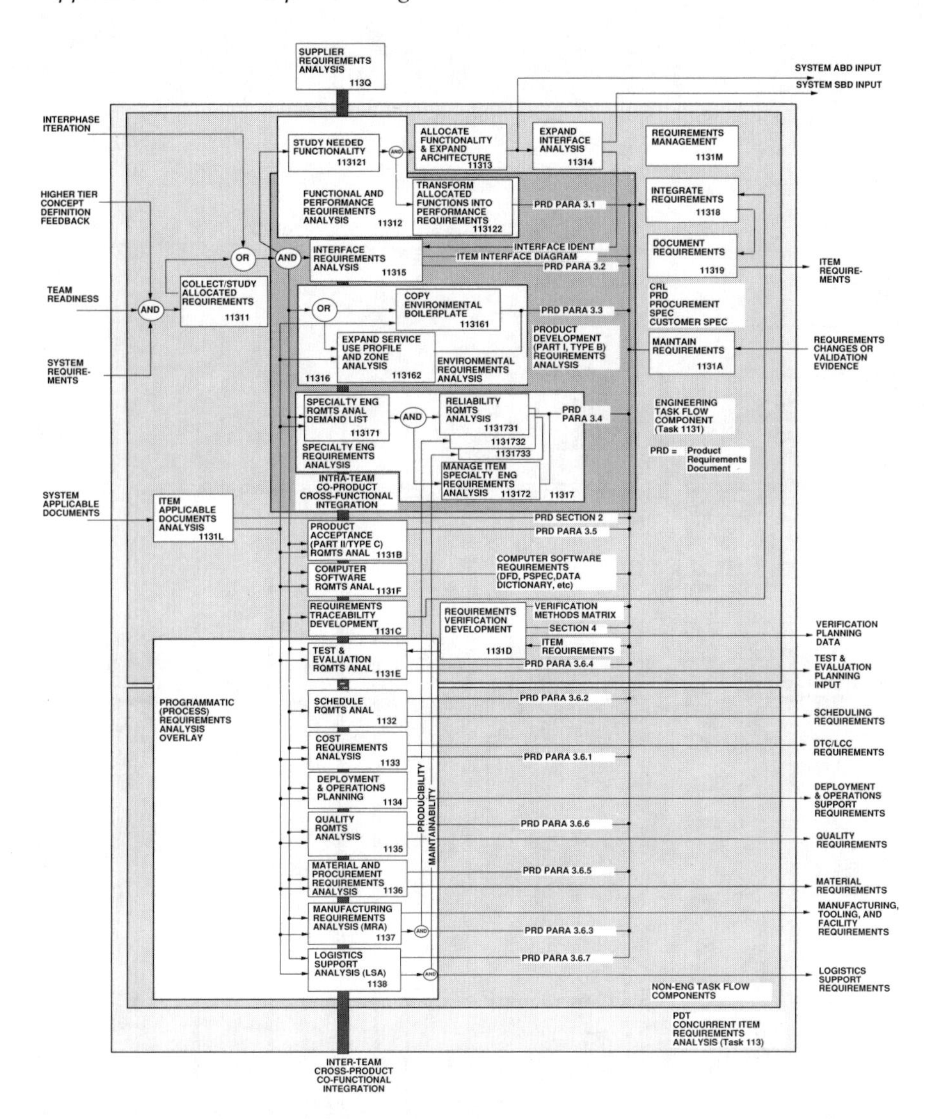

Figure C-7 PDT item concurrent requirements analysis (Task 113).

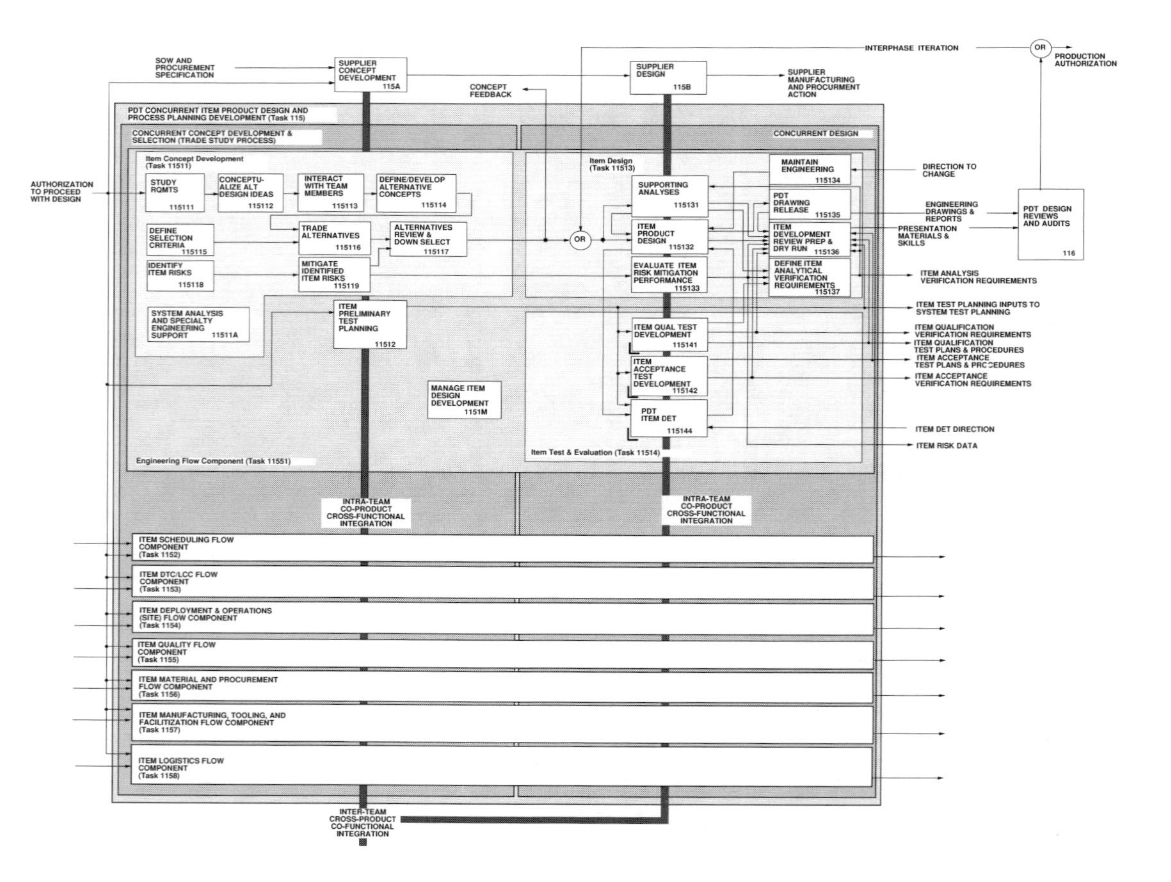

Figure C-8 PDT concurrent item product and process design (Task 115).

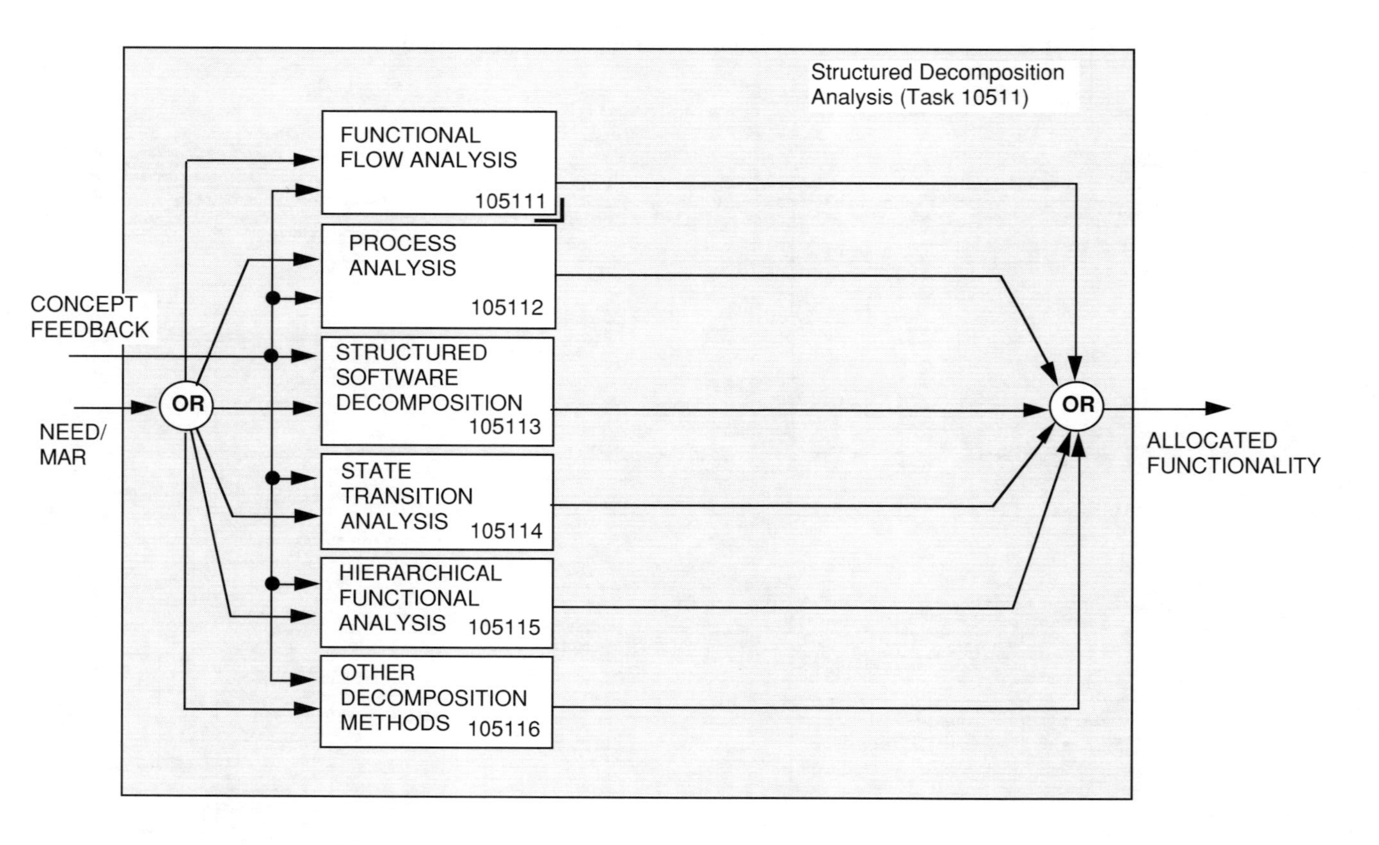

Figure C-9 Structured decomposition analysis (Task 10511).

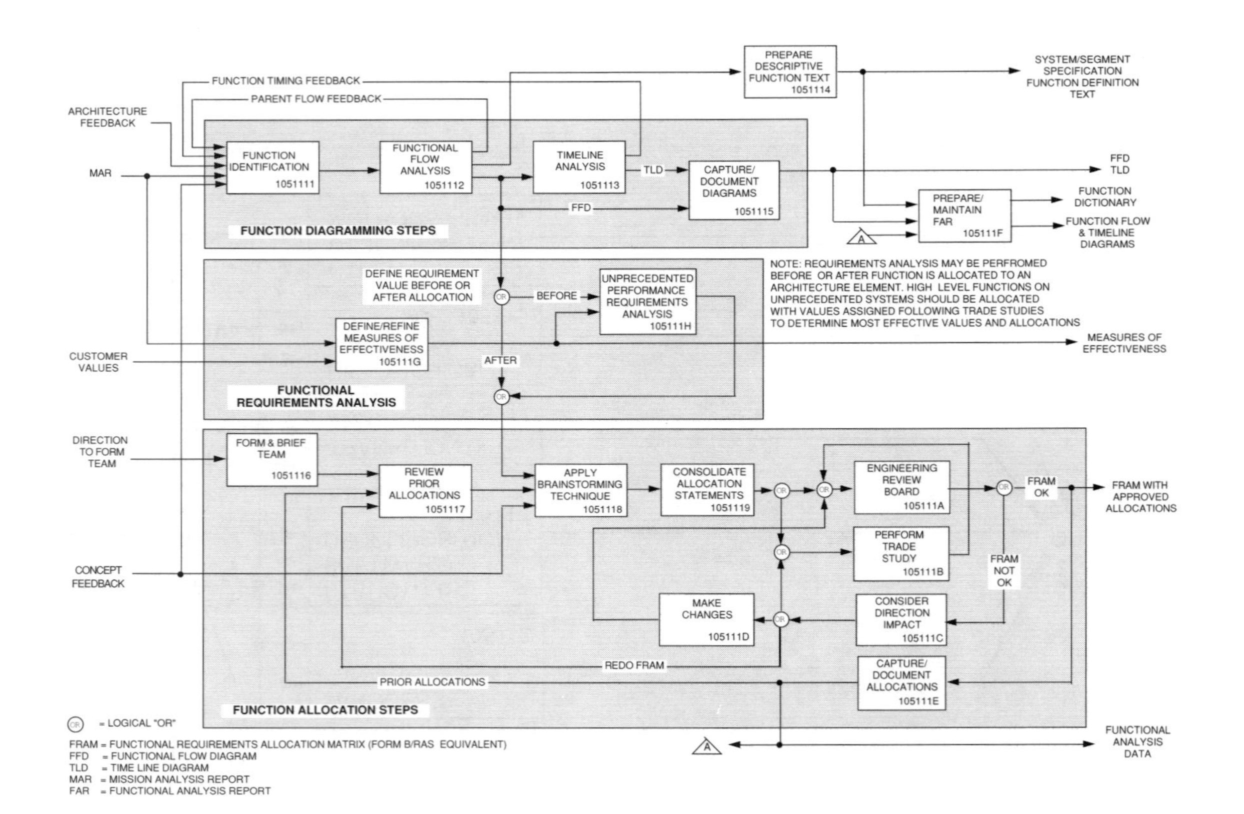

Figure C-10 Functional flow analysis (Task 105111).

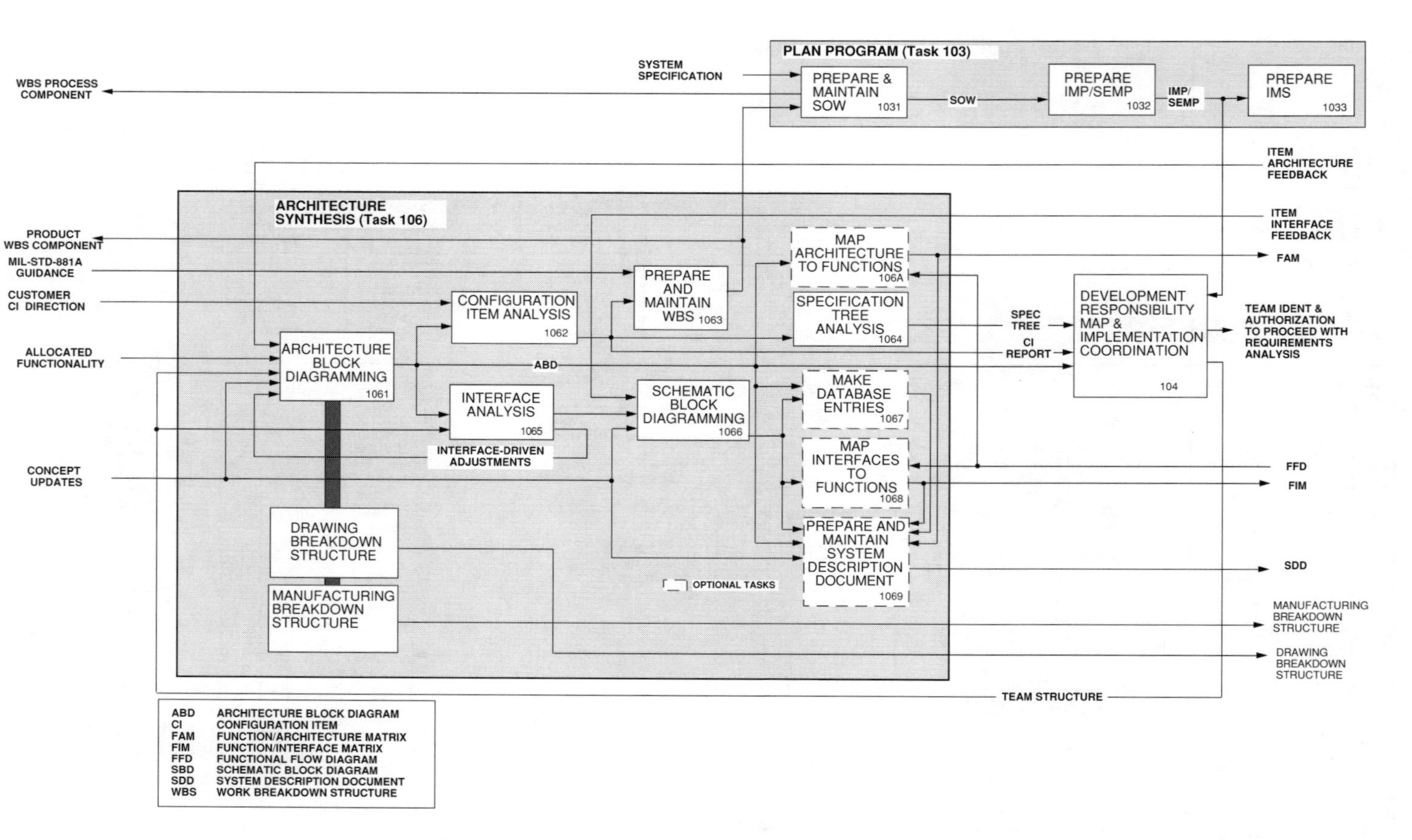

Figure C-11 Architecture synthesis (Task 106).

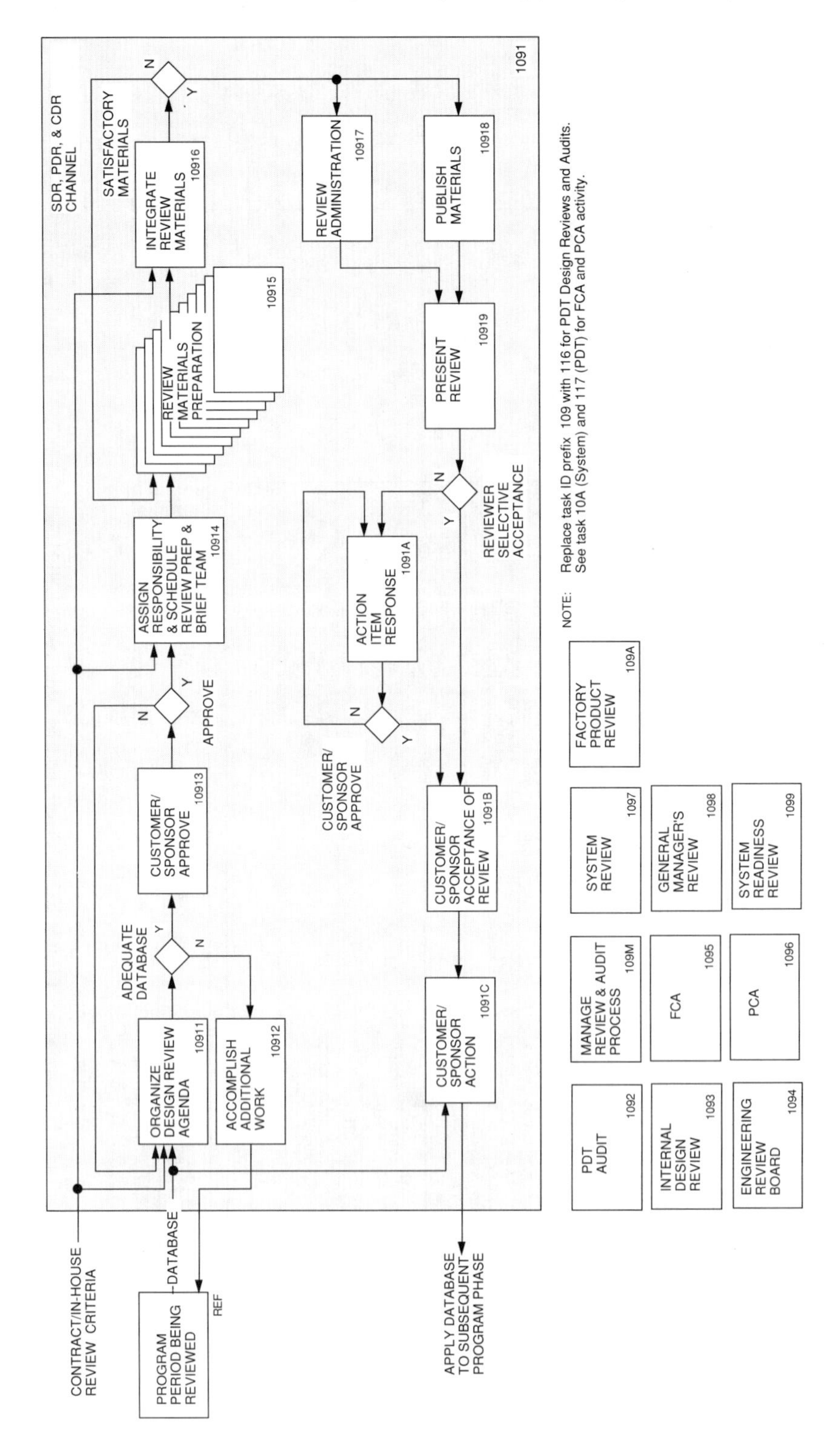

Figure C-12 System level design reviews and audits (Task 109).

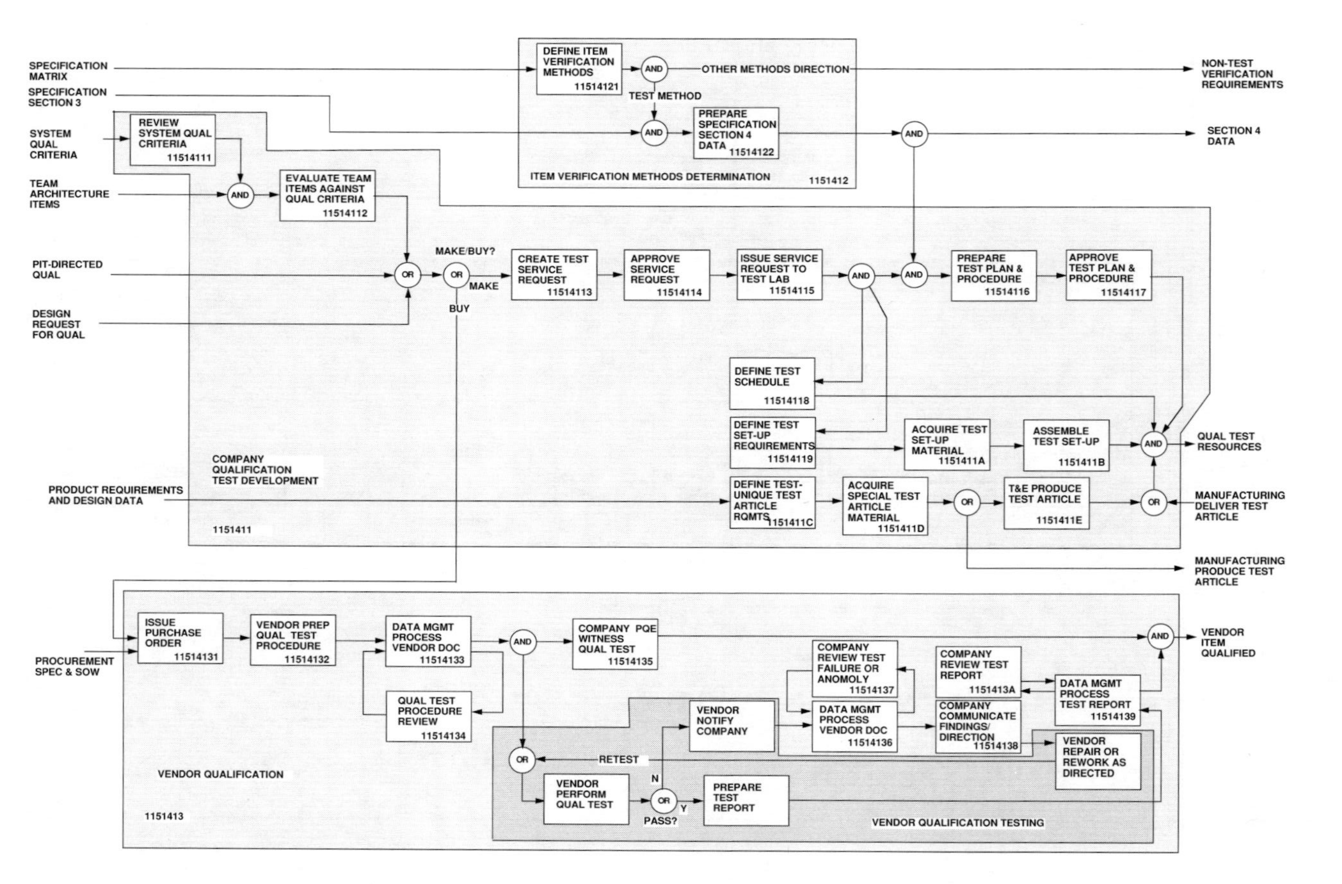

Figure C-13 Item qualification text development (Task 115141).

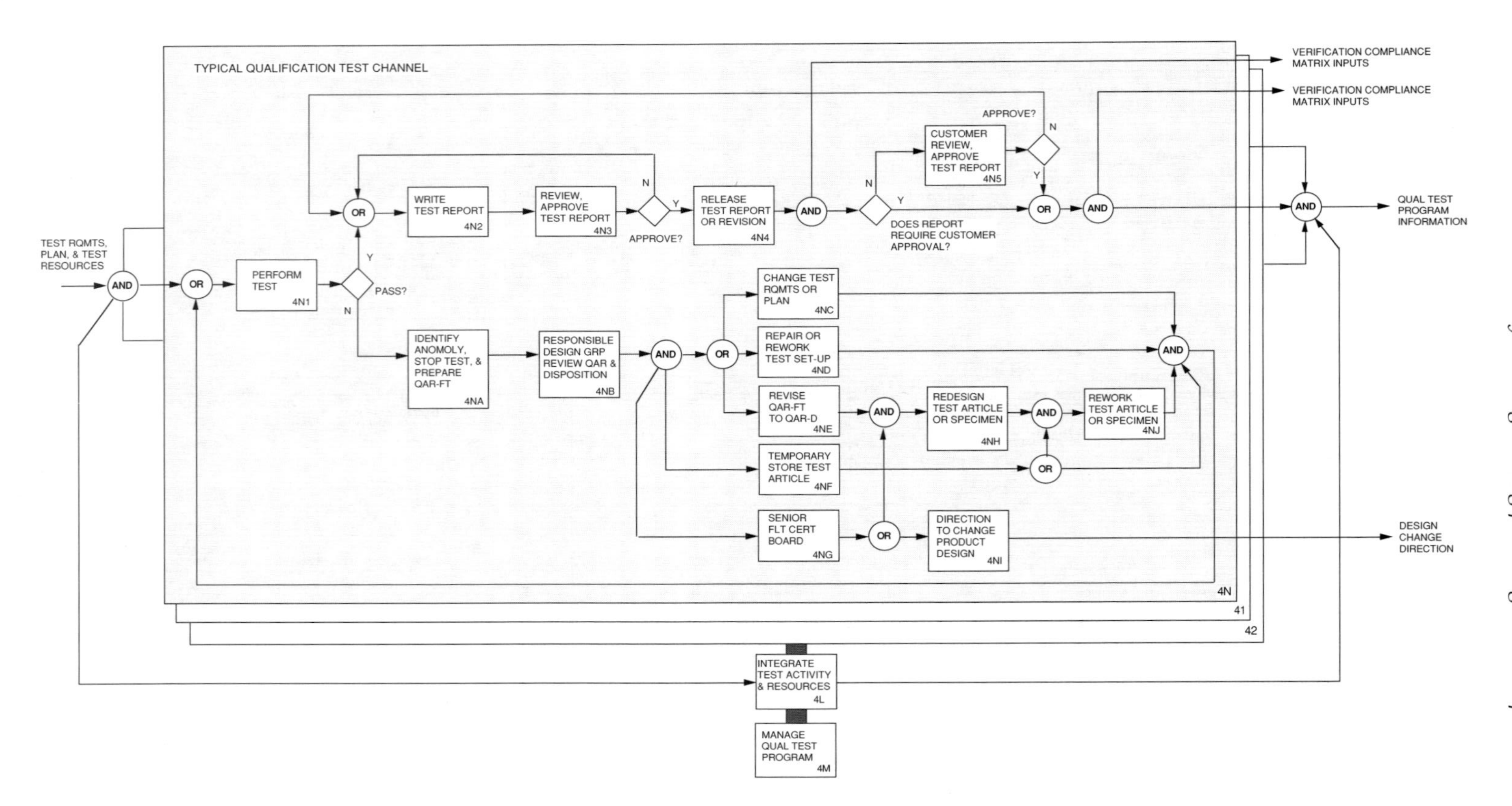

Figure C-14 Qualification testing (Task 4).

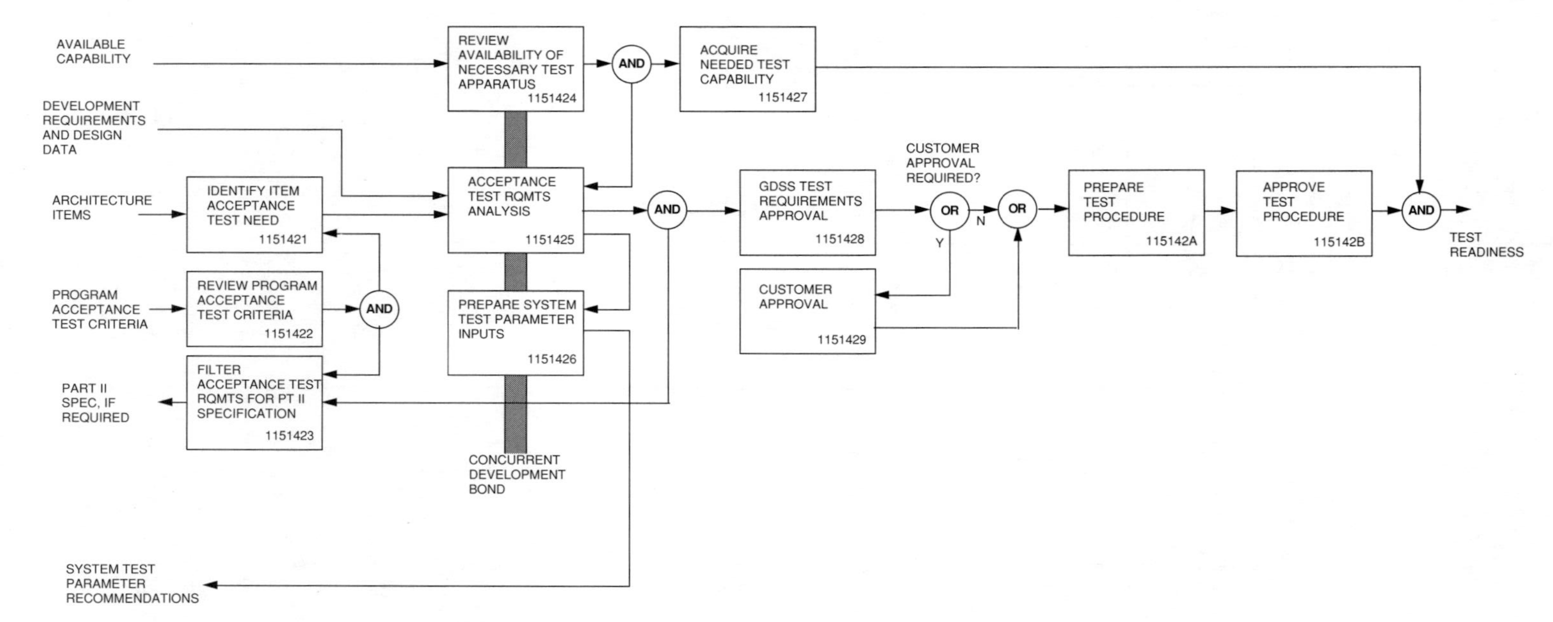

Figure C-15 Item acceptance test development (Task 115142).

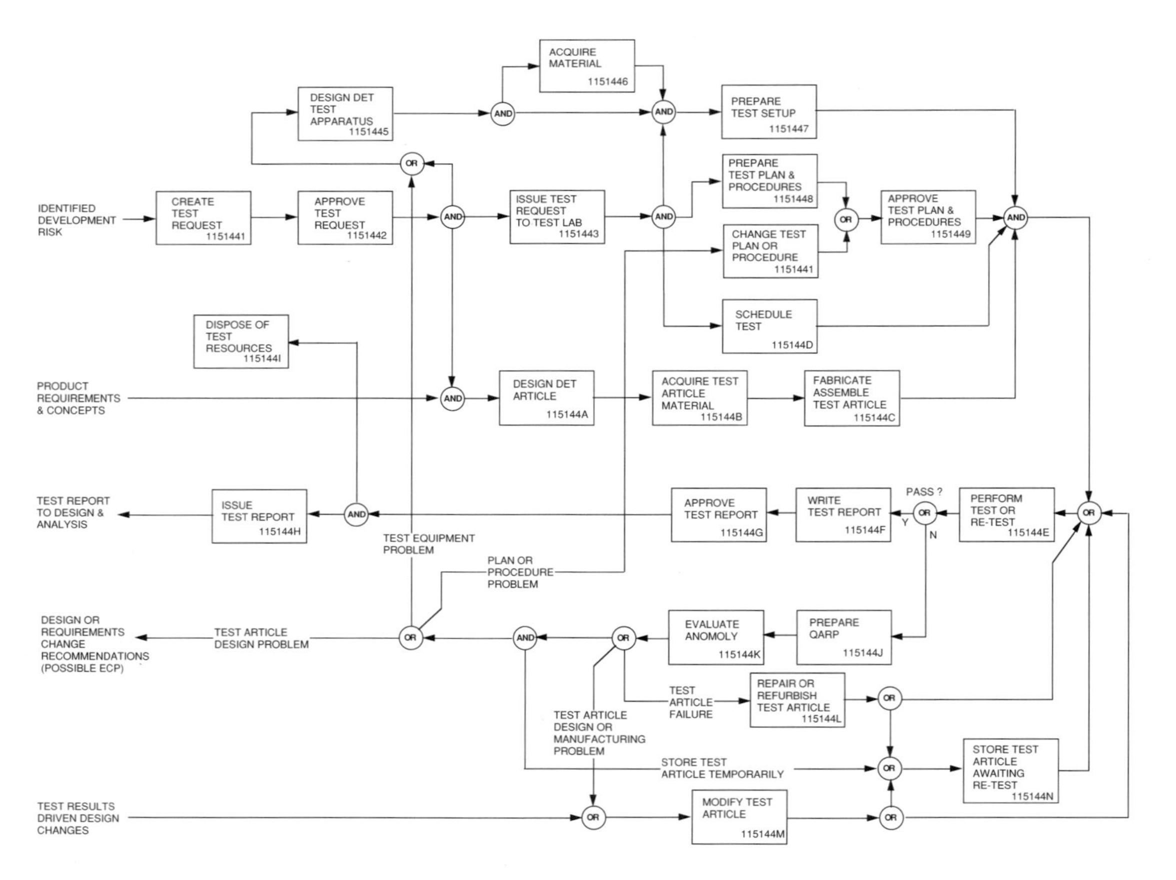

Figure C-16 Item design evaluation testing (Task 115144).

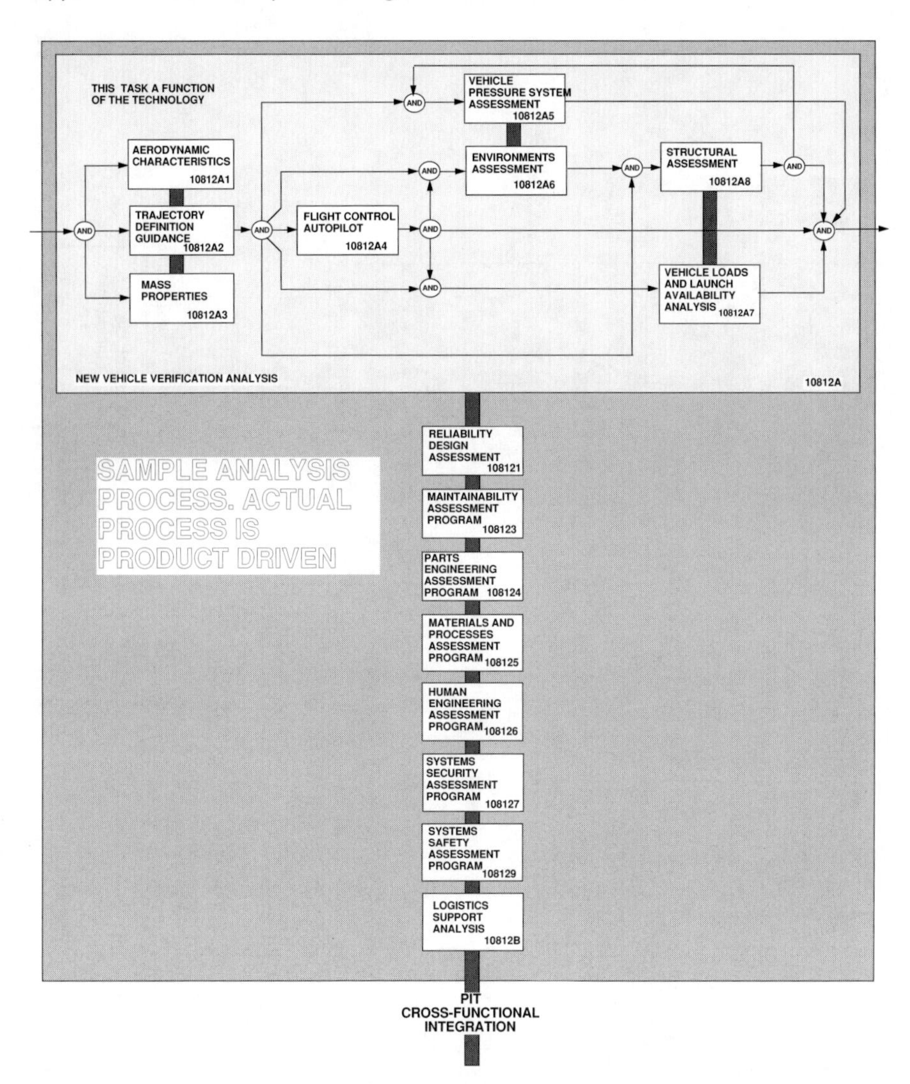

Figure C-17 Systems analysis and specialty engineering assessment (Task 10812).

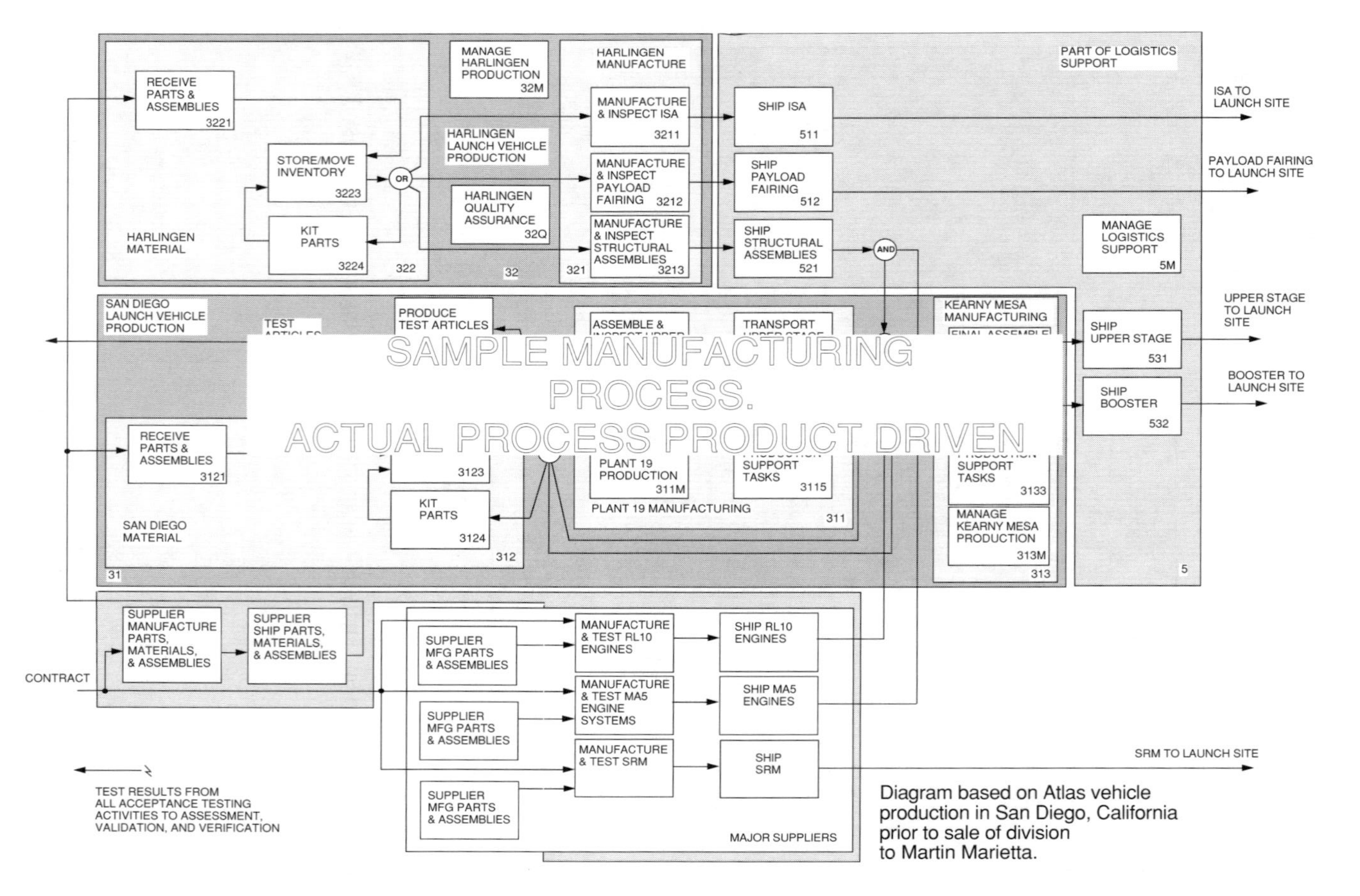

Figure C-18 Production (Task 3).

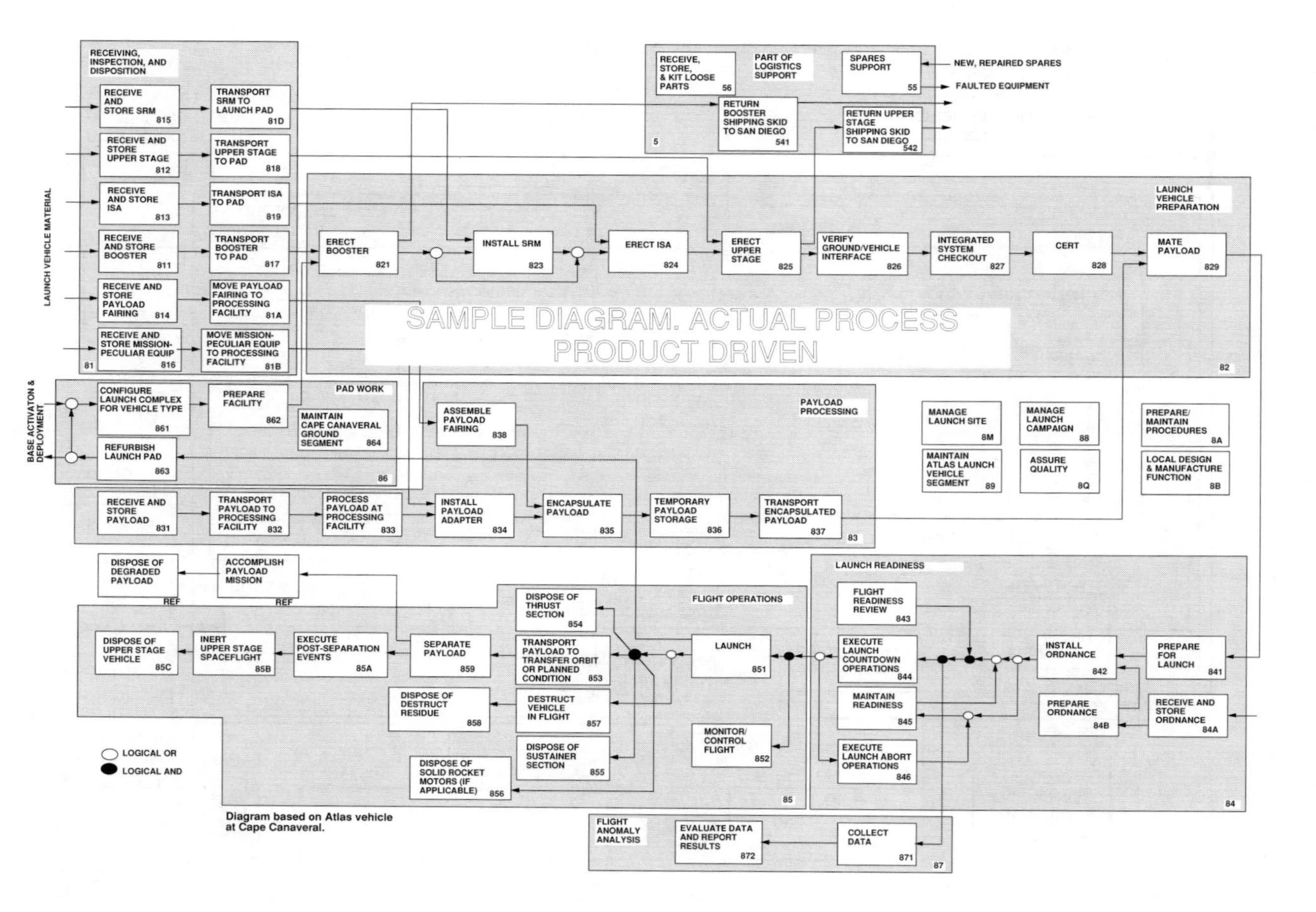

Figure C-19 Operations (Task 8).

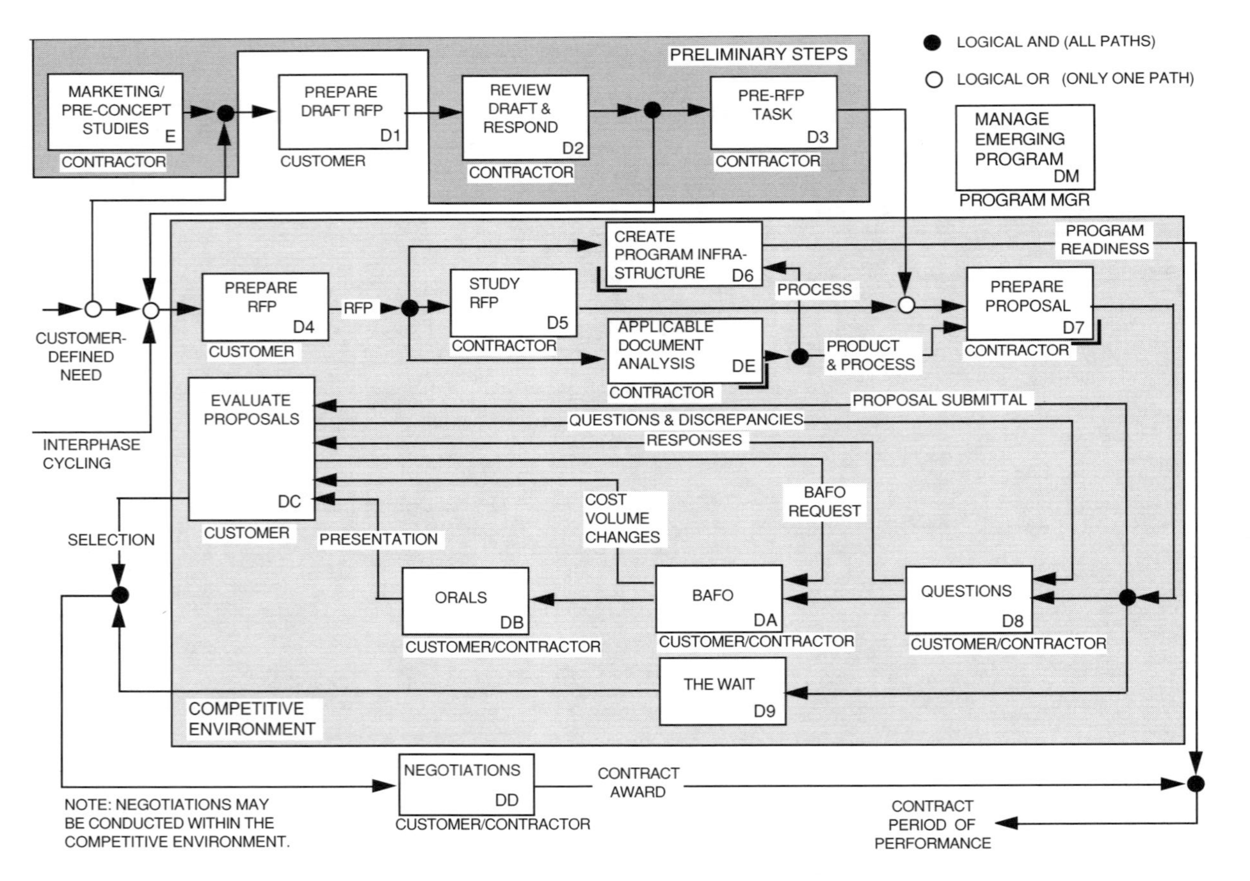

Figure C-20 RFP-proposal cycle (Task D).

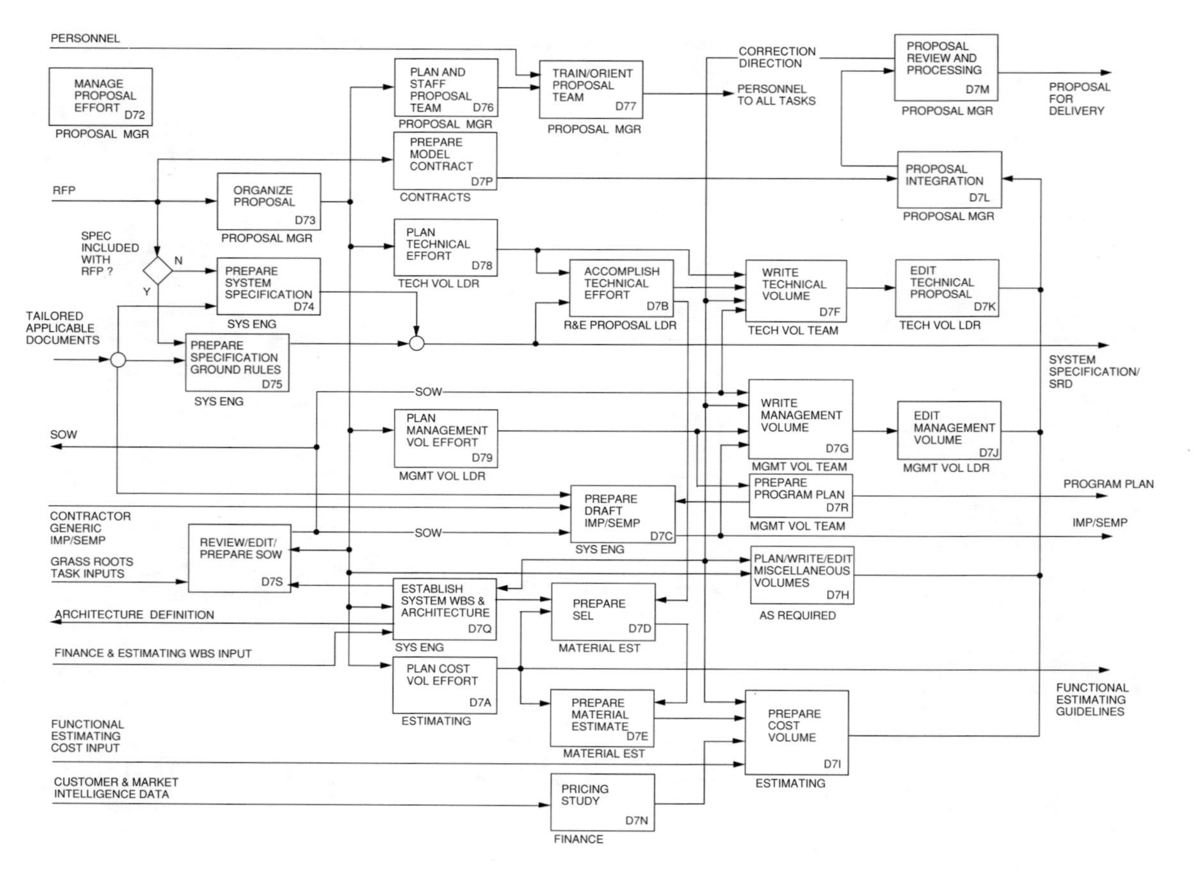

Figure C-21 Prepare proposal (Task D7).

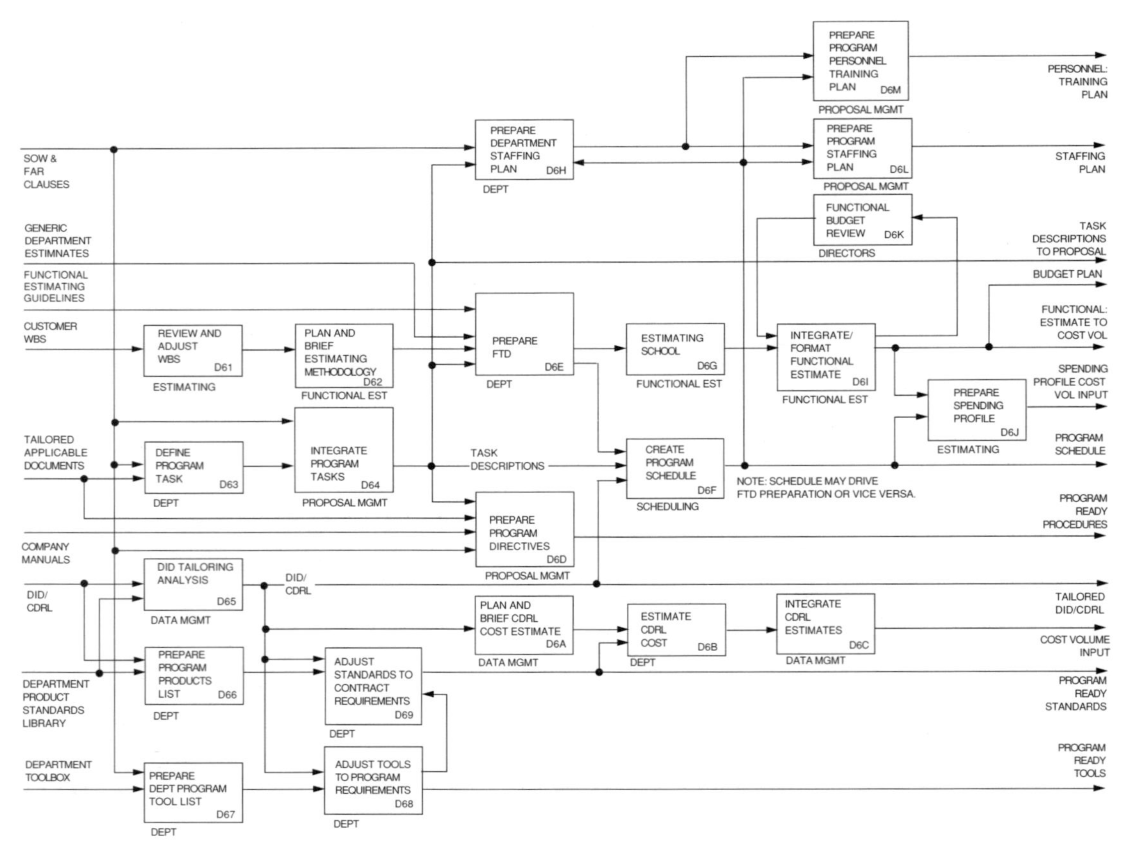

Figure C-22 Create program infrastructure (Task D6).

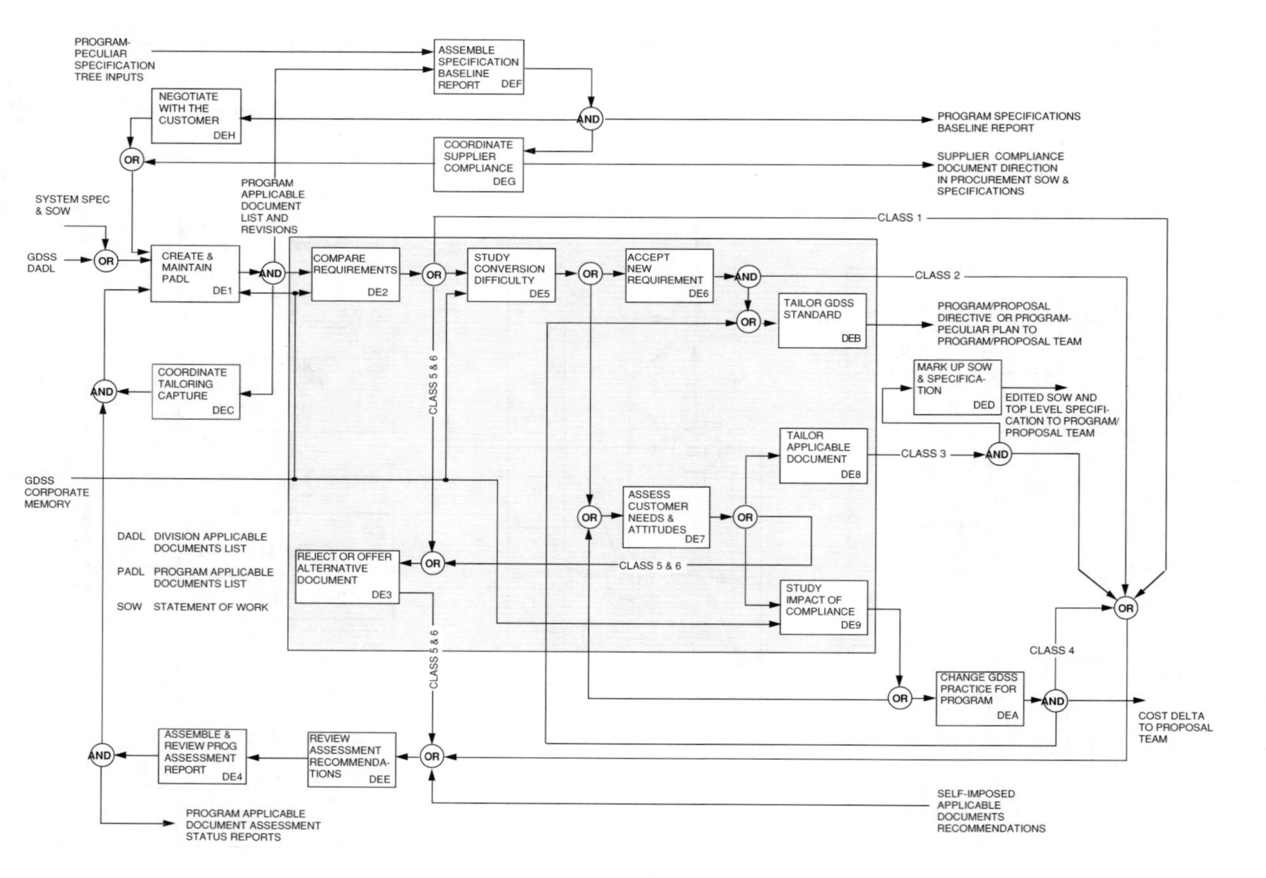

Figure C-23 Applicable document assessment flow.

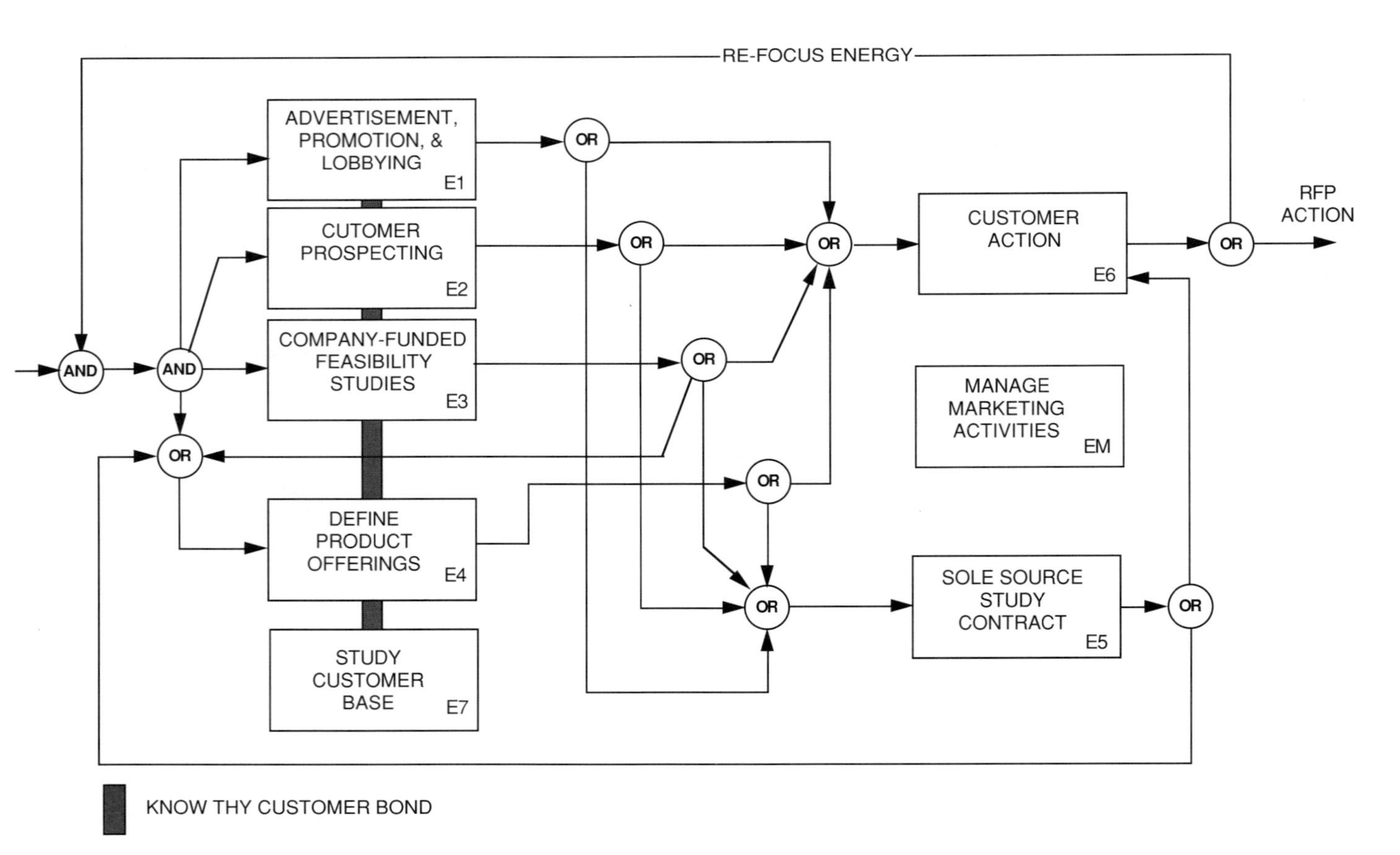

Figure C-24 Marketing and pre-contract studies (Task E).

COMPANY XYZ

FUNCTIONAL
SYSTEM ENGINEERING MANUAL
(SEM)

AND

GENERIC PROGRAM
SYSTEMS ENGINEERING
MANAGEMENT PLAN
(SEMP)

APPENDIX D

TASK DESCRIPTIONS

Appendix D

Task descriptions

D.1 Task list purpose

Table D-1 columns provide information expanding on the identification of tasks shown as blocks in the process diagram included in Appendix C. Each block on a diagram in Appendix C appears as a set of records in Table D-1. Each task requires one or more functional department activities to complete. Each combination of a task and a department constitutes one record in the listing.

D.2 Column headings

The column headings in Table D-1 are explained below:

Task number	The task ID number assigned in the lower right corner of the corresponding task block in Appendix B process diagrams. Acts as a unique identifier of the task.
Dept nbr	Company functional department number responsible for supplying personnel to perform the task on programs. Tasks will be accomplished by PDT on programs but should be accomplished on those teams by persons from the indicated department because they are properly trained in the process. If more than one department is required to accomplish a task, the other department(s) is(are) listed separately and the task descriptions will generally be different to some degree.
Task name	Brief name of the task that appears in the process block included in Appendix C.
Company document	Reference to the company procedure or practice where performance of this task is covered in detail. The document acronym is followed by a section (S) and/or paragraph (P) number. An example of a reference is SEM/SEMP; P3.4.

Source reference	Source references currently listed are as follows: SEM/SEMP Systems Engineering Manual/Systems Engineering Management Plan
Task description	A brief description of the task.
Generic phasing	Identifies the phases shown in Figure 3-1 to which you would commonly apply the task. The Figure 3-1 tasks are listed left to right 1 through 9 in the column. An"X" means it applies and an "O" means it does not.

Table D-1 Enterprise Charter Task List

Task nbr	Dept nbr	Task name	Company document/ Source reference	Task description	Generic phasing
0		Enterprise Management	SPM	The top officer of the enterprise leads and manages the enterprise through a staff composed of leaders of the following functions: (1) business acquisition, (2) Enterprise Integration Team, (3) program management, and (4) functional departments.	XXXXXXXXX
1		Product and Process Development	SEM/SEMP	An organized process for translating a customer need or internally agreed upon commercial initiative into a detailed definition of the product in terms of engineering drawings and reports and a companion process for production and test of that product in terms of plans, procedures, production and quality planning data, initial facilities and materials acquisition, and preparations for deployment, operations, and continued logistic support.	XXXXXXOOO
10		Program Integration Team Activities	SEM/SEMP	Throughout the development period the PIT assigns PDT responsibilities, audits team performance, and manages overall technical effort.	XXXXXXOOO
101		Mission Analysis	SEM/SEMP	Complete mission and operations analysis that might have been begun in proposal period. This includes a logistics basing analysis and a system environment definition as well as a clear definition of one or more design reference mission profiles that will be used as a basis for initial development work.	XXXOOOOOO
103		Plan Program	SPM	Prepare initial WBS, SOW, program plans including SEMP or IMP and IMS or other scheduling media.	XXXXXXOOO
1031		Staff Program	SPM	The PIT coordinates with functional management to identify personnel for assignment to the program.	XXXXXXOOO
1032		Prepare Program CAPs	SPM	The PIT develops cost account plans for PIT and PDTs.	XXXXXXOOO
1033		Define Product Team Structure	SEM/SEMP	The PIT studies the product architecture and determines how best to associate team responsibilities with the architecture. Efforts are made to minimize inter-team interfaces.	XXXXXXOOO
1034		Prepare and Maintain a Program IMP/SEMP	SPM	Adjust planning data created during proposal or create new planning data in the absence of a proposal. Maintain development planning data over the development span. On a plan employing integrated management system approach, planning documentation includes an IMP and IMS referring to a generic SEM/SEMP and other generic planning documents. On a program requiring a program-peculiar SEMP, the generic SEM/SEMP is modified to the minimum extent necessary.	XXXXXXOOO
104		Development Responsibility Map & Implementation Coordination	SEM/SEMP	The PIT publicly identifies items in system architecture that will be assigned to development teams and makes those assignments to initial team leaders. PIT assists the team leaders to staff the team.	XXXXXXOOO
105		System Requirements Analysis	SEM/SEMP	A structured, top-down approach for identifying, specifying, and conceptualizing system elements in a fully coordinated effort.	XXXXXOOOO
10511		Structured Decomposition Analysis	SEM/SEMP	A group of several structured decomposition methodologies that may be used depending on the nature of the system element and the skill and knowledge of the element Principal Engineer. Traditional Functional Analysis and software structured analysis are two methodologies included. The objective of this activity is to understand what the system should consist of in order to achieve its purpose and satisfy the customer need (the ultimate system function). Organized methods are employed to ensure that the system will have all of the functionality needed and that the system development process will create or acquire the physical resources to accomplish that functionality. These processes also develop insights into performance requirements for the elements the functionality is allocated to.	XXXOOOOOO

Table D-1 (continued) Enterprise Charter Task List

Task nbr	Dept nbr	Task name	Company document/ Source reference	Task description	Generic phasing
105111		Functional Analysis	SEM/SEMP	A structured, top-down method for gaining insight into the elements of a system architecture and the performance requirements appropriate for those elements. Analyst lists functions the system must perform to satisfy the customer need and allocates them to architecture elements. Each allocation becomes the attribute fragment for one or more primitive requirement statements for the element to which the function is allocated. This process is repeated downward through the architecture with knowledge of prior allocations. Function identification should be done concurrently with allocation rather than completing the identification process before starting the allocations.	XXXOOOOOO
1051111		Function Identification	SEM/SEMP	A creative activity of thinking about the customer need and functions identified earlier to conceive new lower tier or other functions. This is a listing process and does not attempt to structure the functions in sequence (see 12112). Existing, approved architecture is used as a marker of the current reality.	XXXOOOOOO
1051112		Functional Flow Analysis	SEM/SEMP	Functions are strung into sequences and relative time dependencies explored in an effort to force the systematic identification of all relevant system functions.	XXXOOOOOO
1051113		Timeline Analysis	SEM/SEMP	A time is assigned for each function identified on the functional flow diagram where time criticality is revealed to the analyst. The product may be reported in tabular or diagrammatic form coordinated with the function dictionary. The times developed become timing requirements for the system elements to which the functions are allocated.	XXXOOOOOO
1051114		Prepare Descriptive Function Test	SEM/SEMP	Write descriptive text for function that will become part of the system specification section 3.1 and the Function Dictionary.	XXXOOOOOO
1051115		Capture/Document Function Diagrams	SEM/SEMP	Prepare forma FFD and TLD in the proper reporting context as defined in a CDRL-referenced DID or GDSS procedures.	XXXOOOOOO
1051116		Brief Team	SEM/SEMP	The first time a team applies itself to this functional analysis process, they are briefed by the team leader or a system engineer in how to gain the most benefit and avoid common pitfalls. At each meeting of the team, the team leader briefs the team on the planned activity for that session or series of sessions.	XXXOOOOOO
1051117		Review Prior Allocations	SEM/SEMP	Team members review prior allocations and the concept description applicable to that portion of the system they are about to work on. Functional analysis must be conducted with a progressive approach with increasing knowledge of the architecture and associated concepts to avoid divergence between the product of the functional analysis process and the work of the design community working on concept development. This task provides those doing the functional analysis with an opportunity to refresh themselves on the current concept.	XXXOOOOOO
1051118		Apply Brainstorming Technique	SEM/SEMP	The brainstorming technique calls for two distinctly different phases: (1) idea generation and (2) consolidation. First the team leader states the object of the work to be accomplished. In the idea generation phase, the leader must encourage participation by all members, slavishly capture all inputs from team members, and discourage critical comment from any team member. In the consolidation phase, the leader encourages members to consider the list of ideas, without reference to who originated each, to simplify and organize the list, eliminate conflicting ideas, and evaluate priorities. Applied to function allocation, this technique takes advantage of the different experiences of several people to identify alternative allocations of functions.	XXXOOOOOO

1051119	Consolidate Allocation Statements	SEM/SEMP	The team reaches agreement on the allocations of functions being worked and places them on a Functional Requirements Analysis Matrix (FRAM), or equivalent.	XXXOOOOOO
105111A	Development Review Board	SEM/SEMP	The work of the team responsible for functional analysis is periodically briefed to the Chief Engineer, PIT Leader, or PDT Leader at a DRB for review and approval. The decision-maker selectively approves allocations and directs the system architecture be changed to reflect additions, deletions, and changes. The decision-maker may direct some allocations be re-studied or changed. Where trade studies have been accomplished (10511B) to select a preferred allocation, the trade study recommendations are brought before a DRB for approval.	XXXOOOOOO
105111B	Perform Trade Study	SEM/SEMP	Many function allocations can be accomplished based on good engineering judgment, Division product line experience, historical precedents, or available resources. Where the allocation cannot be easily made, a trade study is commissioned to resolve a preferred allocation from considered alternatives.	XXXOOOOOO
105111C	Consider Company Impact	SEM/SEMP	Given direction by the Chief Engineer, PDT Leader, or PIT Leader at a DRB to reconsider or change briefed allocations, the team considers alternatives and re-enters the analysis process accordingly.	XXXOOOOOO
105111E	Capture/Document Allocations	SEM/SEMP	Incorporate the approved Function Requirements Allocation Matrix (FRAM) content into the program database or functional report.	XXXOOOOOO
105111F	Prepare and Maintain Functional Analysis Report (FAR)	SEM/SEMP	Collect functional flows, timelines, descriptive data, and allocations into a format defined in a CDRL-referenced DID or GDSS standards. Document provides a formal release of a baselined functional analysis product.	XXXOOOOOO
105111G	Define/Refine Measures of Effectiveness (MOE)	SEM/SEMP	Identify and select top level requirements and characteristics for use in making trade study decisions between alternatives.	XXXOOOOOO
105111H	Unprecedented Performance Requirements Analysis	SEM/SEMP	The option is provided to define performance requirements associated with functions prior to their allocation. This option is most appropriate in the case of unprecedented systems to avoid a leap to point designs.	XXXXOOOOO
105112	Process Analysis	SEM/SEMP	The analyst creates an analytical model of system operation and ground processing in block diagram form where the blocks denote processes needed for proper system operation. The lines interconnecting the blocks indicate sequence of operation and flow of principal hardware elements. Existing architecture is mapped to the process blocks and the analyst determines if the processes can be accomplished with those resources. If not, new resources are added to the architecture. This structured decomposition approach is very effective in defining the ground segment after the basic structure of the segment is defined through functional analysis. Process analysis is also a component of LSA, a performance requirements analysis methodology, where the analyst is more interested in human task performance requirements.	OOXXXXOOO
105113	Structured Software Decomposition	SEM/SEMP	Flight Software applies a variation of the Yourdon-Demarco structured software process to flight software needs.	OOXXXOOOO
105114	State Transition Analysis	SEM/SEMP	This is a useful approach for understanding the functionality of a system or element that is not associated with a physical dynamic involving position changes. It encourages identification of specific states and/or modes that are characterized by particular conditions and links these states in possible sequences. State machine or automata theory is useful in applying.	XXXOOOOOO
105115	Hierarchical Functional Analysis	SEM/SEMP	A sub-function list is developed for each function beginning with the customer need statement, sub-functions allocated to architecture, and this pattern is carried downward to lower levels of detail. Functional hierarchy diagram formed from lists. Process tends to develop a one-to-one relationship between a functional hierarchy (problem domain) and an architecture hierarchy (solution domain).	XXXOOOOOO
10512	Mission & Operations Analysis	SEM/SEMP	Operations Analysis leads the effort to expand the customer's need statement into a clear understanding of the effects the system will have on its environment including cooperative, non-cooperative, and hostile systems and the natural environment in the process of	XXXOOOOOO

Table D-1 (continued) Enterprise Charter Task List

Task nbr	Dept nbr	Task name	Company document/ Source reference	Task description	Generic phasing
				accomplishing a specific system mission that supports the customer need statement. The mission analysis report should be stated in the context of system effects and not in terms of the system composition. Subsequent structured analysis will decompose the customer need into an architecture that satisfies the need within the context of the mission definition. Initially, depending on system past history, there may or may not be prior partial knowledge of system architecture. As the system definition process proceeds, the mission analysis may require update within the context of a known system architecture. This task should lead the SRA activity but may continue during the initial SRA implementation. Operations Analysis performs computer simulations to develop figures of merit.	
10513		Basing and Logistics Concepts	SEM/SEMP	Early in the development of the system, alternative concepts are studied for physical and functional location of bases from which to operate and maintain the system and alternative logistics support concepts. Alternatives are evaluated and preferred concepts selected.	XXXOOOOOO
10514		Constraints Allocation	SEM/SEMP	Constraints allocation involves creating a list of constraints that will or may influence one or more system architecture elements and assigning them to the elements. Each allocation (assignment) creates a place holder, or puts a demand on a specialty engineer, for subsequent specialty engineering requirements analysis to identify a constraint of that kind for that element. The constraints allocation mechanism commonly is a design constraints scoping matrix (DCSM).	XXXOOOOOO
10515		Quality Function Deployment (QFD)	SEM/SEMP	A structured methodology for gaining insight into an optimum relationship between customer wants and needs and the requirements for an item. It is an effective method, especially when the voice of the customer comes from several sources, for flushing out customer input on their needs as a part of the process of building the system specification. When applied in its total depth, it also offers a set of features more comprehensive than this involving a series of matrices completed by teams. This additional capability is more finely tuned to product improvement than new product development, however.	XXXOOOOOO
10516		Make Requirements Changes	SEM/SEMP	Alter the requirements previously identified for an item in order to balance requirements across two or more items, respond to higher level requirements, or for other reasons.	OOXXXXOOO
10517		Requirements Flowdown and Margin Management	SEM/SEMP	Manage the identification of lower tier requirements based on higher tier requirements values across several requirements types. Identify requirements to be managed through margins, coordinate assignment of principal engineers for each requirement, coordinate periodic review of margins accounts, and rebalance accounts for optimum condition.	OXXXXOOOO
10518		Product Requirements Analysis	SEM/SEMP	A structured process for identifying the appropriate minimum set of performance requirements and constraints for a system element.	OOXXXOOOO
105181		Performance Requirements Analysis (PRA)	SEM/SEMP	The actual performance of this step is a function of the requirements identification methodology employed. The preferred approach is to follow the discipline of translating each functional requirement allocated to an element into one or more performance requirements for that element. This entails expanding the brief function statement into a complete thought expressed in one or more proper English sentences within the guidelines established by the requirements quality checklist given in the referenced document. These guidelines include the characteristics of quantification, verifiability, and traceability, for example. The process may entail application of Quality Function Deployment at the system level involving consultation with customer and user personnel.	OOXXXOOOO

			The element principal shall lead this effort with support from Operations Research, System Analysis, and other departments.	
105182	Design Constraints Analysis	SEM/SEMP	The identification of appropriate constraints for system elements. This process is accomplished by the analysis and specialty community using specialized techniques. Constraints include environmental, interface, specialty engineering, and physical characteristics requirements.	OOXXXOOOO
1051821	Interface Requirements Analysis (IRA)	SEM/SEMP	For each line on the Schematic Block Diagram the analyst will determine appropriate corresponding requirements that will be levied against the elements on the two ends of the interface.	OOXXXOOOO
1051822	Environmental Requirements Analysis (ERA)	SEM/SEMP	System level natural environmental requirements are determined and then they are allocated to lower tier elements. Commonly, the principal operating system element (vehicle, aircraft, or whatever) is partitioned into environmental zones within which specific environmental requirements apply. When the physical location of a lower tier element is established, the environment corresponding to the zone within which it is located applies to the element. As the nature of the inter-relationship between the evolving system concept and the system natural environment is understood, induced environments are studied and defined by understanding how the system interacts with the natural environment to create other environmental effects that influence the system in ways that the natural environment alone would not.	OOXXXOOOO
1051823	Specialty Engineering Requirements Analysis	SEM/SEMP	Product requirements work accomplished by a predetermined group of specialty engineering personnel based on the product characteristics and customer requirements. Some of these specialists may also have a responsibility to define product development and manufacture process requirements, and this work is covered elsewhere.	OOXXXOOOO
10518231	Reliability Requirements Analysis		A series of tasks that collectively act to define quantitative hardware (and software if covered in the contract) reliability requirements based on a given system level reliability figure of merit.	OOXXXOOOO
105182311	Reliability Modeling (Task 201)		The purpose of this task is to develop a reliability model to be used for making numerical apportionments and reliability predictions from the system through component levels. Reliability personnel create a reliability block diagram from the architecture block diagram and a companion reliability mathematical model reflecting the mission sequences and modes of operation defined on the functional flow diagram or process diagram and system interfaces defined on the schematic block diagram. The reliability block diagram includes all system elements satisfying a reliability criterion for criticality, also established in this task.	OOXXXOOOO
105182312	Reliability Allocations (Task 202)		The purpose of this task is to create quantitative system reliability requirements and allocate or apportion them to lower levels of indenture. Reliability figures for system elements shall be derived from the reliability model by allocating/apportioning parent element reliability to subordinate elements using available reference materials and sound engineering judgment subject to review.	OOOXXXOOO
10518232	Producibility Engineering Requirements Analysis		Manufacturing engineering personnel identify needed product characteristics preferred manufacturing concept and adjust manufacturing concept for operational imperatives.	OOXXXOOOO
10518233	Maintainability Requirements Analysis		Maintainability personnel create a maintainability model selecting system elements from the architecture based on a selection criterion determined in this task. The maintainability figure(s) of merit is (are) determined based on an allocation process from system level figures. If the figure of merit is, for example, Mean Time to Repair (MTTR), the MTTR figures are allocated to lower tier elements in step with the controlled advance of requirements and concept development work regulated by the Chief Engineer. This model is maintained throughout the program phase with adjustments based on trade studies and other sound knowledge of system needs. As the development reaches the third program phase, the principal focus is on predicting maintainability figures of merit based on actual design figures, but there remains some continuing need to readjust the maintainability model.	OOXXXOOOO

Table D-1 (continued) Enterprise Charter Task List

Task nbr	Dept nbr	Task name	Company document/ Source reference	Task description	Generic phasing
10518234		Parts Engineering Requirements Analysis		The number of different parts is minimized and parts standardized. Parts specifications are prepared to control parts to be procured where no current standard exists.	OOXXXOOOO
10518235		Materials and Processes Requirements Analysis		Materials and processes specification content is determined to assure procurement of items appropriate to product needs.	OOXXXOOOO
10518236		Human Factors Requirements Analysis		Product features are defined for optimum human interface.	OOXXXOOOO
10518237		Systems Security Requirements Analysis		The requirements for physical, communications, information, and other forms of security appropriate for the system and elements thereof are identified and quantified.	OOXXXOOOO
105182371		Security Requirements Definition		Study customer-provided threat documentation, operational concepts, mission statements, customer-referenced security requirements documentation, and other source materials and define general and specific system requirements that may be required to adequately secure the system. Study and define needs for security requirements for lower tier system elements that will assure that required system level security is achieved.	OOXXXOOOO
105182372		Prepare Preliminary System Security Concept (PSSC)		Normally the Government customer will perform the analytical work and prepare this document, but GDSS may be called upon to do so. The document describes mission task, operational systems, the operational environment, and personnel/manpower, equipment, and employment issues as they influence security concerns. The security concept that addresses the identified security threats and vulnerabilities is described.	XXXOOOOOO
105182373		Security Threat Definition and Analysis		Specific threats to system security are identified and characterized.	XXXOOOOOO
105182374		Security Training Requirements Analysis		Based on identified threats and concerns, persons who will require training to ensure security success are identified and plans prepared to provide that training is made.	OOXXXOOOO
105182375		Prepare and Maintain a System Security Management Plan		Prepare a formal document that fully describes the planned security tasks required to meet system security requirements, including organizational responsibilities, methods of accomplishment, milestones, depth of effort, and integration of accomplishment with other program engineering, design and management activities, and related systems.	OOXXXOOOO
105182376		Prepare Security Appendix to the Logistics Support Plan		Identify management objectives and procedures associated with security systems and sub-systems.	OOXXXOOOO
10518238		Environmental Impact Requirements Analysis		Any adverse effects of the product system on the natural environment are identified and mitigated to the extent possible. Where appropriate, system characteristics are altered to reduce the environmental impact.	XXXXOOOOO
10518239		Systems Safety Requirements Analysis		Potential risks to life, health, and property damage are identified and evaluated relative to product features. Requirements are identified to preclude adverse safety hazards.	OOXXXOOOO
1051823A		Contamination Requirements Analysis		Potential sources of contamination of the product that may result in decreased performance or failure are identified and ways to preclude their occurrence are studied. Requirements for protection of the product from contamination are defined.	OOXXXOOOO
1051823B		Corrosion Control Requirements Analysis		Potential sources of corrosion effects on product elements are studied and requirements identified to preclude occurrence.	OOXXXOOOO
1051823C		Damage Tolerance Requirements Analysis		Based on the results from hostile action, corrosion control, security, safety, maintainability, and reliability analyses, sources of potential damage to the product are defined and requirements to preclude it are defined. Design features are identified that will permit	OOXXXOOOO

		continued operation of the product in the presence of damage through redundancy and other features.	
1051823D	Health & Safety Requirements Analysis	Adverse impacts on personnel health and physical safety are identified as a result of operation of the product and preventive steps or design features defined to mitigate or preclude their occurrence.	OOXXXOOOO
1051823E	Vulnerability Requirements Analysis	Sources of vulnerability to hostile efforts are studied and requirements identified to reduce risks from these impacts.	OOXXXOOOO
1051823F	Transportability Requirements Analysis	Based on the evolving definition of the employment process, requirements for transportability of product items are defined to ensure the planned process can be performed.	OOXXXOOOO
1051823G	Electromagnetic Compatibility (EMC) Requirements Analysis	Requirements are defined to ensure freedom from interference with system operation from externally created electromagnetic interference and prevention of interference with other systems from internally generated signals.	OOXXXOOOO
1051823H	Mass Properties Requirements Analysis	Weight and center of gravity (CG) requirements for product items are defined and balanced against other item mass properties requirements. The CG of items requiring handling is determined under predetermined anticipated conditions.	OOXXXOOOO
1051823I	Thermodynamics Requirements Analysis	Thermal effects to which product items are exposed and corresponding requirements are studied and defined to protect the product from damage and failure from the presence or absence of heat energy.	OOXXXOOOO
1051823J	Structural Dynamics Requirements Analysis	Strength of materials and the effects of stresses anticipated in different deployed situations are evaluated requirements identified to prevent product damage.	OOXXXOOOO
1051823K	Guidance Accuracy Requirements Analysis	The degree of precision in movement or placement of a product element is defined.	OOXXXOOOO
1051823L	Aerodynamics Requirements Analysis	Items exposed to atmospheric relative motion at significant velocity are studied to define optimum shape for performance and to preclude damage.	OOXXXOOOO
10519	Test Planning Analysis (TPA)	Test and Evaluation integrates the test requirements driven by requirements verification (qualification testing), design evaluation testing, reliability testing, safety testing, and acceptance test needs into an efficient integrated test plan requiring minimum test resources, time, and budget. Special test and flight test article needs are identified and coordinated with the program plans and schedules. This task includes launch site checkout for launch vehicles where GDSS is the system operator, but it does not include operation and maintenance procedures development for systems to be operated by customer personnel. The latter is a logistics technical manual function within task 12612.	OXXXXOOOO
1051H	Integrate Product Requirements	The requirements for an element are developed by many people working somewhat independently in their own field of specialization. The assigned principal engineer must coordinate the set of requirements prepared by supporting analysts to ensure a condition of balance and appropriateness.	OOOXXOOOO
1051I	Requirements Capture and Publication	The traditional specifications group tasks of preparing specification standards (boilerplates), publishing specifications, applicable documents analysis, and requirements baseline maintenance.	XXXXXOOOO
1051I1	Manage Specification Development	Specification types are defined for items in architecture. Specification development and maintenance responsibilities are made clear. All specifications are placed on the program schedule and their timely development tracked.	OOXXXOOOO
1051I2	Requirements Baseline Maintenance	The lead requirements engineer maintains a list of all requirements documents planned for the program and their status. This list acts as an inventory for all of the documents contained in the working electronic media and includes the name and department of the responsible engineer and other information. This task also includes publishing this list together with the list of all of the program applicable documents and associated tailoring in the form of a program Specification Baseline Report.	OOXXXOOOO

Table D-1 (continued) Enterprise Charter Task List

Task nbr	Dept nbr	Task name	Company document/ Source reference	Task description	Generic phasing
1051I3		Prepare Specification Standards		Prepare boilerplate including common material that will be found in every specification. Can be used as basis for creating a program specification through editing and adding item-peculiar requirements.	OOXOOOOOO
1051I4		Generate Specifications		Use one of two methods to generate a specification: (1) text-processing or (2) computer-generation. In the former, a boilerplate is presented to the principal engineer who marks it up for the specific requirements for the element. These changes are entered through word processing and the product reviewed by the analysis and specialty community. Accepted changes are made and the document presented for approval. In the latter, the requirements entered into the computerized requirements system are automatically transferred into a desktop publishing environment where they are combined with boilerplate material. The requirements may be reviewed in the database prior to publication. This task includes the formal review of the completed document in an ERB by competent authority and obtaining approving signatures.	OOXXXOOOO
1051I5		Release, Archive, and Distribute Specifications		Formally approved paper copies (with actual signatures) are formally released by Configuration Management and distributed and retained by the Engineering Vault. Thereafter the documents are available for additional copies from the Vault. As computerization of the specifications process becomes more pervasive, these activities will begin to fade away having been replaced by electronic equivalents.	OOOXXOOOO
1051J		Requirements Audit and Traceability Audit		PIT studies the relationships between the lower tier requirements and system requirements to ensure that all lower tier requirements were properly derived from and traceable to system requirements. Untraceable requirements suggest unnecessary cost. Voids in lower tier requirements suggest unfulfilled customer needs.	OOOXXOOOO
1051L		Applicable Documents Analysis		A team, led by PIT, studies the set of customer-defined applicable documents or develops a GDSS list of applicable documents and tailors them for mutual agreement with the customer to ensure both minimum cost and customer performance requirements are satisfied.	OXXXOOOOO
1051L1		Create & Maintain PADL		A formal list of applicable documents called by program specifications and statements of work is created as the program master from which common tailoring is derived for inclusion in all specifications and SOWs. The list is initially prepared during a proposal effort and thereafter maintained. The list is reviewed and approved as are changes to it.	OXXXOOOOO
1051L2		Compare Requirements		The requirements contained in an applicable document are compared against the corresponding requirements in in-house documentation for the purpose of determining the degree of agreement.	OXXXOOOOO
1051L3		Reject or Offer Alternative Document		Customer proposed applicable documents that are inappropriate are rejected from the list of approved applicable documents and alternative documents identified. Some motivations for this action are: document no longer active replaced by more recent document, document content in conflict with adequate internal practices, and document content not related to product features.	OXXXOOOOO
1051L4		Assemble & Review Program Assessment Report		Study conclusions are collected, integrated into a common format, and prepared in the planned document format. The final report is offered for internal review by functional and program management.	OXXXOOOOO
1051L5		Study Conversion Difficulty		In cases where the customer requirements are valid and inconsistent with company capabilities, the cost and schedule impacts of satisfying the requirements are determined.	OXXXOOOOO

1051L6	Accept New Requirement	Where valid customer requirements are in conflict with internal practices but the internal impact is not excessively adverse, they are accepted into the list of applicable documents.	OXXXOOOOO
1051L7	Assess Customer Needs & Attitudes	Where compliance with customer requirements results in adverse cost and schedule impacts, the basis for customer interest in the requirements is studied to understand if there are suitable alternatives.	OXXXOOOOO
1051L8	Tailor Applicable Documents	Determine how specific applicable documents will be altered to bring about alignment with desired content. Content is deleted, edited, and new material added to bring about alignment.	OXXXOOOOO
1051L9	Study Impact of Compliance	Where customer requirements are valid and not in agreement with internal practices, the precise impact of compliance is determined as a basis for estimating cost and determining program impacts.	OXXXOOOOO
1051LA	Change Company Practice for Program	Standard company practices are changed to agree with necessary customer requirements. Cost and schedule impacts are also determined as a basis for ensuring compatibility between program needs and plans.	OXXXOOOOO
1051LB	Tailor Company Standard	Where company practices must be changed to reflect necessary customer requirements, changes are made to internal practices either for all programs or applicable only to the specific program where the problem is identified.	OXXXOOOOO
1051LC	Coordinate Tailoring Capture	Identify persons responsible for writing tailoring instructions, assign responsibility, and track development of tailoring.	OXXXOOOOO
1051LD	Mark-Up SOW & Specification	Where final applicable document conclusions are in conflict with customer supplied SOW and system specification listings, those documents are edited for agreement with final conclusions.	OXXXOOOOO
1051LE	Review Assessment Recommendations	Final recommended applicable document list and tailoring is reviewed for approval internally.	OXXXOOOOO
1051LF	Assemble Specification Baseline Report	Prepare final report.	OXXXOOOOO
1051LG	Coordinate Supplier Compliance	Interact with planned and potential suppliers to gain agreement on customer requirements and tailoring thereto.	OXXXOOOOO
1051LH	Negotiate With Customer	Discuss differences between customer list of documents and tailoring and the final or interim list developed internally.	OXXXOOOOO
1051M	Manage Requirements Analysis	This task provides leadership and management oversight over the complete SRA process including schedules, budgets, and product quality. PIT will be assigned to perform this task.	OXXXXXOOO
1051N	Verification Integration	Integrate the verification efforts of the PDTs and accomplish system level verification planning activities. Includes development and maintenance of a verification compliance matrix.	OOOXXXOOO
1052	Schedule Requirements Analysis	Establish timing for beginning requirements analysis work, preparation of specifications, and completion of specifications.	OOXXOOOOO
1053	Cost Requirements Analysis	Identify cost goals for product items in the context of a design-to-cost and/or life cycle cost structure. Coordinate changes in the cost distribution to balance requirements burdens between teams and across system items.	XXXXOOOOO
10531	Design To Cost Requirements Analysis	The Design To Cost (DTC) requirements are determined by Economic Analysis. This is a process of decomposing the system DTC goals into appropriate values for system elements while setting aside margins to accommodate development risk abatement.	OOXXOOOOO
10532	Life Cycle Cost Requirements Analysis	The cost of acquisition is combined with the Operations and Maintenance (O&M) cost to project the total cost of the system over the life of the system. This cost figure is maintained as system characteristics change in the early system development phases.	XXXOOOOOO
1054	Operations & Deployment Planning Analysis (DPA)	On systems operated by the customer (blue suiters) involving deployment of the system to multiple sites, DPA provides a disciplined approach to understanding deployment requirements and what elements they influence (are allocated to). The products are deployment performance requirements fed into the requirements sets for the appropriate architecture elements and program planning reports pertaining to deployment scheduling, site	OOXXOOOOO

Table D-1 (continued) Enterprise Charter Task List

Task nbr	Dept nbr	Task name	Company document/ Source reference	Task description	Generic phasing
				preparation planning and work, customer operations and maintenance training scheduling, and supply support planning. Operations requirements are identified based on development and analysis of process flow diagrams.	
1055		Quality Requirements Analysis		Identify quality product and corresponding process requirements. Plan quality assurance program and identify needed process characteristics.	XXXXXOOOO
1056		Material Requirements Analysis		Identify requirements for raw and processed materials needed for development and manufacturing of product. Coordinate product requirements with material related processes. In cooperation with design and manufacturing personnel on PIT define material sources (make/buy plan).	OOXXXOOOO
1057		Manufacturing Requirements Analysis (MRA)		The requirements for the manufacturing process are developed simultaneously with the requirements for the product. These requirements sets should be interactively developed to introduce the most cost-effective features on the two sides of the system development process: product and process. Where manufacturing advantages encourage adoption of operations system components features and it can be shown to be cost effective and not detract adversely from system performance, they may be accepted by the Chief Engineer. Likewise, advantageous operating system features may drive process design in terms of tooling selection, facility station sequence, and exert other influences.	OOXXXOOOO
1058		Logistics Support Requirements Analysis (LSRA)		LSA is a systematic and comprehensive analysis on an iterative basis through all phases of the system life cycle to satisfy supportability objectives. The complete program is composed of requirements analysis included under this heading, planning activity (see task 1MM), design assessment, and design verification activity.	XXXXXOOOO
10581		Logistics Program Planning and Control (Task 100)	ILSP	Accomplish logistics program planning and set strategy for program development.	XXXOOOOOO
105811		Development of an Early LSA Strategy (Task 101)	ILSP	Prepare potential supportability objectives for the new system/equipment and identify proposed LSA tasks and subtasks to be performed early in the acquisition program.	XXXOOOOOO
105812		Logistics Support Analysis Plan (Task 102)	LSP	Prepare an LSA Plan which describes how the LSA program will be conducted to meet program requirements. The LSA Plan may be part of the Logistics Support Plan or a separate document.	XXXOOOOOO
105813		Program and Design Reviews (Task 103)	LSP	Establish and document design review procedures for logistics reviews and provide logistics inputs to planning for major general reviews.	XXXOOOOOO
10584		Mission & Support Systems Definition (Task 200)		Establish supportability objectives and supportability related design goals, thresholds, and constraints through comparison with existing systems and analyses of supportability, cost, and readiness.	XXXOOOOOO
10585		Use Study (Task 201)		Identify and document the pertinent supportability factors related to the intended use of the new system. Factors to be considered include mobility requirements, deployment scenarios, mission frequency and duration, basing concepts, anticipated service life, interactions with other systems, operational environment, and human capabilities and limitations.	XXXOOOOOO
10586		Mission H/W, S/W, & Support Standardization (Task 202)		Identify existing and planned logistic resources which have potential benefits for use on each system/equipment concept under consideration.	XXXXOOOOO
10587		Comparative Analysis (Task 203)		Identify existing systems and subsystems (hardware, operational, and support) useful for comparative purposes with new system/equipment alternatives. Select or develop a baseline	OOXXOOOOO

		comparison system (BCS) for use in comparative analyses and identifying supportability, cost, and readiness drivers of each significantly different new system/equipment alternative. Determine the O&S costs, logistics support resources requirements, reliability and maintainability (R & M) values, and readiness values of comparative systems identified.	
10588	Technological Opportunities (Task 204)	Identify and evaluate design opportunities for improvement of supportability characteristics and requirements in the new system/equipment.	OOXXOOOOO
10589	Supportability and Supp. Related Design Factors (Task 205)	Establish quantitative supportability goals resulting from alternative design and operational concepts and establish supportability and supportability related design objectives, goals, and thresholds, and constraints for new system/equipment for inclusion in program approval documents, system/equipment specifications, other requirements documents, or contracts as appropriate.	OOXXOOOOO
1058A	Preparation and Evaluation of Alternatives (Task 300)	Optimize the support system for the new item and develop a system which achieves the best cost, schedule, performance, and supportability.	OOXXOOOOO
1058A1	Functional Requirements Identification (Task 301)	Logistics participation in the functional analysis process focusing on supportability characteristics.	OOXXOOOOO
1058A2	Support System Alternatives (Task 302)	Alternative logistics concepts are developed and refined supportive of defined functional needs.	OOXXOOOOO
1058A3	Evaluation of Alternatives and Tradeoff Analysis (Task 303)	Logistics participation in trade studies and performance of logistics trades to gain insight into preferred solutions from among alternatives.	OOXXOOOOO
106	Architecture Synthesis	This is a structured process of identifying the architectural relationships of system elements identified in the process of allocating system functions. A block is added to the architecture block diagram, it is determined whether the element is a configuration item and what kind of specification (including none) will be prepared. The element is also assigned to a specific R&E design function as the responsible design agent determines whether it is to be internally designed or procured.	XXXXOOOOO
1061	Architecture Block Diagramming	System elements are assembled into a hierarchical block diagram based on their functional relationships and the organization assigned development responsibilities. This process is interactive with the WBS creation as well as specification and configuration item identification in early stages.	XXXXOOOOO
1062	Configuration Item Analysis	Each system element is screened against a set of selection criteria to determine if it should become a configuration item. The customer will manage the development program through the item selected as configuration items. On commercial programs, this process identifies internal GDSS End Items through which our Program Office will manage the program.	XXXXOOOOO
1063	Prepare and Maintain WBS	Normally the customer will define the WBS based on their view of the application of MIL-STD-881 guidelines to the anticipated program and their program office's needs for visibility. The WBS includes both product and task components and we must attempt to steer the WBS product component into reflecting our view of the system architecture. This is not always successful due to inflexible computer programs used to implement MIL-STD-881 guidelines and it may become necessary to carry two different structures related by a mapping.	XXXXOOOOO
1064	Specification Tree Analysis	Each element on the architecture block diagram is screened for suitability for formal requirements capture in a specification. If an element matches one of the several criteria sets, it is added to the specification tree and a responsible party identified to lead requirements analysis and concept development activities.	XXXXOOOOO
1065	Interface Analysis	The need for interfaces between the system elements is studied to determine the most advantageous allocation of system functions in order to minimize system interfaces. The principal tool used is N-Square diagrams. This task is interactive with schematic block diagram construction.	XXXXOOOOO

Table D-1 (continued) Enterprise Charter Task List

Task nbr	Dept nbr	Task name	Company document/ Source reference	Task description	Generic phasing
1066		Schematic Block Diagramming		The elements of the architecture block diagram are illustrated so as to identify their interface relationships. Each line on the schematic block diagram identifies an interface between two elements. Later steps in the SRA process identify the requirements and concepts for these interfaces. This task is interactive with interface analysis.	XXXXOOOOO
1067		Make Database Entries		If a database is employed to retain functional and architecture definition data, entries are made in that database and thereafter maintained in accordance with decisions reached during continuing analysis.	XXXXOOOOO
1068		Map Interfaces to Functions		The interfaces defined on the schematic block diagram are mapped to the functions depicted on the functional flow diagram by virtue of the architecture elements shown to be effective in each function on the Function/Architecture Map (FAM). The result is a Function/Interface Map (FIM). This product provides insights into interface criticality across the system functions.	XXXXOOOOO
1069		Prepare & Maintain System Description Document (SDD)		Edit the Division SDD boilerplate into a program-peculiar document, establish content needed for the program, and acquire and create initial document content. As the system concept matures, document content is refined and expanded. Minimum content includes: Customer Statement of Need, Architecture Block Diagram (ABD), Schematic Block Diagram (SBD), System Functional Flow Diagram (FFD), Architecture and Interface Dictionaries, and design concept data on each element of the system down to the configuration item or end item level. Concept data should include one or more sketches of the element, a list of salient features and quantified values for key parameters. Data is accepted into the SDD as a result of an ERB approving concepts or optimization data. The SDD concept data in combination with the requirements defined for the element, and the results of any design evaluation tests and analyses or trade studies, should be sufficient input to satisfy a designer's needs for information in preparation for preliminary design. The SDD is maintained only until CDR at the latest. Thereafter, adequate descriptive data is available in drawings and reports to clearly define the system. The program may choose to maintain a subset of the document as a high level information package for new program personnel and other briefing purposes.	XXXXOOOOO
106A		Map Architecture to Functions		The analyst studies the functional flow diagram and determines what architecture elements are required to accomplish each one. A matrix is formed called a Function/Architecture Matrix (FAM). This is a cross check to ensure that all functions have been allocated correctly. The analyst may find it necessary to stimulate re-evaluation of functions based on findings here.	XXXXOOOOO
107		System Requirements Review		Customer is presented with system level requirements developed during initial program period. Approval of requirements is needed before the development team undertakes to develop one or more suitable concepts and selection of a preferred concept for review by the customer.	OXOOOOOOO
108		System Integration		Cooperative interaction between design principal engineer and the specialty engineering and analysis community during early program phases to evolve architecture element concepts compliant with requirements. This is accomplished through product development teams for selected architecture elements and principal engineer led efforts otherwise to coordinate the evolving concept and specialty views of the concept. Where performance, cost, schedule, or technology risks are identified, they are acted upon through a risk management process which may entail conducting of technology searches or demonstration.	OOXXXOOOO

1081	Engineering Integration Component		Engineering functional department component of PIT activity. Personnel assigned to these tasks normally drawn from the engineering department.	OOXXXOOOO
10811	System Optimization and Interface Integration		Generally, the detailed work of a small team focusing on developing an acceptable concept for one system element will focus too narrowly resulting in possible sub-optimization. This possibility is overcome by studying the resulting overall system composition periodically to ensure that it is optimum.	OOXXXOOOO
10812	System Analysis & Specialty Engineering Assessment		The analysis and specialty engineering communities review preliminary and detailed design products (drawings and analysis) to determine if the performance, specialty requirements, and other constraints have been satisfied. Reports are prepared to record the results of these analyses.	OOXXXOOOO
108121	Reliability Design Assessment	LSP	During preliminary and detailed design phases, reliability personnel monitor design progress as directed by the Chief Engineer and work closely with designers developing element designs from element concept or preliminary design and requirements data. This is an interactive process with the reliability engineer assertively ensuring that the designer understands reliability requirements and provides, where necessary, sound examples of satisfactory synthesis of reliability requirements. As the design matures, the reliability engineer studies design features from several perspectives to assure that it does satisfy reliability requirements. Cases of failure to meet requirements are reported back to the designer for resolution, or the Chief Engineer if not acted upon.	OOXXXOOOO
1081211	Reliability Predictions (Task 203)		The purpose of this task is to estimate the reliability of the system and to determine if contractual reliability requirements can be achieved with the proposed design. The preliminary or detailed design of each mission-critical element identified in the mathematical model is predicted based on its design, components/parts used, and environmental factors. Trade studies may be required to study possible reallocation of reliability, cost, and/or maintainability figures to provide a mix that will yield to design solution and support system requirements.	OOXXXOOOO
1081212	Failure Modes, Effects, and Criticality Analysis (Task 204)		The purpose of this task is to determine all possible failure modes and their effects on mission success through a systematic analysis of the design. The analysis is intended to identify needed reliability improvements in a timely manner and to foster interchange of design information with other program activities such as system safety, instrumentation, test, and other reliability analyses. The task is accomplished by analyzing in an organized way each element of a design for possible failure modes and then to evaluate system performance for each one of those failure eventualities. Where part failures lead to mission failure, alternative design solutions, redundancy, procedure changes, and other techniques are reviewed for applicability. It is common to design for single-point failure tolerance in unmanned systems. Man-rated systems may require two-fault tolerant design greatly complicating this reliability analysis.	OOXXXOOOO
1081213	Sneak Circuit Analysis (Task 205)		The purpose of this task is to identify latent paths which may cause occurrence of unwanted functions or inhibit desired functions, assuring all components are functioning properly. The analytical process may be applied to hardware, software, or combinations of both.	OOXXXOOOO
1081214	Electronic Parts/Circuits Tolerance Analysis		The purpose of this task is to examine the effects of part and circuit parameter tolerance and parasitic parameters over the range of specified operating life and conditions and to ensure compliance to approved parts derating criteria.	OOXXXOOOO
1081215	Reliability Critical Items (Task 208)		The purpose of this task is to identify and control those items which require special attention because of complexity, application of state-of-the-art techniques, anticipated reliability problems, or the impact of potential failure on safety, readiness, and mission success. An item is identified as a critical item if it contains one or more single point failure modes. Other criteria are established based on program, product, and customer concerns.	OOXXXOOOO

Table D-1 (continued) Enterprise Charter Task List

Task nbr	Dept nbr	Task name	Company document/ Source reference	Task description	Generic phasing
1081216		Environmental Stress Screening (Task 301)		Parts not complying with customer requirements are subjected to special testing that stresses parts to a level that marginal parts will fail and be replaced prior to delivery avoiding a not uncommon higher failure rate due to infant mortality early in the life of an item.	OOXXXOOOO
1081217		Reliability Development Growth Test Program (Task 302)		Conduct pre-qualification testing (also called test analyses and fix or TAAF) to resolve the majority of reliability problems early in the development phase prior to the start of qualification testing.	OOXXXOOOO
1081218		Reliability Demonstration or Qualification Test (Task 303)		Special tests are planned and implemented to demonstrate that planned reliability is achieved.	OOXXXOOOO
1081219		Production Reliability Acceptance Test (PRAT) Program		Conduct testing in the production environment to assure that the reliability of the hardware is not degraded as a result of changes in tooling, processes, work flow, design, parts quality, or other characteristics.	OOXXXOOOO
108121A		Effects of Normal Use (Task 209)		Task formally called "Effects of Functional Testing, Storage, Handling, Packaging, Transportation, and Maintenance". Determine the effects of these activities on item reliability.	OOOOOXXOO
108123		Maintainability Assessment Program		The maintainability consequences of design solutions are studied for compliance with maintainability requirements.	OOXXXOOOO
108124		Parts Engineering Assessment Program		The parts consequences of designs are reviewed for compliance with parts engineering standards created for the program and other parts requirements. Designs are screened for non-standard parts that can be replaced from the standard parts list.	OOXXXOOOO
108125		Material & Processes Assessment Program		The materials and processes consequences of design solutions are studied for compliance with requirements. Designs are screened for use of non-standard materials and processes that can be replaced by standard ones.	OOXXXOOOO
108126		Human Engineering Assessment Program		The human engineering consequences of design solutions are studied and compared with requirements and human capabilities. Alternative design solutions better matched to human features and capabilities are proposed.	OOXXXOOOO
108127		Systems Security Assessment Program		The safety consequences of design solutions are studied and compared with safety requirements.	OOXXXOOOO
1081271		Design Security System		Synthesize the security system requirements resulting in a preliminary design of the subsystems that comprise the overall security system. This may include component screening in association with Parts Engineering, component response analysis, engineering test planning support, and subsystem verification analysis planning. The design work for the actual physical system elements will be accomplished by the design departments taking advantage of System Security work identified here. Many system elements will include security components that the responsible design function may not fully comprehend the security significance of.	OOXXXOOOO
1081272		Subsystem and System Response Analysis		This analysis consists of threat rejection logic, detailed adversary modeling, subsystem response modeling, subsystem qualification and verification testing planning support, system verification analysis, system response modeling, and system response analysis.	OOXXXOOOO
1081273		Conduct a Preliminary Security Vulnerability Analysis		Study the preliminary baseline design in early program phases including: a. Identify logical security vulnerabilities of the system in its projected operational environments and address general threats stated in the SON and threat definition documents.	OOXXXOOOO

			b. Define security system functional requirements which may effectively secure the system from exploitation. Choose candidate safeguard configurations to mitigate or reduce identified vulnerabilities.	
1081274	Threat Assessment and Advisory Mission Analysis		Identify and evaluate threats to the system achieving its mission purpose and provide mission characteristics inputs based on this assessment.	OOXXXOOOO
108129	Systems Safety Assessment Program		The safety consequences of design solutions are evaluated and compared with safety requirements.	OOXXXOOOO
10812A	New Product Validation Analysis		Highly technical analytical disciplines are applied to evolving design solutions in support of assuring that those design solutions are consistent with requirements and the laws of nature and that they are feasible.	OOXXXXOOO
10812A1	Aerodynamics Characteristics		Aerodynamic characteristics of the evolving design solution are compared with aerodynamic requirements and guidelines through analysis and possible wind tunnel experimentation. Undesired features are discouraged and better solutions offered.	OOXXXXOOO
10812A2	Trajectory Definition Guidance		The optimum and possible paths to be followed by the evolving product design during operations are determined and compared with mission scenarios.	OOXXXXOOO
10812A3	Mass Properties Assessment		The predicted mass properties characteristics of the evolving design are determined through analysis and estimation and comparisons made with required characteristics. Where excess weight is detected, a flag is raised to respect required values. Where weight savings can be made, it is placed in margins or re-distributed to other designs to solve problems. The weights statement is maintained.	OOXXXXOOO
10812A4	Flight Control Autopilot			OOXXXXOOO
10812A5	Vehicle Pressure System Assessment		Evolving product item pressurization needs and capabilities are studied and compared with pressurization requirements and with other product features.	OOXXXXOOO
10812A6	Environments Assessment		The evolving design and its environmental interface is studied for compliance with environmental requirements and environmental impact requirements.	OOXXXXOOO
10812A7	Product Loads			OOXXXXOOO
10812A8	Structural Assessment			OOXXXXOOO
10812B	Logistics Support Analysis		Product-oriented logistics analysis is conducted to assure that design solutions satisfy logistics requirements.	XXXXXOOOO
10812B4	Determination of Logistic Sppt Resource Rqmts (Task 400)		Summary level task. See lower tier tasks.	XXXXXOOOO
10812B41	Task Analysis (Task 401)	LSM	Analyze each operation and maintenance task identified on the process flow diagram. Determine procedural steps required, logistics resources required, task parameters (frequency, interval, elapsed time, and manhours), and maintenance level. Analysis results prepared in logistics support analysis reports (LSAR).	OOXXXOOOO
10812B42	Early Fielding Analysis (Task 402)	LSM	Analyze the effect of introducing the new system into the environment defined by existing systems. Determine effects on customer manpower and how to acquire additional personnel or discontinue support for existing staff. Define the impact of failing to acquire the necessary logistics support resources as an input to program risk assessment. Develop plans to mitigate identified risks.	OOXXXOOOO
10812B43	Post-Production Support Analysis (Task 403)	LSM	Assess the expected life of the system/equipment. Identify support items associated with the system/equipment that will present potential problems due to inadequate sources of supply after shutdown of production lines. Develop a plan that assures effective support during the life of the system.	OOXXXOOOO
10812B5	Supportability Test, Evaluation, and Verification (Task 501)	LSM	Formulate a test and evaluation strategy to assure that specified supportability requirements are achieved for introduction into verification plans. Provide logistics specialty component for verification program.	OOXXXOOOO

Table D-1 (continued) Enterprise Charter Task List

Task nbr	Dept nbr	Task name	Company document/ Source reference	Task description	Generic phasing
10815		Risk & Technology Management		An organized process for constantly or periodically reviewing the program for the existence of risk in terms of cost, schedule, technical (ability to satisfy technical requirements), and technology concerns. Each potential concern is evaluated to determine if a risk condition exists. If a risk is identified, it is assigned to a principal engineer to develop an abatement action plan. Program status with respect to the action plan is reviewed periodically and direction given by management on completing risk abatement actions.	XXXXXXOOO
108151		Technical Performance Measurement		A specific list of requirements against a specific set of architecture items is selected using good engineering judgment and program office and customer guidance and direction. Key parameters are selected that have a lot of leverage across the system and in which system problems will invariably be detected. Each of these parameters is assigned to a principal engineer who is required to maintain track of the value for this parameter and future plans related to causing it to remain within or converge to a required value range. The current value, planned future actions, and trends for each parameter are reviewed periodically and action taken to understand and correct out-of-bounds conditions or trends. Where serious problems are identified, the parameter is established as a technical risk and managed within that environment until the risk is abated and the parameter is no longer a program risk.	OOXXXXOOO
108152		Technology Management		Efforts to understand needed technologies and acquisition of those technologies are coordinated.	XXXXXOOOO
108153		Technology Availability Assessment		Maintain contact with authoritative sources of technology information. Determine availability of specific technologies of potential or real value in association with development of product items.	XXXXXOOOO
108154		Risk Analysis		A continuous process of looking for program risks in several areas produces potential risks or concerns. Each of these items is investigated to determine if a valid risk condition exists.	XXXXXOOOO
108155		Risk Mitigation Evaluation		Alternative risk mitigation strategies are evaluated and a specific path or paths is/are selected.	XXXXXOOOO
108156		Risk Tracking & Documentation		The status of all active risks are tracked and risk principal engineer progress compared with his/her action plan. Where lagging performance is detected, causes are identified and corrective action implemented. Periodic Risk Management Board meeting reviews status and performance of principal engineers in abating risks in accordance with their plans.	XXXXXOOOO
108157		Technology Demo		Where it is uncertain whether the technology is available for a given design concept, a technology demo may be conducted to demonstrate that one or more technologies are available to satisfy the need. This may involve original research, analytical work, design, fabrication, and testing of physical hardware items, development and testing of computer code, or combinations of these activities. Alternatively, the team may discover from the technology demo that a technology is not available for the concept leading to alternative concepts, technology search, or technology development.	XXXOOOOOO
108158		Technology Search		An engineer or analyst familiar with a particular technology field searches available sources for a technology of potential application to satisfy an unfulfilled need on the program.	XXXXXOOOO
1081M		Manage Integration Work		PIT manages PIT integration work and coordinates PDT integration work to achieve program integration goals.	OXXXXOOOO
1081M1		Systems Integration Planning Activities		Prepare integration plans.	OXXXOOOOO

1081M2	Prepare and Maintain TPM Plan	Identify candidate TPM parameters and encourage closure on a specific list of parameters. Assign parameters to parameter principals. Track development and maintenance of TPM data. Schedule TPM reviews and coordinate reviews.	OXXXOOOOO
1081M3	Prepare Program Trade Study Plan	Prepare a list of needed trade studies and define how those studies will be conducted. Prepare document.	OXXOOOOOO
1081M4	Interface Management Planning Activities	Prepare interface management plan. Gain agreement with associates and customer on methods. Prepare memo of understanding. Schedule interface development activities.	OXXXOOOOO
1081M5	Engineering Decision Support	PIT personnel maintain a decision rationale database and provide the PIT with an organized decision-making forum where decisions made are captured, needed actions identified and assigned, and minutes captured for future reference.	XXXXXXOOO
1081M51	Action Item Tracking and Control	An action item database is maintained for formally assigned actions outside of the normal planned work. Reports are published to support PIT management showing status of formal action items.	OOXXXXOOO
1081M52	Development Review Board (DRB)	A DRB is called by the Chief Engineer (PIT Leader) to review an engineering analysis, conclusions, and recommendations about a technical problem encountered in the development of a system or element thereof. It is a formal problem solving environment providing a cross-functional team review and formal decision and action item assignment by the Chief Engineer (PIT Leader).	OXXXXXOOO
1081M6	Manage Trade Study Program	Assign trade study responsibilities, assist in staffing teams, track performance of studies. Facilitate reviews of results.	XXXXXOOOO
1082	Program Schedule Development	Program level schedule development and integration of lower tier schedules developed by PDTs.	XXXXOOOOO
1083	DTC/LCC Integration	The PIT coordinates the DTC/LCC work taking place on PDT, integrates and balances lower tier DTC/LCC conclusions across the system, and balances DTC/LCC requirements with other requirements to achieve the best system-optimized distribution of cost and other characteristics.	XXXXXOOOO
1084	Deployment and Operations Plans and Procedures Integration	Deployment plans and requirements are compared with evolving product features and changes coordinated in plans and features to assure agreement. Actions are taken to acquire access to resources needed for deployment and operational activities.	OXXXXOOOO
1085	Quality Planning Integration	Prepare plans for quality practices and implementation. Coordinate across PDT. Compare evolving design features with pre-determined quality requirements and planning.	XXXXXXOOO
1086	Material and Procurement Integration	Identify material sources, coordinate sources with specific needs, and monitor selection of suppliers.	OOXXXXOOO
1087	Production Process Integration	Production personnel interact with design and specialty engineering personnel to simultaneously develop both the product design concept and the production process concept in terms of tooling concepts, facilitization, production flow, and source alternative exploration. This activity includes producibility input to the product design and design input to the production process definition in a concurrent environment.	OOXXXXOOO
10871	Preparation Steps	Continue previous work to prepare for production work.	OOXXXOOOO
108711	Update Make-Buy Plan	Prepare or update source decisions on product items. Determine whether each product item will be purchased or internally fabricated.	OOXXXOOOO
108712	Update Manufacturing Process Definition	Maintain manufacturing plans consistent with evolving design solution.	OOXXXOOOO
10872	Tooling Acquisition	Partition all tooling into those items that will be internally fabricated and those that will be purchased. Seek out suppliers and select suppliers for needed tooling. Arrange availability timing and track schedule compliance.	OOOXXOOOO
108721	Map Tooling Needs to Process	Coordinate tooling needs identified for items with production process. Integrate PDT tooling recommendations at system level.	OOOXOOOOO

Table D-1 (continued) Enterprise Charter Task List

Task nbr	Dept nbr	Task name	Company document/ Source reference	Task description	Generic phasing
108722		Study Tooling Availability		Determine of tooling appropriate to specific manufacturing tasks is available.	OOOXOOOOO
108723		Procure Tooling		Arrange for purchase of tooling in accordance with specific requirements.	OOXOXOOOO
108724		Acquire Existing Tooling		Arrange for assignment of existing tooling for program purposes and installation in planned facility.	OOOXXOOOO
108725		Acquire Tooling Material		Purchase material needed to fabricate new tools.	OOOXXOOOO
108726		Design Tooling Modification		Accomplish tooling design work necessary to characterize new tooling needed.	OOOXXOOOO
108727		Design New Tooling		Accomplish tooling design work assuring mutual consistency with product design features.	OOOXXOOOO
108728		Manufacture New Tooling		Manufacture new tools that will be provided through in-house building.	OOOXXOOOO
108729		Modify Existing Tooling		Accomplish tooling design changes in accordance with predetermined design work.	OOOXXOOOO
10873		Design and Acquire Production Facility		Manufacturing Engineering and Facilities engineers select machine tools, allocate or design and contract development of needed facilities, plan floor layouts, and model production capability in support of planned production concurrently with the development of the design concept and preliminary design for the production system. These tools and facilities may satisfy the needs partially or completely for special test articles and low rate production. See task 10877 for rate production preparations.	OOXXXOOOO
108731		Map Facility Needs to Process		Ensure that adequate facilities are available for each step in the planned production process.	OOXXOOOOO
108732		Study Facility Availability		Review the features of existing facilities and compare with needed features. Determine availability of useful facilities.	OOXXOOOOO
108733		Acquire Facilities		Make arrangements for availability of needed facilities.	OOOXXOOOO
108734		Design Facility Changes		Design modifications to existing facilities to satisfy program needs.	OOOXXOOOO
108735		Acquire Facility Material		Procure and take delivery of the necessary material to build or modify needed facilities.	OOOXXOOOO
108736		Modify Facilities		Accomplish planned modifications of available facilities.	OOOXXOOOO
10874		Prepare Production Planning	MM	Prepare the detailed planning instructions to be used in manufacture and assembly of product.	OOOXXOOOO
108741		Map Planning Needs to Process	MM	Integrate PDT planning needs and map the resultant system planning needs to process steps.	OOOXXOOOO
108742		Copy/Edit Planning	MM	Planning data prepared for other purposes is edited for applicability to a new or changed product or production situation.	OOOXXOOOO
108743		Prepare New Planning	MM	Write and gain approval for new product manufacturing planning instructions that are coordinated with product requirements and features.	OOOXXOOOO
108744		Prepare Test Article Planning	MM	Write and gain approval for special manufacturing planning for special test articles. These are normally one-of-a-kind items.	OOOXXOOOO
10875		Personnel Acquisition		Acquire needed manufacturing personnel to accomplish planned manufacturing tasks.	OOOXXOOOO
108751		Map Manufacturing Personnel Needs to Process		Coordinate process steps with personnel needs in terms of numbers and specialties in order to refine the definition of needed personnel.	OOOXXOOOO
108752		Study Personnel Availability	MM	Review the availability of manufacturing personnel from existing employee's base and define any need for hiring new personnel. Define required skills and knowledge for any new personnel required.	OOOXXOOOO
108753		Acquire/Train Needed Personnel	MM	Where new personnel are needed, Human Resources must acquire them and the enterprise must train them for any differences between employee qualifications and process requirements.	OOOOXOOOO
10876		Maintain Process Definition		Edit process in keeping with improvements and corrections.	OOOXXOXOO
10877		Design and Acquire Rate Production Capability		Manufacturing Engineering, Manufacturing Technology, and Facilities Engineers design, develop, inspect facilities, tooling and equipment to accomplish low rate and high rate	OOOXXOXOO

			production. These actions may entail updating and increasing the capacity of test article production facilities or complete replacement of them.	
1087M	Manage Production Process Design	MM	Lead and manage the development of a manufacturing design solution for the evolving product design such that it is in mutual agreement with planned product features.	OOOXXOXOO
1088	Logistics Integration		All system logistics activities are coordinated between PDTs and PIT including both product features supportive of sound logistics capabilities and logistics processes that will support the product subsequent to delivery.	OOOXXXXXO
109	System Level Design Reviews and Audits		At appropriate times in the development process, the following reviews and audits are implemented: SDR, PDR, CDR, FCA, and PCA (SRR covered under task 107). Each of these activities involves preparing materials for presentation to the customer that reflect a prior work phase, presentation, and response to customer-identified areas of further work through action items. Customer approval of the review generally signifies approval to move to a subsequent program phase.	XXXXXXXOO
1091	SDR, PDR & CDR Channel		Accomplishment of these reviews is a fairly generic process including planning, presentation, and action item resolution actions.	XXXXXOOOO
10911	Organize Design Review Agenda		Review customer requirements in MIL-STD-1521B, MIL-STD-499B or other compliance document or internal guidance, and create an initial list of topics that must be covered at the review. Subject this list to critical comment by PIT, IPD, and program management and respond to critical input. Determine appropriate presenters for listed items and ascertain presenter readiness to cover these topics to the planned depth. Review prior activity schedule and tasks performed to ensure that work related to review topics has been accomplished and produced adequate material reflecting that work.	XXXXXOOOO
10912	Accomplish Additional Work		Where it is determined that work has not been completed to the degree necessary to support the review being planned, the PIT Manager directs appropriate IPD teams, PIT, and specific persons to complete additional work to fill out the presentation plan.	XXXXXOOOO
10913	Customer/Sponsor Approve		Either the external customer or an internal sponsor must review the planned agenda and approve same.	XXXXXOOOO
10914	Assign Responsibility, Schedule Review Prep, & Brief Team		Map the approved agenda to the program organization and specific principal engineers. Schedule the review preparation period showing when internal reviews/dry runs will be accomplished. Define review materials formats and make boilerplate electronic media available. Determine how the materials will be integrated into the final package and ensure resources are available to support the plan. Prepare and deliver a briefing on the preparation process to the review team including all presenters/principal engineers. Maintain contact with those responsible for preparing materials and ensure guidelines are being followed.	XXXXXOOOO
10915	Review Materials Preparation		The PIT coordinates the preparation of presentation and supporting materials required for review. Where proper equipment exists, presentations are encouraged from existing materials stored in an organized fashion in a development information grid on a networked computer system serving the program.	XXXXXOOOO
10916	Integrate Review Materials		Collect presentation materials, review them for format, content, and completeness. Coordinate with presenters where materials are not totally satisfactory. Assemble materials into a complete review presentation book in accordance with the plan and gain approval of the package from the PIT Manager.	XXXXXOOOO
10917	Review Administration	SEM/SEMP	Organize and facilitate meeting environment; attendance; copy, materials, computer, and secretarial needs; presentation aids (screens, projectors, etc.); and facilities.	XXXXXOOOO
10918	Publish Materials		Arrange for reproduction of adequate number of copies to be available in accordance with the customer's needs. Arrange for shipment of required copies to customer offices in accordance with CDRL where appropriate. Distribute copies to management and presenters. Acquire presentation viewgraph materials where that media is used. Where computer video	XXXXXOOOO

Table D-1 (continued) Enterprise Charter Task List

Task nbr	Dept nbr	Task name	Company document/ Source reference	Task description	Generic phasing
				presentation is used, organize the collected electronic files for easy access and brief presenters on that arrangement.	
10919		Present Review		Actual review event.	XXXXXOOOO
1091A		Action Item Response		Where the customer identifies an enterprise action item accepted by the review leadership, responsibility is assigned for the item and a response coordinated in accordance with a due date.	XXXXXOOOO
1091B		Customer/Sponsor Acceptance of Review		The customer or sponsor of the review signals acceptance of the review generally in the form of a letter or statement to be included in review minutes signed by the customer.	XXXXXOOOO
1091C		Customer/Sponsor Action		A major review is held to determine the status of the work accomplished to that time based on a need to move on to a new program plateau and product baseline working toward overall program goals. Therefore there should be some significant action resulting from the review such as approval for movement to the next phase or release of funding.	XXXXXOOOO
1092		Product Development Team Audit		PIT audit of PDT activity intermediate to major or internal reviews.	OOOXXOOOO
1093		Internal Design Review		An intermediate review between major reviews or in preparation for a major review organized by the PIT and generally accomplished by a PDT.	OOXXXOOOO
1094		Development Review Board		The Board Integration Team (PIT) uses the Development Review Board (DRB) as a means to hear and respond to PDT concerns and recommended changes in current plans, requirements, and design concepts based on a perceived problem that must be solved. The DRB at this level also deals with PIT problems.	OOXXXXOOO
1095		Functional Configuration Audit		A major customer review where the customer reviews available evidence that their requirements defined in specifications have been fully satisfied by the design. The essential question the customer is trying to answer at an FCA is, "If an article is produced faithfully to the drawings will it satisfy the requirements expressed in the system and configuration item specifications?" Another way of saying this is, "Does the current design satisfy the requirements?"	OOOOOXOOO
1096		Physical Configuration Audit		A major customer review of the first production article to verify that it accurately reflects the requirements defined in drawings and planning.	OOOOOOXOO
109M		Manage Review and Audit Process		Identify needed reviews, determine participants for each review, schedule reviews, and facilitate reviews and post-review actions.	XXXXXXXOO
10A		System Assessment, Validation, and Verification		PIT evaluates evolving designs for compliance with requirements, coordinates analyses focused on requirements verification, collects and organizes test and analysis evidence of compliance, and coordinates test and analysis efforts needed to validate that current requirements will yield to design efforts within the current state of the art.	OOOXXXOOO
10A1		Specialty Engineering Verification		Work collection heading. See lower tier tasks.	OOOXXXOOO
10A11		Reliability Verification		Conduct reliability assessment of evolving design and assure that quantitative reliability model and other reliability requirements have been satisfied. Reliability testing and demonstrations may also be involved.	OOOOOXXOO
10M		Manage Program Integration Team (PIT)		Manage the activities of the PIT both internally and its influences on PDTs.	XXXXXXOOO
11		Integrated Product Development		The cross-functional process for development of new product and modification of existing products.	OOXXXOOOO

111	Form Product Development Team	An organized process for placing a PDT in operation.	OOXXOOOOO
113	Item Requirements Analysis	Perform analyses to define the requirements for PDT item requirements.	OOXXXOOOO
1131	Engineering Task Flow Component (Item Requirements Analysis)	The Research & Engineering contingent on the assigned PDT leads the identification of product requirements for item assigned.	OOXXXOOOO
11311	Collect/Study Allocated Requirements	PDT members review allocations of parent requirements made by others or by the team and reach consensus that these are correct allocations. The team interacts with other PDT or the PIT to discuss alternative values or requirements.	OOXXOOOOO
11312	Functional and Performance Requirements Analysis	The assigned PDT identifies needed functionality, allocates it to elements of their assigned item (in task 11313), or associated team items, and transforms allocated functionality into appropriate performance requirements. Also, the team transforms functions allocated by other teams to their item into performance requirements.	OOXXOOOOO
113121	Study Needed Functionality	Functions are strung into sequences and relative time dependencies explored in an effort to force the systematic identification of all relevant system functions.	OOXXOOOOO
113122	Transform Allocated Functions Into Performance Requirements	The actual performance of this step is a function of the requirements identification methodology employed. The preferred approach is to follow the discipline of translating each functional requirement allocated to an element into one or more performance requirements for that element. This entails expanding the brief function statement into a complete thought expressed in one or more proper English sentences within the guidelines established by the requirements quality checklist given in the referenced document. These guidelines include the characteristics of quantification, verifiability, and traceability, for example. The process may entail application of Quality Function Deployment at the system level involving consultation with customer and user personnel. The element principal shall lead this effort with support from Operations Research, System Analysis, and other departments.	OOXXOOOOO
11313	Allocate Functionality and Expand Architecture	The team allocates functionality defined in task 113121 to team items, or items for which other teams are responsible, and coordinates with PIT to expand the system architecture to include these items where they were not previously defined.	OOXXOOOOO
11314	Expand Interface Analysis	As team expands item functionality it encounters expanded insight into needed interfaces between items subordinate to their item and between their item and others. The team is fully responsible for identification of internal interface, and jointly responsible with other teams for cross-team interfaces. Team members interact with PIT to expand the system interface definition within their item and with interface persons on other teams through a PIT-chaired interface working group to resolve inter-team interfaces. Item interfaces that cross the boundaries for which GDSS is responsible are flagged for PIT coordination with associates or customer through a ICWG or equivalent meeting structure.	OOXXOOOOO
11315	Interface Requirements Analysis (IRA)	For each line on the Schematic Block Diagram the analyst will determine appropriate corresponding requirements that will be levied against the elements on the two ends of the interface.	OOXXOOOOO
11316	Environmental Requirements Analysis (ERA)	System level natural environmental requirements are determined and then they are allocated to lower tier elements. Commonly, the principal operating system element (vehicle, aircraft, or whatever) is partitioned into environmental zones within which specific environmental requirements apply. When the physical location of a lower tier element is established, the environment corresponding to the zone within which it is located applies to the element. As the nature of the inter-relationship between the evolving system concept and the system natural environment is understood, induced environments are studied and defined by understanding how the system interacts with the natural environment to create other environmental effects that influence the system in ways that the natural environment alone would not.	OOXXOOOOO

Table D-1 (continued) Enterprise Charter Task List

Task nbr	Dept nbr	Task name	Company document/ Source reference	Task description	Generic phasing
113161		Copy Environmental Boilerplate		If the PIT has established a boilerplate environmental definition for the team item, the team simply copies that boilerplate to their requirements set. Some editing may be required.	OOXXOOOOO
113162		Expand Service Use Profile and Zone Analysis		If the team item is an end item, the team must cooperate with PIT to expand the service use profile as it relates to the item and conduct a zoning analysis for the item to define appropriate environmental requirements for components located within the item.	OOXXOOOOO
11317		Specialty Engineering Requirements Analysis		PDT specialty engineering requirements analysis work for the items under their responsibility.	OOXXOOOOO
113171		Create Specialty Engineering Rqmts Analysis Demand List		Constraints allocation involves creating a list of constraints that will or may influence one or more system architecture elements and assigning them to the elements. Each allocation (assignment) creates a place holder, or puts a demand on a specialty engineer, for subsequent specialty engineering requirements analysis to identify a constraint of that kind for that element. The constraints allocation mechanism commonly is a design constraints scoping matrix (DCSM).	OOXXOOOOO
113172		Manage Item Specialty Engineering Requirements Analysis		The team must coordinate the product of the several specialty engineering disciplines to ensure they respond to the demand created in task 113171 and that the result is free of internal inconsistencies.	OOXXOOOOO
1131731		Reliability Requirements Analysis		A series of tasks that collectively act to define quantitative hardware (and software if covered in the contract) reliability requirements based on a given system level reliability figure of merit.	OOXXOOOOO
1131732		Producibility Engineering Requirements Analysis		Coordinated requirements work between production process requirements and product feature requirements to assure an optimum manufacturing process.	OOXXOOOOO
1131733		Maintainability Requirements Analysis		Maintainability personnel create a maintainability model selecting system elements from the architecture based on a selection criterion determined in this task. The maintainability figure(s) of merit is (are) determined based on an allocation process from system level figures. If the figure of merit is, for example Mean Time to Repair (MTTR), the MTTR figures are allocated to lower tier elements in step with the controlled advance of requirements and concept development work regulated by the Chief Engineer. This model is maintained throughout the program phase with adjustments based on trade studies and other sound knowledge of system needs. As the development reaches the third program phase, the principal focus is on predicting maintainability figures of merit based on actual design figures, but there remains some continuing need to readjust the maintainability model.	OOXXOOOOO
1131734		Parts Engineering Requirements Analysis		Prepare list of standard, approved, or preferred parts for use on the program.	OOXXOOOOO
1131735		Materials and Processes Requirements Analysis		Define standard, preferred, or approved materials and processes for use on the program. Prepare formal list and gain approval. Maintain list.	OOXXOOOOO
1131736		Human Factors Requirements Analysis		Define human factors requirements, generally in terms of a reference to a commonly accepted standard of human capabilities.	OOXXOOOOO
1131737		Systems Security Requirements Analysis		The requirements for physical, communications, information, and other forms of security appropriate for the system and elements thereof are identified and quantified.	OOXXOOOOO
1131739		System Safety Requirements Analysis		Identify safety requirements, commonly through reference to a safety standard with possible tailoring. Interpret these requirements in the context with the specific system and situation for development teams.	OOXXOOOOO

113173A	Contamination Requirements Analysis		Identify any needed protections from contamination that will be required for the system or items thereof. This may entail use of certain materials, use of clean rooms with specified degree of cleanliness, or protection of items potentially damaged by contamination.	OOXXOOOOO
113173B	Corrosion Control Requirements Analysis		Identification of required or approved methods for preventing corrosion of system surfaces. May entail protection of metallic surfaces, elimination of materials that may cause corrosion, or use of non-corrosive materials.	OOXXOOOOO
113173C	Damage Tolerance Requirements Analysis		Identify requirements to prevent system or item failure as a result of specific damage. This may be referred to as robustness by the customer.	OOXXOOOOO
113173D	Health & Safety Requirements Analysis		Identify specific product requirements to encourage a healthy and uninjured system operating and maintenance workforce.	OOXXOOOOO
113173E	Vulnerability Requirements Analysis		Identify specific list of sources of energy or action that may adversely influence system operation. Define specific actions the product must be designed to survive.	OOXXOOOOO
113173F	Transportability Requirements Analysis	LSM	The need for items to be moved from place to place is studied and corresponding requirements identified where appropriate.	OOXXOOOOO
113173G	Electromagnetic Compatibility (EMC) Requirements Analysis		EMC requirements are defined for equipment that may generate interference or be subjected to adverse effects from interference. Commonly these requirements are stated in terms of a reference to a standard on the subject with possible tailoring.	OOXXOOOOO
113173H	Mass Properties Requirements Analysis		A system or item mass properties (weights) model is created and weights allocated down through the architecture as it is defined. Where appropriate, center of gravity requirements are also defined.	OOXXOOOOO
113173I	Thermodynamics Requirements Analysis		Requirements involving heat or the absence of it are defined for items that have temperature sensitivities.	OOXXOOOOO
113173J	Structural Dynamics Requirements Analysis		Planned structures are analyzed to ensure they have sufficient strength to sustain expected physical, thermal, vibration, and other loads in static and dynamic situations encountered during the item life cycle.	OOXXOOOOO
113173K	Guidance Accuracy Requirements Analysis		The degree of accuracy needed for present position identification, future (target) position identification, and movement between them is determined.	OOXXOOOOO
11318	Integrate Product Requirements		The requirements for an element are developed by many people working somewhat independently in their own field of specialization. The assigned principal engineer must coordinate the set of requirements prepared by supporting analysts to ensure a condition of balance and appropriateness.	OOXXOOOOO
11319	Document Requirements		The traditional specifications group tasks of preparing specification standards (boilerplates), publishing specifications, applicable documents analysis, and requirements baseline maintenance.	OOXXOOOOO
1131A	Maintain Requirements		Control changes to requirements.	OOXXXOOOO
1131B	Product Acceptance Requirements Analysis		A structured process for identifying the appropriate minimum set of performance requirements and constraints for a system element.	OOOXXOOOO
1131C	Requirements Traceability Development		The team cooperates with PIT to ensure that item requirements are traceable to parent requirements and that evidence of that traceability can be produced on demand.	OOXXOOOOO
1131D	Requirements Verification Development		Each item requirement shall be studied by the team to determine verification needs for inclusion in Section 4 and/or in test or analysis planning. Team works with PIT to expand verification management database content.	OOXXXOOOO
1131E	Test and Evaluation Requirements Analysis		Test and Evaluation integrates the test requirements driven by requirements verification (qualification testing), design evaluation testing, reliability testing, safety testing, and acceptance test needs into an efficient integrated test plan requiring minimum test resources, time, and budget. Special test and flight test article needs are identified and coordinated with the program plans and schedules.	OOXXXOOOO

Table D-1 (continued) Enterprise Charter Task List

Task nbr	Dept nbr	Task name	Company document/ Source reference	Task description	Generic phasing
				This task includes launch site checkout for launch vehicles where GDSS is the system operator, but it does not include operation and maintenance procedures development for systems to be operated by customer personnel. The latter is a logistics technical manual function within the task 12612.	
1131L		Item Applicable Documents Analysis		The element Principal compares specification section 2 documents listed with those called out in sections 3, 4, and 5 and ensures that all documents called in text are listed in section 2 and no documents are listed in section 2 that are not called in text. In addition, the Principal ensures that all document tailoring is consistent with program tailoring or deviates only in ways that have been reviewed and accepted by engineering management for the specific case.	OOXXOOOOO
1131M		Manage Item Requirements Analysis		This task provides leadership and management oversight over the complete SRA process including schedules, budgets, and product quality. On NASA Phase A/B and DoD Phase 0/1 programs 877–1 may be assigned to perform this task.	OOXXXOOOO
1132		Schedule Requirements Analysis		PDT schedule requirements analysis.	OOXXOOOOO
1133		Cost Requirements Analysis		PDT cost analysis from DTC and/or LCC perspective.	OOXXOOOOO
1134		Deployment & Operations Planning Analysis		On systems operated by the customer (blue suiters) involving deployment of the system to multiple sites, DPA provides a disciplined approach to understanding deployment requirements and what elements they influence (are allocated to). The products are deployment performance requirements fed into the requirements sets for the appropriate architecture elements and program planning reports pertaining to deployment scheduling, site preparation planning and work, customer operations and maintenance training scheduling, and supply support planning. Operations requirements are identified based on development and analysis of process flow diagrams.	OOXXXOOOO
1135		Quality Requirements Analysis		Item product quality and quality process requirements are defined based on system level requirements.	OOXXXOOOO
1136		Material Requirements Analysis		Item materials are specified based on the program material standards.	OOXXXOOOO
1137		Manufacturing Requirements Analysis (MRA)		The requirements for the manufacturing process are developed simultaneously with the requirements for the product. These requirements sets should be interactively developed to introduce the most cost-effective features on the two sides of the system development process: product and process. Where manufacturing advantages encourage adoption of operations system components features and it can be shown to be cost effective and not detract adversely from system performance, they may be accepted by the Chief Engineer. Likewise, advantageous operating system features may drive process design in terms of tooling selection, facility station sequence, and exert other influences.	OOXXXOOOO
1138		Logistics Support Requirements Analysis (LSRA)		LSA is a systematic and comprehensive analysis on an iterative basis through all phases of the system life cycle to satisfy supportability objectives. The complete program is composed of requirements analysis included under this heading, planning activity (see task 1MM), design assessment, and design verification activity.	OOXXXOOOO
113Q		Supplier Requirements Analysis		Cooperative work by supplier representatives as participant of company PDT.	OOXXXOOOO
114		PDT Requirement Review and Audit		A major customer review to ensure that the system requirements are understood and agreed upon by customer and contractor.	OOOXXOOOO

115	PDT Concurrent Item Product Design and Process Planning	PDT process for development of design solutions to item requirements previously developed.	OOOXXOOOO
1151	Engineering Flow Component	A work grouping for all Engineering Department work. See lower tier details.	OOOXXOOOO
11511	Trade Study Methodology	An organized method for reaching sound decisions within the context of a difficult question with insufficient information for easy judgment. Alternatives are evaluated against predefined criteria and requirements and relative or absolute merit determined as a prerequisite to selection of the most advantageous alternative.	OOOXOOOOO
115111	Study Element Requirements	As a prerequisite to defining and evaluating alternatives, the trade study team studies the requirements for the item to ensure that all alternatives conceived will respect those requirements.	OOOXOOOOO
115112	Develop Alternative Concepts	The trade study team conceives two or more alternative solutions to the requirements.	OOOXOOOOO
115113	Interact With Team Members		OOOXXOOOO
115114	Define/Develop Alternative Concepts	The concepts are initially conceived using a very simple description. Each must thereafter be developed sufficiently to ensure feasibility and to provide enough detail for specialized team members to evaluate their features from their perspective.	OOOXOOOOO
115115	Define Selection Criteria	A value system is defined as a basis for selecting the preferred concept. All alternatives must satisfy the requirements but each will be more or less advantageous than others with respect to the selection criteria.	OOOXOOOOO
115117	Alternatives Review and Down-Select	The PIT convenes a review of the team concept(s). The team offers the alternatives they considered, the selection criteria, requirements, results of their evaluation, and a recommendation. The PIT meeting chairman and evaluation team members ask questions and evaluate presented material. The PIT Chairman makes a final selection, provides guidance and direction on future preliminary and detail product and process design work, and assigns action items to account for additional work required to complete concept development activities.	OOOXOOOOO
115118	Identify Item Risks	PDT define risks associated with product item and related processes.	OOXXOOOOO
115119	Mitigate Identified Item Risks	Study ways to mitigate risks within the resource limits available to the team. Notify PIT of risks beyond PDT limits.	OOXXXOOOO
11511A	System Analysis & Specialty Engineering Support	Specialty engineering and analysis groups interact with the design principal engineer or staff to ensure that design personnel understand the specialty requirements and have access to good examples of compliance, to support trade studies evaluating alternative design concepts, and to evaluate particular concept features for compliance. Systems Analysis personnel support design with required analyses.	OOXXXOOOO
11512	Item Preliminary Test Planning	Define which items under team responsibility will have to be tested for development, qualification, and acceptance purposes, when that testing must take place, and the resources needed to accomplish needed testing.	OOXXXOOOO
11513	Item Design	The design and analysis activity that translates the system and system element concepts and associated requirements into a preliminary design in terms of engineering drawings and analysis reports that support the suitability of the design.	OOOXXOOOO
11514	Item Test and Evaluation Development	This is a high level work category. Refer to lower tier details.	OOOXXOOOO
115141	Item Qualification Test Development	Develop qualification test plans including what items must be subjected to testing, test scheduling, requirements, procedures, and test article development.	OOOXXOOOO
151411	Company Qualification Test Development	Qualification program for team items is developed. PIT evaluates and integrates work of all PDT in this area which may change some of the plans and results for system testing optimization.	OOXXXOOOO
11514111	Review System Qualification Criteria	The PIT must develop a criterion for selection of items for qualification testing by the product teams. The product team reviews this criterion to ensure they understand it in preparation for identification of team items that will require qualification.	OOXXXOOOO

Table D-1 (continued) Enterprise Charter Task List

Task nbr	Dept nbr	Task name	Company document/ Source reference	Task description	Generic phasing
11514112		Evaluate Team Items Against Qual Criteria		The PDT studies each team item against the test criteria and selects or rejects items for testing.	OOXXOOOOO
11514113		Create Test Service Request		For each item that will require qualification testing, a formal statement is created to signal that selection and start the qualification test planning process.	OOOXOOOOO
11514114		Approve Service Request		The PDT Leader, and possibly the PIT Leader, must approve each test service request as a management cross check.	OOOXOOOOO
11514115		Issue Service Request to Test Lab		The approved test request is given to the agency responsible for developing the test.	OOOXOOOOO
11514116		Prepare Test Plan & Procedure		The test is planned based on item verification requirements developed during item requirements analysis and a responsive test procedure is created.	OOOXOOOOO
11514117		Approve Test Plan & Procedure		The PDT Leader reviews plans and procedures for compliance with item verification requirements and program requirements.	OOOXOOOOO
11514118		Define Test Schedule		The item test schedule is created based on FCA timing and the schedule for higher level qualification actions dependent upon completion of lower tier qualification actions.	OOOXOOOOO
11514119		Define Test Set-Up Requirements		The test apparatus needed to accomplish the planned test is defined in terms of test and support equipment, instrumentation, and services.	OOOXOOOOO
1151411A		Acquire Test Set-Up Material		Acquire test equipment and materials needed to accomplish the test.	OOOXXOOOO
1151411B		Assemble Test Set-Up		Assemble and install test set-up in preparation for test run.	OOOXXOOOO
115141C		Define Test-Unique Test Article Requirements		The majority of test article requirements should be derived from the item requirements in order that the test article test results will provide credible evidence of having satisfied those requirements. But, there may be test-specific requirements that must be satisfied as well in the development of the test article.	OOOXOOOOO
1151411D		Acquire Special Test Article Material		Materials needed to create the special test article are acquired and made available for test article fabrication or assembly.	OOOXXOOOO
1151411E		T&E Produce Test Article		The test article may be created by the test and evaluation community rather than manufacturing.	OOOOXOOOO
1151412		Item Verification Methods Determination		The PDT must determine whether it is necessary to verify compliance with item requirements through testing or if other methods (including no verification requirement) can or should be employed.	OOOXXOOOO
11514121		Define Item Verification Methods			OOOXXOOOO
11514122		Prepare Specification Section 4 Data		A paragraph is written for section 4 of each team item specification where the verification method is a test that prescribes what test activity shall be done to prove that the design satisfies the requirement.	OOOXOOOOO
1151413		Vendor Qualification		Items produced by suppliers may be qualified by the supplier using the supplier's test and evaluation process. This task defines a generic supplier process that includes steps that must be included. This includes both in-house activity and supplier activity.	OOOXXXOOO
11514131		Issue Purchase Order		A purchase order is prepared that includes vendor qualification testing.	OOOXOOOOO
11514132		Vendor Prepare Qual Test Procedure		The vendor prepares a qualification test procedure that is responsive to item test requirements in the procurement specification.	OOOXXOOOO
11514133		Data Management Process Vendor Documentation		Vendor test procedure is routed through specific reviewers within the PDT. Data Management ensures that reviews and response to review comments are completed within schedule requirements resulting in an approved procedure.	OOOXXOOOO

11514134	Qual Test Procedure Review	Appropriate members of the responsible PDT and PIT review the vendor document and provide Data Management with responses. Where there is considerable negative comment, this step may require a review meeting to determine the best course of action to satisfy potentially disparate concerns.	OOOXXOOOO
11514135	Company PQE Witness Qual Test	Company PQE may have to witness vendor tests to ensure tests are accomplished in accordance with approved plans and procedures and that reports accurately reflect results.	OOOOXXOOO
11514136	Data Management Process Vendor Documentation	Test failure status is communicated to the PDT and PIT.	OOOXXXOOO
11514137	Company Review Supplier Status Notification	The PDT must review a vendor test failure notice to determine the most appropriate response (retest, requirements change, design change).	OOOXXXOOO
11514138	Company Communicate Findings/ Direction	Results of PDT review of test failure notification is communicated to the vendor.	OOOXXXOOO
11514139	Data Management Process Test Report	The formal test report is received by Data Management and provided to the persons who must review and approve it. Ideally, the test reports should be placed on line for access during verification work leading up the FCA.	OOOOXXOOO
1151413A	Company Review Test Report	The PDT reviews the test report and either approves it or tells Data Management to accomplish additional work coordinated through contracts.	OOOOXXOOO
115142	Item Acceptance Test Development	Identify all of the items that must be subjected to an acceptance test and in each case develop the procedure, resources definition, and process design; acquire the needed resources; and create the testing capability using those resources.	OOOXXOOOO
1151421	Identify Item Acceptance Test Need	Evaluate the need for an acceptance test for specific items using a criteria developed for the program. All items in the architecture must be filtered through this activity and a decision made to test or not to test for acceptance purposes.	OOOXXOOOO
1151422	Review Program Acceptance Test Criteria	The team developing the acceptance test must be familiar with program criteria for testing relative to the content of test procedures and Part II specifications. Part II specification content need not include requirements for every test activity. They should include the requirements corresponding to the principal tests that must be accomplished to assure the customer that the article complies with the previously approved design.	OOOXXOOOO
1151423	Filter Acceptance Test Requirements for Part II Spec	Review all test requirements defined for the test activity. Partition those requirements into those top level requirements that will provide the customer with clear knowledge of the degree of compliance of the product with approved engineering and manufacturing planning.	OOOXXOOOO
1151424	Review Availability of Necessary Test Apparatus	Determine if the resources exist and are available to support acceptance test for a given item.	OOOXXOOOO
1151425	Acceptance Test Requirements Analysis	Determine all requirements for acceptance test. What parameters must be tested or measured to what accuracies? What combinations of stimuli are required in combination with what measurements?	OOOXXOOOO
1151426	Prepare System Test Parameter Inputs	While defining item test parameters, the IPD Team may gain insight into needed system parameters. This is a two way street coordinated by the PIT in task 10813.	OOOXXOOOO
1151427	Acquire Needed Test Capability	Given that adequate resources are not available for a particular test, these resources are either purchased or designed and built within the enterprise.	OOOXXOOOO
1151428	Company Test Requirements Approval	Review and approve test requirements. PIT should review all PDT prepared test requirements.	OOOXXOOOO
1151429	Customer Approval	Where required, the test requirements are approved by the customer prior to procedure development.	OOOXXOOOO
115142A	Prepare Test Procedure	Prepare a test procedure based on the approved requirements that accomplishes the goals defined in the requirements.	OOOXXOOOO
115142B	Approve Test Procedure	Test procedures are approved internally and possibly by the customer. The PIT should review and approve all PDT prepared procedures.	OOOXXOOOO

Table D-1 (continued) Enterprise Charter Task List

Task nbr	Dept nbr	Task name	Company document/ Source reference	Task description	Generic phasing
115144		Item Design Evaluation Testing		The need for all development testing is defined based on PDT needs. These needs are driven by anticipated risks in the team's ability to successfully synthesize the item requirements or to aid in defining requirements values that are both consistent with the customer's need and possible to achieve.	OOXXXOOOO
1151441		Create Service Request		Each development test should be defined in a test request as a prerequisite to developing the test. These tests may include both PDT item level tests and tests at higher levels to support the needs of the PDT. The PIT must approve all test requests requiring cooperation beyond the control of the PDT.	OOXXXOOOO
1151442		Approve Service Request		Tests that can be conducted within the span of control of the PDT may be approved by the PDT Leader. All tests that will require customer or Enterprise resources beyond the control of the PDT must be approved by PIT.	OOXXXOOOO
1151443		Issue Service Request to Test Lab		The approved test request is provided to the test agency responsible for the test.	OOXXXOOOO
1151445		Design DET Test Apparatus		The test apparatus suitable for accomplishing needed testing is developed or purchased and assembled in preparation for the test.	OOXXXOOOO
1151446		Acquire Material		Needed material for the test set-up is acquired.	OOXXXOOOO
1151447		Prepare Test Set-Up		Acquired test apparatus and materials are organized and assembled into the required test apparatus in preparation for the test.	OOXXXOOOO
1151448		Prepare Test Plan and Procedures		A test plan/procedure is prepared to accomplish the requirements and goals expressed in the test request.	OOXXXOOOO
1151449		Approve Test Plan and Procedures		Prepared plan/procedures is/are reviewed and approved.	OOXXXOOOO
115144A		Design DET Article		If required, the design of one or more test articles is prepared as a prerequisite to acquiring conforming articles and materials.	OOXXXOOOO
115144B		Acquire Test Article Material		Materials defined in the test article design are acquired.	OOXXXOOOO
115144C		Fabricate/Assemble Test Article		The test article is assembled/fabricated from acquired materials in accordance with the prepared design.	OOXXXOOOO
115144D		Schedule Test		The test is placed on the test schedule so as to avoid conflicts with other testing and resources.	OOXXXOOOO
115144E		Perform Test or Re-test		Accomplish tests in accordance with test procedures and make available test results and measurements.	OOXXXOOOO
115144F		Write Test Report		Prepare a written report of test results.	OOXXXOOOO
115144G		Approve Test Report		Review and approve the test report.	OOXXXOOOO
115144H		Issue Test Report		The approved test report is made available for use on the program.	OOXXXOOOO
115144I		Dispose of Test Resources		All resources used in testing are disposed of. Some resources may be useful in later planned tests involving qualification of items.	OOXXXOOOO
115144J		Prepare QDR		Where test results are not fully satisfactory, a quality deficiency report (QDR) is prepared describing the nature of the problem with a recommendation for subsequent actions needed.	OOXXXOOOO
115144K		Evaluate Anomaly		Study the cause of the observed anomaly and determine an appropriate response to the identified cause.	OOXXXOOOO
115144L		Repair or Refurbish Test Article		Where the test anomaly resulted in damage to the test article or redesign or repair is necessary, it is accomplished as a prerequisite to returning to testing.	OOXXXOOOO
115144M		Modify Test Article		Where the anomaly has resulted in a conclusion that the test article does not properly represent the needed characteristics, it is modified as a prerequisite to returning to testing.	OOXXXOOOO

115144N	Store Test Article Awaiting Test	If conflicts exist precluding immediate return to testing, the test article is stored until a return to testing can be arranged.	OOXXXOOOO
1156	Item Material and Procurement	Make or Buy decisions are made for all system elements identified in the design concept. Make items are fed to the production planning functions (1Q and 1R) while procurement items are associated with available sources within this task. Statements of work, procurement specifications, and model contracts are developed and competitive bids evaluated followed by selection of sources. Correct materials are ordered and stocked consistent with production needs and made available to the production process in a timely way.	OOXXXOXOO
115A	Supplier Concept Development	Supplier work to develop a design concept for an item defined in a procurement specification.	OOXXOOOOO
115B	Supplier Item Design	The supplier translates procurement specification requirements into a design concept, preliminary design, and detailed design with interaction between the responsible PDT and the supplier development team. In-process and major reviews are accomplished to ensure that the supplier is proceeding on a course that will meet with success.	OOOXXOOOO
116	Design Review	PDT design review process.	OOXXXOOOO
1161	Item Concept Design Review	A major PDT item level internal review independent of the major customer review called the System Design Review or called as a preparatory step to the SDR. Team presents their concept in terms of sketches and analysis results. PIT reviews data to ensure that the team concept is acceptable and responsive to system requirements. Commonly the requirements for this review for internal purposes are satisfied by task 115117.	OOXXOOOOO
1162	Item Preliminary Design Review	When this review is called as an incremental customer review, the customer reviews all work and products during the preliminary design period to reach a decision whether the activity has progressed satisfactorily to justify commitment to detailed design and development of any needed test articles. The same review may be called for internal purposes in the absence of a customer requirement.	OOOXOOOOO
1163	Item Critical Design Review	When this review is called an incremental CDR by the customer, the customer reviews the essentially complete design for readiness to manufacture the initial system elements. Normally this point is characterized by 95% of all planned drawings complete. This review may be called for internal purposes when not mandated by the customer at this level.	OOOOXOOOO
1164	Item Internal Design Review	A review of a vehicle or ground subsystem concept or preliminary or detailed design by the Chief Engineer or responsible design department Director for the purpose of establishing an internal design baseline. This activity may also be used as a dry run for a formal SDR, PDR, or CDR.	OOXXXOOOO
117	PDT Assessment, Validation, and Verification	The PDT is responsible for accomplishing assessment, validation, and verification work within the bounds of their authority and resources.	OOOXXOOOO
1171	Requirements Verification Management	Throughout this phase, concurrently with specialty assessment and component and full scale testing activity, the verification function coordinates the assignment of verification evidence development responsibilities, tracks the status of verification activity, and maintains verification records. This activity is in preparation for a successful FCA.	OOOXXXOOO
118	Release PDT Product Design Package	Original release of design package or changes thereto. Package includes product design as well as coordinated material, manufacturing, quality, tooling, and logistics designs.	OOOOXOOOO
1M	Manage Development Program	Program management of the complete program. Entails intense interaction with customer management and leadership of the program team effort through team leaders. Program cost and schedule as well as the evolving design solution are reviewed against planned activity and adjustments made to ensure success.	XXXXXXOOO
2	Procurement	Identify sources of required material, interact with those sources to acquire material, store and transport that material to manufacturing sites that will use that material.	OOXXXXOXO
3	Production	The enterprise shall provide the capability to manufacture, assemble, and test those items identified for in-house production.	OOOOOOOXO

Table D-1 (continued) Enterprise Charter Task List

Task nbr	Dept nbr	Task name	Company document/ Source reference	Task description	Generic phasing
32Q		Plant Quality Assurance		A quality assurance process shall be applied to all manufacturing activity performed within the control of the enterprise.	OOOOOOXOO
4		Qualification Testing		Qualification testing must be applied to all items where it is necessary to verify that the development requirements have been satisfied in the design and no other verification method is appropriate in establishing convincing evidence of compliance.	OOOOOXOOO
4L		Integrate Test Activity and Resources		At any one time there are many qualification tests ongoing. These tests may require use of shared resources and facilities. Management skills must be applied to coordinate overall test achievement.	OOOOOXOOO
4M		Manage Program Qualification Test Program		Overall management of the qualification test activity through the PIT.	OOOOOXOOO
4N		Typical Qualification Test Channel		This task represents one of many qualification tests conducted on a program. Each test follows the same pattern of activities, with the specific path within dictated by test results (fail or pass for example). If the test is successful, a test report describes the results. If a test fails for any reason, the reason is determined and an appropriate response undertaken until such time as the test results are accepted or the test is deleted.	OOOOOXOOO
4N1		Perform Qualification Test		Planned test activities are accomplished in accordance with approved procedure and results made available.	OOOOOXOOO
4N2		Write Test Report		The test results are collected and published with the results of any needed analysis applied to present the results in a useful way.	OOOOOXOOO
4N3		Company Review, Approve Test Report			OOOOOXOOO
4N4		Release Test Report or Revision			OOOOOXOOO
4N5		Customer Review, Approve Test Report			OOOOOXOOO
4NA		Identify Anomaly, Stop Test, and Prepare QDR		Where an anomaly occurs in testing, it must be analyzed and a course of action taken to return to testing. Generally, this will require suspension of testing while a quality deficiency report (QDR) prepared to describe the anomaly is evaluated.	OOOOOXOOO
4NB		Responsible Design Group Review QDR and Disposition		The QDR is assigned to the appropriate team of design function for development of an appropriate response. Dispose of the QDR so as to enable a return to testing.	OOOOOXOOO
4NC		Change Test Requirements or Plan			OOOOOXOOO
4ND		Repair or Rework Test Set-up			OOOOOXOOO
4NE		Revise QDR			OOOOOXOOO
4NF		Temporary Store Test Article			OOOOOXOOO
4NG		Senior Flight Certification Board		Board composed of senior management personnel reviews test results and intended resolution and determines whether or not product design must be changed as suggested by test results.	OOOOOXOOO
4NH		Redesign Test Article or Specimen		Redesign the test article to solve problems identified in testing and from other sources.	OOOOOXOOO
4NI		Direction to Change Product Design		Where test results show that the design does not satisfy requirements, direction is issued to change the design to comply and to change the test article(s) accordingly.	OOOOOXOOO
4NJ		Rework Test Article or Specimen		Changes are made to the test article in preparation for a return to testing.	OOOOOXOOO
5		Logistics Support		The product is supported logistically based on logistics planning work accomplished concurrently with the product development.	OOOOOXXXO

6	Base Activation & Deployment	Where appropriate to the product, customer or enterprise facilities are established in the field as preparation made for receiving the product in preparation for a planned initial operating capability.	OOOOOXOXO
7	Operations Testing	Plan, coordinate, implement, and report upon tests conducted at operational sites (such as launch sites) that seek to verify that the system will satisfy customer needs and requirements. These tests employ operational hardware and software in the intended operational environment.	OOOOOXXXO
8	Operations	System operation in accordance with operations, logistics, and maintenance plans and procedures to satisfy the customer's mission needs.	OOOOOOOXO
821	Erect Booster		
842	Install Ordnance		
844	Execute Launch Countdown Operations		
9	Program Management	Chief Engineer leads the technical development effort in early program phases as the Program Integration Team (PIT) Leader. Reports to the Program Manager. Subsequent to PCA or first article inspection, a person with manufacturing skills is installed as the PIT Leader and the Chief Engineer remains responsible for leadership of R&E personnel on the PIT.	XXXXXXXXX
A	Quality Assurance	Quality engineering and assurance activities focused on a particular program.	OOOXXXXXO
B	Engineering Change Proposal	Work undertaken on a specific program to change contractually stipulated characteristics of the product (ECP) or process (CCP) based on sound information of a need for change.	OOOXXXXXO
B1	Prepare and Submit ACSN	Where the customer requires submission of advanced change study notices (ACSN) as a precursor of an engineering change proposal, the ACSN is prepared in accordance with customer requirements.	OOOXXXXOO
B5	Change Integration	As part of the ECP development process, the technical aspects of the change are defined by the appropriate team under the guidance of an assigned principal.	OOOXXXXXO
C	Mission Adaptation Process	Where the product line can be applied to many different missions, this activity provides for adaptation of the principal product design to alternative missions. This involves design of special adaptation and interfaces for the mission-peculiar aspects of the application.	XXXXXXXXO
D	RFP-Proposal Process	This process covers the work cycle from receipt of a request for proposal (RFP) through notification of a win or loss.	XXXXOOXXX
D1	Prepare Draft RFP	Customer prepares a draft RFP to draw comment from potential bidders.	XXXXOOXXX
D2	Review Draft & Respond		XXXXOOXXX
D3	Pre-RFP Task	Prior to entering into a competitive environment there may be some preparatory tasks that the enterprise can ethically accomplish to improve competitive position. This is a very sensitive area that must be carefully reviewed with respect to enterprise ethics standards, customer rules, and U.S. law.	XXXXOOXXX
D4	Prepare RFP	Customer prepares the RFP, possibly using inputs from cooperating contractors who submitted responses to the draft RFP (if used).	XXXXOOXXX
D5	Study RFP	The proposal team studies the RFP carefully and meets to share impressions prior to racing into proposal preparation.	XXXXOOXXX
D6	Create Program Infrastructure	Concurrent with the preparation of the proposal it is necessary to prepare for the possibility that we will win the competition. The Proposal manager (or the Program Manager if assigned) will start to create the Program team during the proposal effort. This includes creating an organization and the fundamental resources needed to support proposal preparation and subsequent program award. Assigned personnel, some of whom will remain with the team through the customer selection period, adapt generic tools, procedures, and standards for customer requirements. A Program Directive Manual is begun that contains procedures replacing Division standard practices that do not comply with customer requirements. Customer applicable documents and Data Item Descriptions (DID) are tailored to the	XXXXOOXXX

Table D-1 (continued) Enterprise Charter Task List

Task nbr	Dept nbr	Task name	Company document/ Source reference	Task description	Generic phasing
				maximum extent to reflect GDSS procedures. Functional task descriptions (FTD) are prepared reflecting the tasks required on the program coordinated with cost estimates and schedule needs. Estimates are reviewed in an estimating school and integrated into functional estimate, reviewed by functional management, and translated into a portion of the program spending profile (material spending profile component added in Task 67). Personnel planning is accomplished for the program win and training needs assessed. Facilities needs are coordinated with Facilities contingent on a program win.	
D61		Review and Adjust WBS		Prior to entering the competitive environment, efforts should have been made to find out if the customer will tolerate changes to the Work Breakdown Structure (WBS). Often the customer is wedded to a particular computer implementation of the WBS and is dependent upon that implementation to manage cost. In such cases it is very hard to gain approval of changes to reflect our computer models. Compromises may be possible, but it is necessary to understand if the selection team will consider changes, however well explained, as unresponsive.	XXXXOOXXX
D62		Plan and Brief Estimating Methodology		The Estimating function or Program Business Team (if formed during the proposal) defines the estimating methods to be used (bottom-up or top-down, for example) and briefs the team on how to respond.	XXXXOOXXX
D63		Define Program Task		Team members review the program requirements and determine which generic tasks should be applied to satisfy these requirements. This is principally a matter of mapping detailed generic tasks to the program SOW.	XXXXOOXXX
D64		Integrate Program Tasks		The proposal team members are polled for inputs to the list of tasks that must be accomplished by GDSS departments in response to the Statement of Work (SOW). The Proposal Manager, Program Manager, and/or Chief Engineer must review these inputs for consistency and compatibility with SOW requirements. The resultant tasks are subjected to a cost estimating exercise and proposal writing response subsequently.	XXXXOOXXX
D65		Data Item Description (DID) Tailoring Task		Data Management assigns each CDRL and corresponding DID to a responsible principal engineer from the department responsible for preparing the document covered by them. Wherever possible, within the Proposal/Program Manager's guidelines for RFP responsiveness, the CDRLs and DIDs are tailored for agreement with GDSS practices and formats and these changes coordinated with proposal writers.	XXXXOOXXX
D66		Prepare Program Products List		Based on the tasks that must be accomplished on the program, each functional proposal representative develops a list of products that must be prepared mapped to specific tasks. These inputs are integrated into an overall program list.	XXXXOOXXX
D67		Prepare Program Tool List		Each functional representative identifies tools that will be useful in satisfying program needs. These inputs are combined into a single program list and reviewed for completeness.	XXXXOOXXX
D68		Adjust Tools To Program Requirements		Where absolutely required by customer needs, tools are changed to permit use on the program.	XXXXOOXXX
D69		Adjust Standards To Contract Requirements		Tailor external standards or edit internal standards to comply with program requirements.	XXXXOOXXX
D6A		Plan and Brief CDRL Cost Estimates		Each information product (document) that must be formally delivered to the customer must be estimated based on any extra work required beyond performing the work related to the document.	XXXXOOXXX

D6B	Estimate CDRL Cost	Estimate special CDRL costs not included in cost of work related to the CDRL.	XXXXOOXXX
D6C	Integrate CDRL Estimates	Estimating must review CDRL estimates and ensure there are no omissions or double booking in the estimates.	XXXXOOXXX
D6D	Prepare Program Directives	Where it is necessary to deviate from generic enterprise practices, these cases must be documented in program directives.	XXXXOOXXX
D6E	Prepare Functional Task Description (FTD)	Each unique task on the program is planned in terms of schedule and cost as an input to the total cost estimate.	XXXXOOXXX
D6F	Create Program Schedule	The Integrated Master Schedule is prepared and coordinated with lower tier schedules.	XXXXOOXXX
D6G	Estimating School	Functional lead people bring their completed FTDs to a meeting room and present them to estimators who have to understand them in order to prepare the functional cost estimate. Any unusual cost factors uncommon in past experience are pointed out and explained.	XXXXOOXXX
D6I	Integrate/Format Functional Task Estimate	The functional inputs are reviewed and compared with the estimating expectations. Questionable inputs are reviewed with sources. Care is taken to ensure that no inputs have been overlooked.	XXXXOOXXX
D6J	Prepare Spending Profile	Estimating computes the amount of money obligated as a function of time and seeks to reduce large exclusions through consultation with scheduling and functional planners.	XXXXOOXXX
D6K	Functional Budget Review	Functional management reviews the estimate prior to submission to the customer.	XXXXOOXXX
D6L	Prepare Program Staffing Plan	Personnel needs in time are determined and agreements made with functional management to support these needs.	XXXXOOXXX
D6M	Prepare Program Personnel Training Plan	Special training needs for personnel assigned to the program are identified and a means devised to provide needed training in a timely way.	XXXXOOXXX
D7	Prepare Proposal	The proposal phase for a potential future program activity. Systems Development prepares assigned proposal text, prepares a SEL and draft SEMP, and plans personnel requirements, computer tool needs, products to be produced, and tasks to be accomplished against a schedule.	XXXXOOXXX
D72	Manage Proposal Effort	Proposal manager is responsible for all proposal efforts.	XXXXOOXXX
D73	Organize Proposal	The proposal manager and staff design proposal organization, initial schedule, and work plan.	XXXXOOXXX
D74	Prepare System Specification	Where required, a system specification is either prepared or one supplied by the customer is edited for agreement with enterprise understanding of the customer need and what is possible with the current technology.	XXXXOOXXX
D75	Prepare System Specification Ground Rules	Proposal management must define or approve a strategy for use in preparing or editing the system specification. This may be coordinated with specific enterprise strengths or competition weaknesses.	XXXXOOXXX
D76	Plan and Staff Proposal Team	Personnel needed for the proposal and how they should be organized are determined. Lead persons are selected and installed.	XXXXOOXXX
D77	Train/Orient Proposal Staff & Team		XXXXOOXXX
D78	Plan Technical Effort	The technical volume of the proposal will require a clear description of the proposed product concept. In order to provide this description, some work will be required to develop the concept. Proposal team management must plan this work and assign responsibilities.	XXXXOOXXX
D79	Plan Management Volume Effort	Based on RFP content, the management volume is planned in terms of content, page count, assignments, and so forth.	XXXXOOXXX
D7A	Plan Cost Volume Effort	Customer requirements for documenting cost estimating results are studied and the consequences translated into planned actions and responsibilities.	XXXXOOXXX
D7B	Accomplish Technical Effort	This task entails the engineering work required to develop a sound technical solution to the customer's need. It may involve requirements analysis, risk analysis, trade studies for alternative approaches, pre-design actions, tests and analyses, and model making.	XXXXOOXXX
D7C	Prepare Draft System Engineering Management Plan (SEMP)	Study customer's RFP (SOW in particular) and make minimum adjustments required in generic GDSS SEMP to fully satisfy customer requirements with respect to managing the	XXXXOOXXX

Table D-1 (continued) Enterprise Charter Task List

Task nbr	Dept nbr	Task name	Company document/ Source reference	Task description	Generic phasing
				engineering effort. Assist Proposal Leader, Program Manager, and Proposal R&E Lead Engineer/Chief Engineer in the development of a sound organizational structure and technical management approach. Map the program tasks to the organizational structure. Identify required SEMP sub-plans and coordinate ownership and preparation schedule. Coordinate SEMP content with Management Volume Leader. Organize a review of the draft SEMP and gain approval in accordance with proposal schedule.	
D7D		Prepare System Equipment List		Determine what the delivered system shall consist of and what additional resources are required for development testing, manufacturing, spares, and other purposes. This will be the basis for the material cost estimate.	XXXXOOXXX
D7E		Prepare Material Estimate		Translate the content of the SEL into a material estimate.	XXXXOOXXX
D7F		Write Technical Volume			XXXXOOXXX
D7G		Write Management Volume			XXXXOOXXX
D7H		Plan/Write/Edit Miscellaneous Volumes			XXXXOOXXX
D7I		Prepare Cost Volume			XXXXOOXXX
D7J		Edit Management Volume			XXXXOOXXX
D7K		Edit Technical Proposal			XXXXOOXXX
D7L		Proposal Integration			XXXXOOXXX
D7M		Proposal Review and Processing			XXXXOOXXX
D7M1		Pink Team		A pink team reviews the planned proposal response early in the development process as a means to steer a flawed plan toward a winning direction. Staff the team with functional management personnel.	XXXXOOXXX
D7M2		Red Team Proposal		The Red Team provides an in-house review of the final proposal based on customer requirements defined in the RFP. Functional management should staff the team.	XXXXOOXXX
D7M3		Acquire Enterprise Proposal Approval		Prior to the submission of the proposal to the customer, it must be reviewed by enterprise management and approved or alternative direction given.	XXXXOOXXX
D7M4		Process Proposal for Delivery		The approved proposal is printed and delivered to the customer with great care to ensure that it arrives at the correct address by the need date.	XXXXOOXXX
D8		Questions		The customer may formally ask the contractors questions about their proposal. These questions must be carefully studied for hidden meaning that may provide insights into customer thinking. The question is assigned to a principal engineer who must respond with an answer within the allocated time. This sometimes entails corresponding proposal writeup changes and cost estimate changes in the immediately affected area and elsewhere. This task also covers response to Discrepancies which the customer may submit to the enterprise to notify us of an area in our proposal that they consider a deficiency, that is not responsive to their requirements contained in the RFP.	XXXXOOXXX
D9		The Wait		Throughout the period from proposal completion (Red Team response completion) and selection (or rejection), it will be necessary for the Program/Proposal Manager to retain and support a minimum staff to continue program organization buildup in anticipation of a win. There are a host of useful tasks that these people can accomplish aside from responding to questions, discrepancies, BAFO input requests, and orals presentation materials preparation.	XXXXOOXXX

		This is a very sensitive area that must be carefully evaluated against GDSS ethics rules and customer regulations.	
DA	Best and Final Offer (BAFO)	The customer may request a Best and Final Offer from each competitor. Generally when they do this the suggestion is intended that you are supposed to reduce the cost estimate. The Proposal Manager may at this time call for a scrub of the cost inputs. These changes may be submitted as a letter or a formal change to the proposal depending on customer requirements.	XXXXOOXXX
D8	Orals	The customer may require a brief oral presentation of the GDSS proposal. Generally, this presentation will involve materials already prepared for the proposal, but may require that some of them be adjusted for improved understanding in a meeting environment. A small team will accompany the Proposal/Program Manager to the customer's site to make and support the presentation. Sometimes the customer will require a video tape of the presentation instead of or in addition to the in-person presentation.	XXXXOOXXX
DC	Evaluate Proposals	The customer evaluates the proposals and makes a selection. This process may include calls for responses to questions and discrepancies and an oral presentation or none of these other inputs.	XXXXOOXXX
DD	Negotiations	This process may take place concurrently between the customer and all competitors prior to selection or, more commonly, only with the winner after the selection. In the negotiations, the customer is trying to reach a clear mutual understanding of the cost, schedule, and performance issues with the enterprise. The result is a definitized contract between the enterprise and the customer. The customer may sign a letter contract with the enterprise subsequent to selection and prior to the completion of negotiations to allow the program to begin while the details are resolved in negotiations.	XXXXOOXXX
DE	Applicable Document Analysis	This task is a duplicate of Task 128 except this one takes place during the proposal period and 128 takes place as a part of the contract task. Replace the 128 characters in 128 subtasks with 6E for subtasks. The purpose here is to adjust where possible the customer's applicable document list and tailoring included within the System Specification (if supplied), SOW, and FAR clauses toward agreement with enterprise practices. Principal engineers are identified for each document and they are asked to review the content against enterprise practices. Differences are assessed with respect to cost impacts and potential for customer willingness to change within the Program/Proposal Manager's guideline for RFP responsiveness.	XXXXOOXXX
DF	Manage Emerging Program		XXXXOOXXX
E	Marketing/Pre-Concept Studies	Work is accomplished to develop and maintain an understanding of the customer's needs. Efforts are made to develop concepts that solve anticipated customer needs and to communicate these to the customer.	XXXOOOOOO
E31	Mission Analysis		XXXOOOOOO
E32	Operations Analysis		XXXOOOOOO
E33	Establish Criteria and Measures of Effectiveness	Identify the parameters that will be used in making value judgments in the selection of preferred concepts and the measures of effectiveness of interest to the customer.	XXXOOOOOO
E34	Economic Analysis	Life cycle cost figures are estimated and computed for planned configurations.	XXXOOOOOO
E35	Requirements Definition	Identify and rationalize customer requirements in consultation with customer and program office representatives. Coordinate the requirements analysis work of other departments including establishing a demand for needed work, assertive coordination of supporting analysis departments, review and integration of their work, and coordination of needed changes. Accomplish requirements analysis tasks assigned to PIT. Consolidate requirements into architecture element sets. prepare a system level requirements document, identify any others needed, and assign responsibilities for their development. Ensure availability of requirements document boilerplates, if required. At this stage in a program, only a system level requirements document is generally required. Gain study leader approval of requirements, baseline same, and ensure they are available to study members.	XXXOOOOOO

Table D-1 (continued) Enterprise Charter Task List

Task nbr	Dept nbr	Task name	Company document/ Source reference	Task description	Generic phasing
E36		Design Concepts & Analysis		Synthesize requirements into alternative concepts and accomplish refinement of these concepts in accordance with the established criteria and measures of effectiveness as required to assure an effective evaluation of relative merits. Document concepts in a form consistent with the study System Description Document (SDD).	XXXOOOOOO
E37		Concept Driven Requirements Analysis		Extract requirements from concepts where concept work identifies a need for requirements not foreseen from front end requirements work. Capture these requirements into the baseline as approved.	XXXOOOOOO
E38		System Analysis & Evaluation		The alternative concepts are subjected to evaluation from the perspective of several analytical perspectives as a function of the criteria and MOE established in task 103. This may include relative performance in one or more areas, reliability, maintainability, safety cost, and other factors.	XXXOOOOOO
E39		Study Management		Manage all study activities including: (1) plan study activities, (2) ensure that planned activities occur within schedule and budget constraints, (3) adjust schedules and budgets as required by conditions and team performance, (4) ensure study is staffed with personnel qualified to satisfy study requirements, (5) review and approve study products, (6) monitor and direct study progress, and (7) lead study team.	XXXOOOOOO
EM		Manage Marketing Activities and Pre-Concept Studies		Manage the work accomplished to identify future contract possibilities with present and potential customers.	XXXOOOOOO
F		Manage Business Acquisition			XXXOOOOOO
G		Study Lessons Learned		Cross-functional department team studies results from programs and reaches decisions about the utility of current practices, tools, and training. The team identifies candidate improvements that could be made to correct perceived imperfections in program activity.	
H		Independent Research and Development (IRAD)		Enterprise efforts to develop new technology appropriate to the enterprise product line. Where possible these studies are correlated with urgent customer needs.	
I		Process Benchmarking		A cross-functional management team studies competitor methods and tools and makes comparisons with GDSS methods and tools. Conclusions are reached about the relative ranking of GDSS with others and priorities are determined on which areas GDSS should seek to improve its performance with respect to a company that is accepted to perform at a world class level in that area.	
J		Define Improvement Actions		A cross-functional management team reviews program lessons learned, IRAD results, and benchmarking study results and reaches conclusions about priorities for enterprise improvements. Selected improvement actions are then further defined and responsibility for accomplishment of needed actions is defined.	
K		Implement Improvement Actions		The responsible person (identified in task J) assembles a team (if necessary), makes a detailed plan of approach responsive to the defined actions needed, and accomplishes planned improvement tasks reporting progress and problems periodically to management.	
L		Maintain Division Best Practices		Enterprise procedural controls are maintained to reflect the best methods the functional departments are capable of. As improvement actions are completed that involve practices changes, those changes are made.	
M		Functional Management		Functional management is responsible for maintaining the Division capability to satisfy the needs of its customer base in terms of best practices, effective tools and facilities, and a	

		qualified pool of personnel.	
N	Maintain Facilities and Tools	Test labs and other facilities as well as computer and special purpose tools are maintained in good condition through capital improvements, purchases, maintenance work, and other activities such that they may be used to satisfy needs of programs.	
P	Maintain Personnel Knowledge and Skills Base	Personnel are exposed to new ideas, current practices and tools, and techniques appropriate to current and potential Division work both to ensure they are qualified to perform department tasks on programs and to encourage maximum flexibility of assignment consistent with depth of knowledge needed in specialized areas.	
Q	Acquire/Separate Personnel	Functional management, in concert with HR, acts to understand current and future needs, acquires needed new personnel in a timely way, and reassigns or separates personnel proven to be unsuited for their current assignment.	
R	Personnel Training Program	Functional management collectively, with Human Resources leadership, provides Division personnel with task and product oriented training in off hours (lunch time and evening) and supports local college degree and certificate programs related to Division needs.	
S	Enterprise Integration Team (EIT)	Enterprise management staffs an integration team to study and balance the needs of the several programs within available resource constraints. Team also advises enterprise management of new resource needs based on existing and anticipated commitments to customers and potential customers. Team attempts to discover conditions of overload in enterprise resources and find ways to permit all programs to satisfy their customer commitments. Where this is not possible, compromises are made for the near term and plans made for future resource changes.	XXXXXXXXX
S2	Enterprise Resources Balancing	Enterprise functional management coordinated through Division Staff reviews program needs for personnel, facilities, manufacturing, and operations resources and seeks to satisfy the needs of all programs within established constraints. Where constraints are restrictive and impact customer requirements, management studies alternatives for solution (including re-balancing program requirements) to remove constraints.	XXXXXXXXX
S241	Deployment & Operations Requirements Integration	Program operations leadership integrates the product requirements with deployment and operations requirements at the system and program level with a principal focus on deployment and operations needs.	XXXXXXXOO
S242	Deployment & Operations Process Design Integration	Logistics personnel study deployment and operations designs and determine resultant demands they place on enterprise resources.	OOOXXOXXO
S251	Quality Requirements Integration	Quality specialists on the EIT study the quality obligations and expectations on all programs and work to resolve any conflicts they imply for the enterprise.	OOXXXXXXO
S252	Quality Process Design Integration		OOXXXXXXO
S261	Material & Procurement Requirements Integration	Program material requirements are studied searching for potential conflicts that have to be resolved.	OOXXXXOOO
S262	Material & Procurement Process Design Integration	Program material plans are reviewed and conflicts resolved at the enterprise level.	OOXXXXOXO
S271	Production Requirements Integration	Production requirements for each program are reviewed at the enterprise level and conflicts with enterprise capability resolved.	OOXXXXOOO
S272	Production Process Design Integration		OOXXXXOXO
S281	Logistics Support Requirements Integration		OOOXXXOXX
S282	Logistics Support Process Design Integration		OOOXXXOXX

Main body index

SEM/SEMP index

H

I

L

M

N

O

P

Q

R